Photochemistry for Environmental Remediation

Photochemistry for Environmental Remediation

Dr. Gabor Kovacs

ARCLER
PRESS

www.arclerpress.com

Photochemistry for Environmental Remediation

Dr. Gabor Kovacs

Arcler Press
4164 Lakeshore Road
Burlington, ON L7L 1A4
Canada
www.arclerpress.com
Email: orders@arclereducation.com

ISBN: 978-1-77956-950-9 (Hardcover)

ABOUT THE AUTHOR

Dr. Gábor Kovács is a lecturer at the Sapientia Hungarian University of Transylvania, Faculty of Technical and Human Sciences, Department of Horticulture in Târgu Mureș, Romania. He has been in this role since 2021.

Kovács obtained his B.Sc. and M.Sc. in Chemistry/Applied Electrochemistry from Babeș-Bolyai University in Cluj-Napoca, Romania, in 2007 and 2008, respectively. He later earned his PhD in Chemistry from Babeș-Bolyai University in 2012.

Before his current lecturing position, Kovács had several research roles. He was an assistant Researcher at the Babeș-Bolyai University Faculty of Physics / Institute of Interdisciplinary Bio-Nano Sciences in Cluj-Napoca. Kovács also held postdoctoral research positions, including at the Department of Applied and Environmental Chemistry at the University of Szeged in Hungary, and at the Institute of Environmental Science and Technology of the same university.

Kovács has taught a variety of chemistry-related courses at the B.Sc. and M.Sc. levels, including General/Inorganic Chemistry, Electrochemistry, Corrosion and Anticorrosive Protection, Organic Chemistry, Biochemistry, Phytosanitary Chemistry, and Toxicology.

Kovács has led several research projects funded by the Hungarian Academy of Sciences and the National Research, Development and Innovation Office of Hungary. These projects have focused on areas such as the synthesis, characterization, and applicability of photocatalytic nanostructures, carbon nanotube-based composites, and investigations dealing with the ecotoxicological aspects of these materials.

TABLE OF CONTENTS

List of Figures .... *xi*

Preface .... *xvii*

Chapter 1 **Introduction to Photochemistry and Environmental Importance** **1**

Introduction 1

1.1. Fundamental Principles of Photochemistry 6

1.2. Types of Photochemical Reactions 12

1.3. Photochemistry in the Atmosphere 17

1.4. Photochemistry and Water Quality 22

1.5. Photochemistry in Renewable Energy 27

1.6. Environmental Impacts of Photochemical Processes 33

Summary 37

References 40

Chapter 2 **Principles of Photophysical and Photochemical Processes** **45**

Introduction 45

2.1. Basic Concepts of Light 50

2.2. Photophysical Processes 56

2.3. Photochemical Processes 61

2.4. Spectroscopy Techniques 66

2.5. Applications in Material Science 71

2.6. Environmental Implications 76

Summary 81

References 83

Chapter 3 **Photocatalysis – Mechanisms and Materials** **87**

Introduction 87

3.1. Fundamental Mechanisms of Photocatalysis 91

3.2. Types of Photocatalysts 96
3.3. Factors Influencing Photocatalytic Activity 102
3.4. Applications in Environmental Remediation 107
3.5. Applications in Energy Conversion 112
3.6. Challenges and Limitations 117
Summary 123
References 125

Chapter 4 Advanced Oxidation Processes (AOPs) in Environmental Remediation 133
Introduction 133
4.1. Mechanisms of AOPs 138
4.2. Types of Advanced Oxidation Processes (AOPs) 143
4.3. Applications of AOPs in Water Treatment 148
4.4. AOPs in Soil Remediation 154
4.5. Advantages and Limitations of AOPs 159
4.6. Future Trends and Innovations in AOPs 163
Summary 169
References 171

Chapter 5 Photochemical Applications in Water Treatment 177
Introduction 177
5.1. Principles of Photochemistry 182
5.2. Photocatalysis in Water Treatment 187
5.3. Advanced Oxidation Processes (AOPs) 193
5.4. Solar Water Treatment Technologies 198
5.5. Comparison with Traditional Water Treatment Methods 203
5.6. Challenges and Limitations 208
Summary 213
References 216

Chapter 6 Photochemical Methods in Air Pollution Control 221
Introduction 221
6.1. Understanding Photochemical Reactions 227
6.2. Mechanisms of Photochemical Air Pollution Control 232
6.3. Photochemical Methods in Industrial Applications 237

6.4. Advantages of Photochemical Methods 242
6.5. Challenges and Limitations 246
6.6. Future Directions in Photochemical Research 251
Summary 257
References 259

Chapter 7 Pollutant Degradation Pathways and Kinetics 267
Introduction 267
7.1. Types of Pollutants 272
7.2. Mechanisms of Degradation 278
7.3. Factors Influencing Degradation Rates 283
7.4. Kinetic Models in Pollutant Degradation 288
7.5. Analytical Techniques for Studying Degradation 293
7.6. Case Studies of Pollutant Degradation 298
Summary 303
References 306

Chapter 8 Design and Operation of Photochemical Reactors 313
Introduction 313
8.1. Principles of Photochemistry 317
8.2. Types of Photochemical Reactors 322
8.3. Design Considerations for Photochemical Reactors 327
8.4. Operational Parameters 332
8.5. Applications of Photochemical Reactors 337
8.6. Challenges in Photochemical Reactor Design and Operation 342
Summary 347
References 350

Chapter 9 Environmental, Energy, and Economic Considerations 353
Introduction 353
9.1. Environmental Impact of Energy Production 357
9.2. Economic Implications of Energy Choices 362
9.3. Policy Frameworks for Sustainable Energy 367
9.4. Technological Innovations in Energy Efficiency 372
9.5. Social Considerations in Energy Transition 378

9.6. Global Perspectives on Energy and Environment 383
Summary 389
References 391

Chapter 10 Future Prospects, Innovations, and Case Studies 397
Introduction 397
10.1. Fundamental Principles of Photochemistry 402
10.2. Innovations in Photochemical Techniques 408
10.3. Applications in Environmental Science 413
10.4. Photochemistry in Medicine 418
10.5. Case Studies of Successful Photochemical Applications 423
10.6. Future Prospects in Photochemistry 427
Summary 432
References 435

Index 439

LIST OF FIGURES

Figure 1.1. Introduction to photochemistry and environmental importance
Figure 1.2. Importance of photochemistry in modern science
Figure 1.3. Fundamental principles of photochemistry
Figure 1.4. Photochemical reactions and mechanisms
Figure 1.5. Role of quantum mechanics in photochemistry
Figure 1.6. Types of photochemical reactions
Figure 1.7. Photodegradation of pollutants
Figure 1.8. Role of sunlight in atmospheric chemistry
Figure 1.9. Formation of ozone and its implications
Figure 1.10. Impact of photochemical smog on air quality
Figure 1.11. Photochemistry and water quality
Figure 1.12. Use of photochemical methods for water purification
Figure 1.13. Photochemistry in renewable energy
Figure 1.14. Solar energy conversion technologies
Figure 1.15. Role of photochemistry in hydrogen production
Figure 1.16. Contribution to climate change
Figure 2.1. Principles of photophysical and photochemical processes
Figure 2.2. Photochemical processes
Figure 2.3. Basic concepts of light
Figure 2.4. Interaction of light with matter
Figure 2.5. Quantum theory of light
Figure 2.6. Photophysical processes
Figure 2.7. Energy transfer mechanisms
Figure 2.8. Photochemical processes
Figure 2.9. Mechanisms of light-induced chemical reactions
Figure 2.10. Applications in organic synthesis
Figure 2.11. Spectroscopy techniques

Figure 2.12. Fluorescence spectroscopy for analyzing photochemical reactions
Figure 2.13. Time-resolved spectroscopy for dynamic studies
Figure 2.14. Applications in material science
Figure 2.15. Solar energy conversion technologies
Figure 2.16. Role of photochemical processes in atmospheric chemistry
Figure 3.1. Photocatalysis – mechanisms and materials
Figure 3.2. Photocatalysis
Figure 3.3. Fundamental mechanisms of photocatalysis
Figure 3.4. Reaction pathways and product formation
Figure 3.5. Types of photocatalysts
Figure 3.6. Semiconductor photocatalysts
Figure 3.7. Metal-based photocatalysts
Figure 3.8. Light wavelength and intensity
Figure 3.9. Applications in environmental remediation
Figure 3.10. Water purification techniques
Figure 3.11. Applications in energy conversion
Figure 3.12. CO_2 reduction to fuels
Figure 3.13. Stability and reusability of photocatalysts
Figure 3.14. Scale-up and industrial application issues
Figure 4.1. Advanced oxidation processes (AOPs)
Figure 4.2. Mechanisms of advanced oxidation processes (AOPs)
Figure 4.3. Role of oxidants in AOPs
Figure 4.4. Types of advanced oxidation processes
Figure 4.5. Fenton and Fenton-like reactions
Figure 4.6. Photocatalytic advanced oxidation processes (AOPs)
Figure 4.7. Applications of AOPs in water treatment
Figure 4.8. Removal of organic contaminants
Figure 4.9. Treatment of industrial wastewater
Figure 4.10. AOPs in soil remediation
Figure 4.11. *In situ* versus *ex situ* remediation techniques
Figure 4.12. Benefits over traditional treatment methods
Figure 4.13. Future trends and innovations in AOPs
Figure 4.14. Emerging technologies and research directions

Figure 5.1. Photochemical processes in water treatment
Figure 5.2. Photochemical processes
Figure 5.3. Principles of photochemistry
Figure 5.4. Role of light in chemical processes
Figure 5.5. Photocatalysis in water treatment
Figure 5.6. Mechanism of photocatalysis
Figure 5.7. Advantages of photocatalytic methods
Figure 5.8. Advanced oxidation processes (AOPs)
Figure 5.9. Effectiveness of AOPs in removing contaminants
Figure 5.10. Solar water treatment technologies
Figure 5.11. Benefits and limitations of solar technologies
Figure 5.12. Comparison with traditional water treatment methods
Figure 5.13. Environmental impact considerations
Figure 5.14. Technical challenges in implementing photochemical methods
Figure 5.15. Future research directions in photochemical water treatment
Figure 6.1. Photochemical methods in air pollution control
Figure 6.2. Importance of air quality management
Figure 6.3. Overview of photochemical methods
Figure 6.4. Principles of photochemistry
Figure 6.5. Role of sunlight in photochemical processes
Figure 6.6. Mechanisms of photochemical air pollution control
Figure 6.7. Photochemical oxidation processes
Figure 6.8. Photochemical methods in industrial applications
Figure 6.9. Implementation in power plants and manufacturing
Figure 6.10. Environmental benefits
Figure 6.11. Potential for scalability and adaptability
Figure 6.12. Technical challenges in the implementation of photochemical processes
Figure 6.13. Economic and regulatory barriers
Figure 6.14. Future directions in photochemical research
Figure 6.15. Potential for policy development and funding
Figure 7.1. Pollutant degradation pathways and kinetics
Figure 7.2. Overview of kinetics in pollutant degradation
Figure 7.3. Types of pollutants
Figure 7.4. Inorganic pollutants

Figure 7.5. Emerging contaminants
Figure 7.6. Biodegradation processes
Figure 7.7. Chemical degradation processes
Figure 7.8. Chemical structure of pollutants
Figure 7.9. Kinetic models in pollutant degradation
Figure 7.10. Zero-order kinetics
Figure 7.11. First-order kinetics
Figure 7.12. Chromatography methods
Figure 7.13. Mass spectrometry (MS)
Figure 7.14. Pesticide degradation in soil
Figure 7.15. Heavy metal remediation in water
Figure 7.16. Pharmaceutical degradation in wastewater
Figure 8.1. Design and operation of photochemical reactors
Figure 8.2. Principles of photochemistry
Figure 8.3. Role of energy transfer in photochemical processes
Figure 8.4. Batch photochemical reactors
Figure 8.5. Continuous flow photochemical reactors
Figure 8.6. Design considerations for photochemical reactors
Figure 8.7. Light source selection and placement
Figure 8.8. Operational parameters
Figure 8.9. Influence of solvent choice on reaction efficiency
Figure 8.10. Synthesis of pharmaceuticals
Figure 8.11. Production of renewable energy sources
Figure 8.12. Challenges in photochemical reactor design and operation
Figure 8.13. Photodegradation of materials and catalysts
Figure 8.14. Economic considerations and cost-effectiveness
Figure 9.1. Environmental, energy, and economic considerations
Figure 9.2. Environmental impact of energy production
Figure 9.3. Effects of fossil fuels on air and water quality
Figure 9.4. Renewable energy sources and their ecological benefits
Figure 9.5. Renewable versus non-renewable energy
Figure 9.6. Policy frameworks for sustainable energy
Figure 9.7. National policies promoting renewable energy
Figure 9.8. Technological innovations in energy efficiency

Figure 9.9. Advances in renewable energy technologies
Figure 9.10. Smart grid technology and its economic benefits
Figure 9.11. Social considerations in energy transition
Figure 9.12. Public perception and acceptance of renewable energy
Figure 9.13. Role of education in promoting sustainable practices
Figure 9.14. Comparison of energy policies in developed
Figure 9.15. International cooperation in addressing environmental challenges
Figure 10.1. Photochemistry
Figure 10.2. Scope of photochemistry
Figure 10.3. Importance in modern science and technology
Figure 10.4. Fundamental principles of photochemistry
Figure 10.5. Light–matter interaction
Figure 10.6. Energy transfer mechanisms
Figure 10.7. Advances in photocatalysis
Figure 10.8. Innovations in solar energy conversion
Figure 10.9. Photochemical processes in water treatment
Figure 10.10. Contributions to green chemistry
Figure 10.11. Photodynamic therapy for cancer treatment
Figure 10.12. Innovations in diagnostic imaging
Figure 10.13. Solar fuels production
Figure 10.14. Future prospects in photochemistry
Figure 10.15. Emerging technologies and research trends

PREFACE

Photochemical science has emerged as a critical discipline in solving global environmental issues, providing new solutions to mitigate pollutants and restore ecological balance. Based on light-activated chemical reactions, photochemical processes can break down toxic substances, detoxify toxins, and reduce environmental threats without introducing secondary contaminants. With growing industrial processes, urbanization, and energy generation, the imperative for advanced and sustainable remediation strategies is greater than ever before. Photochemistry differs in this regard because it is both scientifically accurate and ecologically sound, harnessing renewable light sources like the sun to power chemical reactions to be used for environmental good. In addition to solving existing contamination problems, this is also an environmentally friendly direction for the future.

Over the past few decades, environmental engineers and scientists have expanded the application of photochemistry from laboratory experimentation to practical, successful, in-field remediation technologies. To purify contaminated water, to break down airborne toxins, or to treat complicated industrial effluents are but a few examples of the applications of photochemical technologies in remediation. These processes work by generating reactive species that are capable of acting on a wide range of pollutants, often with total mineralization to harmless by-products like water and carbon dioxide. Integration of photochemistry with other environmental engineering technologies is creating advances in energy efficiency, cost-effectiveness, and scale-up application, and it is rapidly emerging as a fundamental technology for modern environmental management.

Photochemistry for Environmental Remediation is designed to provide a comprehensive review of how photochemical principles and technology can be applied to solve environmental problems. The book bridges fundamental science and engineering practice by explaining fundamental principles and the reality of implementation. It addresses photophysics and photochemistry reaction principles, touches on photocatalysis and advanced oxidation processes, and their use in air, water, and industrial waste treatment. Additionally, the book covers reactor design, pollutant degradation mechanisms, and environmental, economic, and energy consequences of photochemical solutions. Interleaving theoretical descriptions with case studies and practical examples, this book gives readers the skills to apply photochemical methods in environmental cleanup projects.

Chapter 1 presents the topic of photochemistry with emphasis on its environmental applications. It explains the scientific justification of using light-induced processes in pollution control and cleanup and indicates the advantages over conventional treatment processes. Significant environmental issues which can be addressed by photochemistry are discussed, leading to the rest of the book.

Chapter 2 covers the principles governing photophysical and photochemical reactions. It delves into light absorption, excited-state behavior, quantum yields, and reaction routes and provides the theory necessary to grasp how light initiates and drives chemical reactions relevant to environmental remediation.

Chapter 3 elaborates on photocatalysis, such as mechanisms, common catalyst materials such as titanium dioxide, and key factors influencing catalytic efficiency. It also deals with applied issues of material choice, surface treatment, and regenerating catalysts and offers insight into maximizing photocatalysis for various remediation technologies.

Chapter 4 also explains AOPs based on photochemical reactions for generating highly reactive oxidizing agents. Chapter 4 explains techniques like UV/H_2O_2, photo-Fenton treatment, and photocatalytic ozonation, explaining how the methods enable rapid and total removal of contaminants in various environmental media.

Chapter 5 reviews photochemical applications to water treatment, e.g., removal of organic pollutants, disinfection, and treatment of novel, emerging pollutants like pharmaceuticals and microplastics. The chapter considers both laboratory-scale processes and actual-scale water treatment plants, with an emphasis on efficiency and regulatory acceptability.

Chapter 6 extends the treatment to photochemical methods of air pollution control. Topics covered include volatile organic compound degradation, photochemical smog reduction, and industrial exhaust gas treatment. The chapter also mentions scaling up problems with air treatment equipment and integrating it with existing pollution control equipment.

Chapter 7 deals with pollutant degradation routes and kinetics, explaining how to degrade pollutants in photochemical conditions and how the degradation is influenced by reaction parameters. The chapter enumerates kinetic models, analytical techniques, and illustrations of degradation tests for various pollutants.

Chapter 8 examines the structure and functioning of photochemical reactors. It provides reactor designs, selection of light sources, optical design, mixing, and flow dynamics and gives useful recommendations to engineers and scientists who are tasked with installing photochemical systems.

Chapter 9 contrasts environmental, energy, and economic considerations, answering the question of the sustainability of photochemical technologies with respect to life cycle assessment, energy input, cost of operation, and potential environmental trade-offs. Chapter 9 also addresses the application of renewable energy resources in the integration of photochemical processes.

Chapter 10 looks to the future with possibilities, innovation, and case studies, which encapsulate new upcoming materials, hybrid technology, and reactor engineering advances. Illustrations from real life are presented to provide successful experience and lessons learned from experience in different environmental conditions.

The book is intended for environmental engineers, chemists, researchers, and graduate students interested in widening their perspective on sustainable pollution control technology. It will also find application for policymakers, industry professionals, and environmental consultants who wish to incorporate photochemical solutions into remediation programs. Blending scientific principles with applied solutions, it is both a reference and a guide to the advancement of the application of photochemistry for the protection and restoration of the environment.

CHAPTER 1

INTRODUCTION TO PHOTOCHEMISTRY AND ENVIRONMENTAL IMPORTANCE

Contents

Introduction 1
1.1. Fundamental Principles of Photochemistry 6
1.2. Types of Photochemical Reactions 12
1.3. Photochemistry in the Atmosphere 17
1.4. Photochemistry and Water Quality 22
1.5. Photochemistry in Renewable Energy 27
1.6. Environmental Impacts of Photochemical Processes 33
Summary 37
References 40

INTRODUCTION

The field of photochemistry is the study of chemical reactions and processes initiated by the absorption of light, essentially combining the science of chemistry and physics. Photochemical reactions are of pivotal importance in a variety of natural phenomena and artificial systems, influencing everything from the process of plant photosynthesis to solar energy technology design. Comprehending photochemistry is important since it opens up the possibilities of how light energy can be used to cause chemical change, resulting in the formation of new materials or energy products. The field analyzes the interactions between matter and light on a molecular basis, enabling researchers to construct definite reactions that tap solar energy effectively. Such advances hold extremely promising potential for environmental utilization, namely in the creation of clean sources of energy and the mitigation of climate change, and therefore further our knowledge of their environmental safety and health implications. The most intriguing application of photochemistry might be in the field of environmental science, namely pollutant degradation. Photochemical reactions play a crucial role in the atmosphere, with sunlight catalyzing the breakdown of some dangerous substances, such as volatile organic compounds and greenhouse gases.

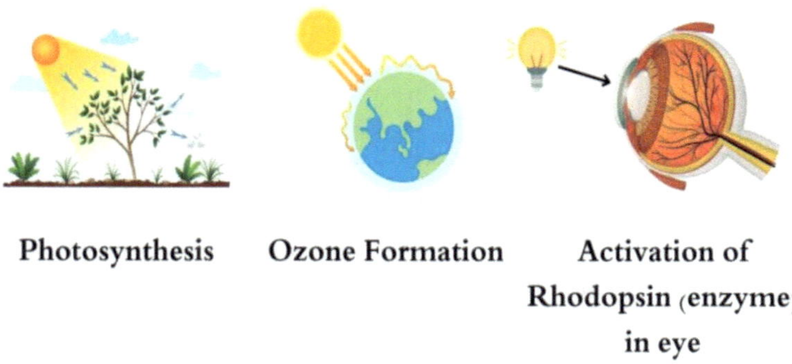

Figure 1.1. *Introduction to photochemistry and environmental importance.*

Source: https://themasterchemistry.com/wp-content/uploads/2023/06/Examples-of-photochemical-reaction-image-1-768x429.jpg.

For instance, the process of photooxidation has the potential to break down harmful chemicals into less toxic products, thus purifying air and reducing the rate of pollution. Moreover, studies attest to the significance of photoresponsive materials, including hydrogels, which have properties that vary with light exposure. These materials stress the synergy between innovation and photochemistry with an unprecedented opportunity to address environmental problems using responsive smart materials. Through the integration of photochemical processes into advanced technology, scientists can potentially facilitate more effective measures to control pollution and conserve the environment and transform scientific research into a sustainable goal. The topic of biomass burning (BB) also illustrates the intricate relation between photochemistry and environmental impacts. BB is a major source of air pollution, affecting air quality, public health, and climate worldwide. Comprehensive studies in nations like China have thoroughly investigated BBs contribution to the chemical dynamics of atmospheric chemicals with regard to photochemical processes involved in secondary pollutant formation. The products emitted from biomass combustion undergo complex photochemical processes, further aggravating air quality conditions and facilitating the transport of harmful substances. Understanding these interactions is crucial in the creation of policies aiming to control BB and its impacts on health and the environment. With increasing awareness of the environmental significance of photochemistry, it is increasingly important to apply such information in the creation of appropriate regulations and interventions in order to create a healthier world and improved public health in the context of current environmental issues (Figure 1.1).

Definition of Photochemistry

Photochemistry as a discipline is concerned chiefly with chemical reactions and processes initiated or preceded by light. Specifically, it is concerned with the interaction of photons and matter that leads to changes in energy states and in molecular structure. Through this interaction, a variety of phenomena are brought about, ranging from the synthesis of complex organic molecules via photosynthesis to the photochemical destruction of atmospheric pollutants. Knowledge of photochemical principles is important, especially in environmental sciences since these events are key players in atmospheric chemistry and the generation of photochemical smog. This knowledge helps scientists forecast the outcome of human activities, such as emissions that have the potential to affect the global climate and air quality in the long run. In environmental relevance, photochemistry is of fundamental importance to comprehend natural and anthropogenic influences on atmospheric composition. For instance, comparative research on photochemical environments, e.g., NASAs PEM-West and TRACE-P field experiments, illustrate the importance of photochemical cycles in comprehending air quality processes in areas like the western North Pacific. The findings of these studies highlight the role of various factors, such as seasonal fluctuation and pollutant concentration, in the creation and annihilation of ozone. In all situations, photochemical activity like that observed is extremely fluctuating based on atmospheric conditions, indicating that permanent anthropogenic activities can lead to crucial variations in photochemical parameters, which further affect climate and environmental health (Anderson et al., 2003). In addition, the quantification of specific compounds such as dimethyl sulfide (DMS) also indicates the prevalent role of photochemistry in the atmosphere. DMS is a natural gas emitted by ocean phytoplankton and functions to nucleate clouds and regulate climate by virtue of its oxidation to sulfate aerosols. Airborne measurements have revealed impressive variability in sea-to-air fluxes of DMS, illustrating the intricate interaction among biogeochemical and photochemical processes in the air. They are concerned with how vertical mixing and *in situ* atmospheric chemistry are major factors in determining emissions and transformation. By investigating these interactions in an orderly fashion, researchers can refine their models of atmospheric behavior, eventually making more sustainable environmental policies (Bandy et al., 1999).

Historical Background and Development

The history of photochemistry dates back to the very first discoveries in chemistry and physics, and with time, the subject evolved through the work of the principal scientists. Around the late 19th century, light's interaction with chemical reaction began receiving good attention due to the pioneering figures Hermann von Helmholtz and Svante Arrhenius. These early ventures betrayed the profound impact that light has on molecular forms, tasting the essence of energy conversions in chemical processes. Continuing work in the 20th century, advances in spectroscopy and the understanding of molecular interactions spawned ever more sophisticated studies of the photochemical reaction mechanisms. This opened the door to uses today, allowing scientists to view light as having a central place not just in the laboratory but also in nature, thereby forming photochemistry as a significant field of environmental science. The 1970s experienced a fundamental change in the understanding of photochemical processes as it became increasingly apparent that atmospheric chemistry was light-dependent. This period witnessed the beginning of research connecting photochemical reactions with environmental phenomena, in particular the depletion of the ozone layer and the effects on human health and ecological processes. It was found from research that ultraviolet (UV) radiation caused reactions that produced noxious pollutants, necessitating the need for metrics to determine these changes, as encapsulated in the Tropospheric Ozone Assessment Report (TOAR) (Allen et al., 2018). As a result, routine procedures came into effect to measure atmospheric concentrations of ozone and their impact on health, opening the door to the regulation necessary to cut down air pollution. This context serves to emphasize mounting realization of the need for interaction between photochemical science and environmental policy to control worldwide problems. During the past few years, the integration of computational models with experiments has further developed the understanding of photochemical processes. Advanced techniques, particularly hybrid quantum mechanics and molecular mechanics (QM/MM) models, have allowed researchers to study ultrafast dynamics in light-absorbing proteins responsible for many biological processes (Gozem et al., 2017). These developments explain light-induced chemical reaction processes at the molecular level, illuminating natural processes as well as theoretical opportunities for solar energy conversion. The continued development of photochemistry, particularly against the backdrop of environmental issues, portends a symbiotic interaction between theoretical advances and pragmatic solutions to world problems such as climate change and energy sustainability. This fluid history of photochemistry is not just a chronicle of past progress but an ever-working road to the application of light for beneficial environmental processes.

Importance of Photochemistry in Modern Science

The importance of photochemistry today is increasing on a daily basis, with its influence extending into various sciences, including environmental science, biochemistry, and astrobiology. Central to the understanding of life on earth, photochemistry describes how light from the sun triggers chemical reactions needed for the formation of organic compounds. The role of photosynthesis, for instance, in taking up carbon dioxide and water to form glucose and oxygen serves to underscore the role of photochemical processes in sustaining life. Furthermore, the data on prebiotic chemistry obtained through the study of nitrogen oxides (NO_x^-) illustrate how light-activated mechanisms influenced early life's evolution on Earth. Current research proposes that these molecules, i.e., nitrate and nitrite, were exposed to photolytic processes that greatly altered their amounts in the ancient environment, showing new information on the chemical pathways that supported life (Babbin et al., 2019). Moreover, photochemistry takes a key role in addressing current environmental challenges, i.e., pollution and climate change. The studies of photochemical reactions help in understanding the mechanisms by which certain pollutants change when exposed to the sun's radiation, producing both detrimental and beneficial effects on ecosystems. For instance, the photodegradation of harmful chemicals can reduce their persistence in the environment and hence mitigate their adverse effects. Other than that, photochemistry is key in the development of new materials and new sources of power, such as solar cells that harness sunlight to produce electricity. Since sustainability is society's issue today, the application of photochemical principles provides a path towards cleaner technology that can be used to counteract environmental degradation, rendering the field more important (Devincenzi et al., 2025). Photochemistry is one of the most important windows in astrobiology that researchers utilize to understand the possibility of life elsewhere in the cosmos. By exploring the interaction among light and other chemical species, scientists can make educated guesses about what kinds of life might possibly exist on other worlds. Photochemical processes are of particular interest to extraterrestrial environment interests as those that govern prebiotic chemistry necessary for life's formation. For instance, the stability and reaction of nitrogen oxides under various conditions indicate that the chemistry in Earth's early environments was also as complicated, maybe just as that of other worlds. Understanding the interactions informs our activities in searching for life beyond Earth and highlights the application of light as a universal catalyst in the chemical evolution of life on the universe (Babbin et al., 2019). This interdisciplinary approach not only advances our understanding of possible extraterrestrial conditions but also consolidates the principles behind photochemistry as a hub of modern scientific inquiry (Figure 1.2).

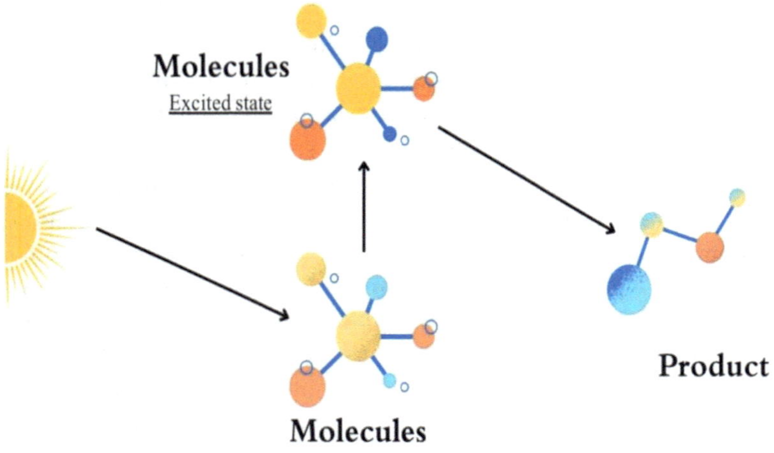

Figure 1.2. *Importance of photochemistry in modern science.*

Source: https://themasterchemistry.com/storage/2023/06/Image-of-Photochemical-reactions-1024x572.jpg.

1.1. FUNDAMENTAL PRINCIPLES OF PHOTOCHEMISTRY

The fundamental principles of photochemistry are based on the study of interactions between light and molecules, which capture how light energy triggers chemical alteration. Photochemistry really studies the excitation of molecules upon photon absorption and subsequent electronic transition. Upon absorption of light by a molecule, it can proceed from a ground state to an excited state, the central transition for most chemical reactions. Such interactions are critical in the explanation of phenomena such as photosynthesis and the degradation of pollutants in ecosystems. From this point of view, researchers can investigate the function of light irradiation in reactivity and stability and highlight the importance of wavelength and energy issues in predicting the behavior of molecules when subjected to UV and visible light. Moreover, excited states play a role beyond that of mere energy absorption; they play a crucial role in the kinetics of chemical reactions. Once excited, a molecule can go through various channels to form products or relax into the ground state with or without energy release.

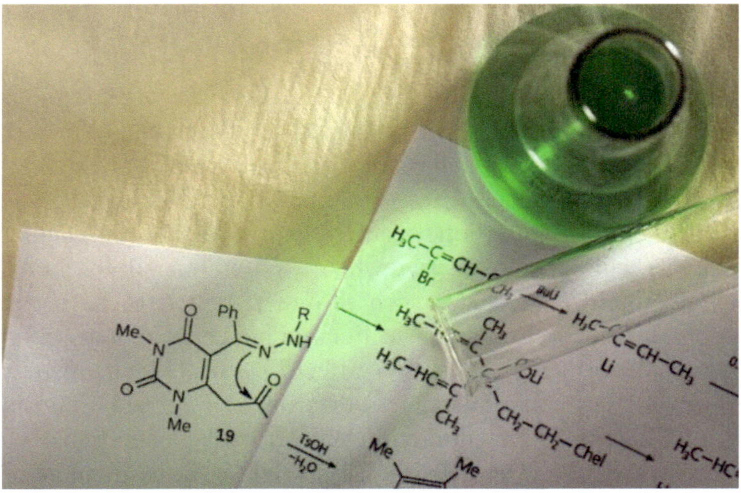

Figure 1.3. *Fundamental principles of photochemistry.*

Source: https://microbiozindia.com/wp-content/uploads/2023/11/photochemistry.jpg.

The efficiency of such routes may be regulated by environmental factors such as temperature and the presence of other chemical entities. Such processes are vital in applications such as atmospheric chemistry, where reactions initiated by sunlight contribute to air quality and climate. The environmental interaction of photochemistry points towards performing ample research to quantify the interactions as they play a crucial role in natural and anthropogenic processes. In the context of a rapidly evolving environment, application of photochemistry principles has grand implications on the understanding of pollution and preventing its effects. For instance, photochemical phenomena over snow cover are processes of intricate interactions between atmospheric chemistry, trace gases, and biogeochemical cycles. Through high-level models, researchers have demonstrated that light-driven chemical transformations play a significant role in controlling the atmosphere's oxidation potential, with snowpack chemistry affecting concentrations of reactive species such as hydroxyl radicals. These findings demonstrate the interlinking of photochemistry and climate dynamics such that knowledge of the underlying mechanisms can inform the crafting of solutions for addressing environmental issues. Conclusions drawn through research conducted via photochemical methods, including computational modeling, are core to developing solutions for enhancing environmental sustainability as well as limiting negative impacts (Assadi et al., 2012; Dibb et al., 2012) (Figure 1.3).

1.1.1. Light Absorption and Energy Transfer

Energy transfer and absorption of light are at the core of the understanding of photochemistry and its implications on the environment. When chemicals and light interact, processes such as absorption lead to the excitation of the electrons, which may cause a series of chemical reactions in response. In the environment, chemical absorption spectra influence whether the chemicals such as air pollutants will become reactive when exposed to sunlight. For example, certain neonicotinoid insecticides such as nitenpyram have been reported to possess decreased absorption cross sections in aqueous solvents compared to non-aqueous solvents. This reduction affects their photochemistry, making their lifetimes shorter in natural water bodies since calculated lifetimes of only 5 minutes were obtained by employing standard solar conditions. Therefore, light absorption properties' knowledge is essential for the assessment of the fate of such pollutants in aquatic systems and thus the relevance of photochemical processes to environmental chemistry (Ezell et al., 2019). The role of light absorption extends far beyond the case of individual pollutants and impacts general ecological processes as well. Oceanic CDOM is a light absorber that can significantly influence the photochemical kinetics of these systems. Photolysis of CDOM yields a number of dissolved gases, including carbon monoxide, through the absorption of light and energy transfer. Modeling simulations have indicated that photochemistry-mediated photolysis of CDOM is a significant source of carbon monoxide in the open ocean, and production rates are extremely variable according to environmental conditions. This production is significant for untangling carbon cycling in marine systems since it connects photochemistry to large-scale biogeochemical cycles. Consequently, light absorption implications for CDOM extend beyond pollution, affirming its pivotal role in marine chemistry (Law et al., 2006). Furthermore, processes of energy transfer through the influence of light absorption identify the complexities involved in photochemical reaction. Compounds can absorb light and transfer energy through various channels such that occurrence of various photoproducts depending on the given conditions is feasible. For instance, the adsorption behavior of certain pollutants would lead to the formation of stable secondary metabolites that may have longer life in the environment. This expresses the interconnectedness of photochemistry wherein not just the changes are promoted by light energy but also the durability as well as toxicity of chemical species. Through the study of such interactions, researchers are able to clarify light absorption's role in environmental quality and guide policy and remediation actions important to ecosystem health. Clarification of these processes is crucial in the prevention of climate change and pollution because they cannot be decoupled from mechanisms of energy transfer within photochemical systems (Ezell et al., 2019; Law et al., 2006).

1.1.2. Photochemical Reactions and Mechanisms

Photochemical reactions play a pivotal role in numerous natural and man-made processes with the utilization of light energy to effect chemical transformations. Photochemical reactions are pivotal in a number of environmental processes, particularly pollutant degradation and the generation of clean energy. For instance, photocatalysis, a process in which light is employed to stimulate a catalyst, has been referred to as a rising method of environmental cleaning. Titanium dioxide (TiO_2) is among the most widely studied photocatalysts, and it is also highly known for efficiently harnessing solar energy. Its applications in water purification and the degradation of organic contaminants demonstrate the power of photochemical reactions in answering pressing environmental concerns. Improvements to enhance the performance of TiO_2 photocatalysts, such as charge separation and widening the light absorption to the visible spectrum, are needed to optimize their efficiency in real-world applications, thus re-establishing the role of photochemical reactions in sustainable environments (Tee et al., 2024). Photochemical reaction processes are complex and primarily dictated by the interaction between photons and molecular species. After light absorption by a molecule, it can excite the electrons to upper energy levels and lead to the production of reactive species such as free radicals. These reactive species, in turn, can trigger additional chemical reactions, including degradation of other pollutants. All these processes need to be understood in order to develop effective photocatalytic systems. Current studies have explained how certain structural modifications to photocatalysts influence their photochemical performance and consequently efficiency and selectivity of target reactions. For example, the refinement in the nanostructuring of TiO_2 illuminates the influence that nanoscale modifications exert on the dynamics of the charge carriers, which is significant for the enhancement of photocatalytic performance overall (Tee et al., 2024). As further research continues, knowledge of these processes will become vital to maximizing photochemical techniques for environmental applications. Photochemical reaction analysis also extends to new avenues such as energy conversion and storage. Photochemical systems can potentially harness solar energy and store it as chemical energy, which is central to mitigating climate change and reducing the reliance on fossil fuels. One application that is especially significant is photochemical conversion of carbon dioxide to useful hydrocarbons, a process that imitates photosynthesis and potentially creates a sustainable carbon cycle. Breakthroughs in light-matter interactions have also brought new opportunities for optimizing photochemical reactions in controlled systems, such as optical cavities. Recent studies of interactions between light intensity and photochemical reaction rates highlight the importance of precise conditions for achieving desired product yields. Through an understanding of these phenomena, scientists can devise new approaches to enhancing the efficiency of photochemical reactions, thus enabling environmentally friendly technologies and energy solutions (Figure 1.4).

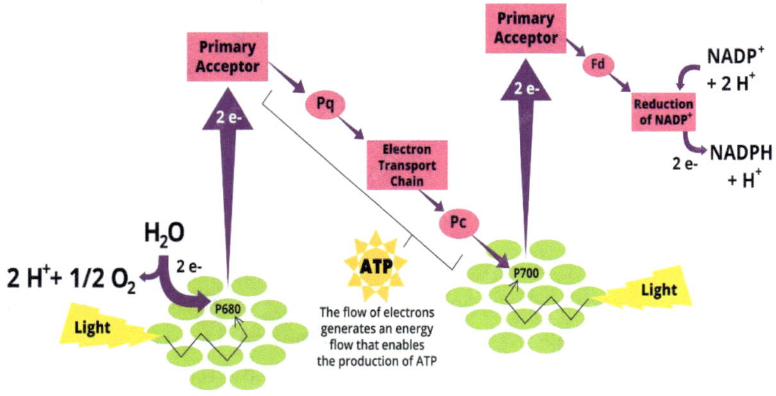

Figure 1.4. *Photochemical reactions and mechanisms.*

Source: https://img.freepik.com/premium-vector/photochemical-phase-photosynthesis-biological-vector-illustration-diagram-scheme_667988-41.jpg?w=2000.

1.1.3. Role of quantum mechanics in photochemistry

Quantum mechanics is included in the fundamental description of light-matter interactions in photochemistry, particularly in photon absorption and subsequent energy transfer. At the atomic scale, quantum mechanics provides the underlying principles that govern the distribution of electrons and energy in molecules. In the case of an interaction between a photon and a molecule, a photon is able to excite the electrons to higher levels of energy that lead to other chemical reactions. It is essential to understand these excitations as they determine the pathways of energy transfer and transformation in most photochemical reactions, which are inextricably tied to environmental processes, such as photosynthesis by plants and degradation of pollutants.

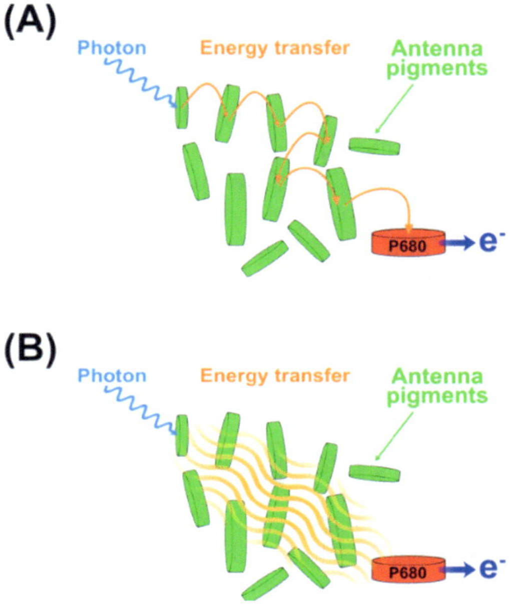

Figure 1.5. *Role of quantum mechanics in photochemistry.*

Source: https://external-content.duckduckgo.com/iu/?u=https%3A%2F%2Fd3i71xaburhd42.cloudfront.net%2F51098a9fbb9f2b50c7f17e6646a7ce3e9b9aa92e%2F3-Figure1-1.png&f=1&nofb=1&ipt=f8d5792f8172ddcba22756ef648f95077784d8d44b875bdc63b40c65a6079d73.

The role of these quantum mechanics principles is not only in their abstract theoretical significance but also provides a basis for the enhancement of photochemical technologies designed to address environmental issues. The molecular orbital theory that stems from quantum mechanical theory also describes the photochemical reactions essential to the environmental sciences. Molecular orbitals define the electron probability distributions in a molecule and its reactivity. For instance, the stability of different polymorphs of titanium dioxide (TiO_2) such as anatase and rutile can be investigated through quantum mechanical computations. Recent ab-initio density functional theory (DFT) computations revealed that some cation dopants are able to favorably stabilize the anatase phase over rutile, modulating the materials efficiency in photocatalysis applications (Assadi et al., 2012). Not only do such findings contribute to the design of more efficient photochemical materials but also identify the role that quantum mechanical principles play in the design of chemical properties in environmental remediation approaches. Also, quantum mechanics is based on knowledge regarding the manipulation of photochemical reactions in order to get them optimally work effectively. Quantum system optimal control theory shows that with the knowledge of quantum states and their interaction with the outside world, one could manipulate these parameters to maximize desirable chemical reactions (Pechen et al., 2010). This ability to adjust responses is crucial in the development of advanced photochemical methods for clean energy technologies, such as solar energy conversion and green compound production. With heightened searching for new solutions to overwhelming environmental challenges, the impact of quantum mechanics in facilitating photochemical application will become ever more central, creating a connection between fundamental science and efficient environmental solutions (Figure 1.5).

1.2. TYPES OF PHOTOCHEMICAL REACTIONS

Study of photochemical reactions plays a significant role in the understanding of diverse environmental events, including those related to air quality and climate change. Among the significant types of photochemical reactions are the reactions involving formation and destruction of reactive intermediates. For instance, in urban areas, nitrogen oxides (NO_x) and volatile organic compounds (VOCs) form complex reactions that lead to the generation of secondary pollutants such as peroxyacetyl nitrate (PAN), a photochemical smog constituent. The role of VOCs is especially significant; certain compounds, such as isoprene and other human-made hydrocarbons, serve as major precursors in PAN formation, which further affects the overall air quality. Interpretation of these answers using mechanisms like the Master Chemical Mechanism (MCM) enhances our comprehension of how single species contribute to photochemical pollution in urban cities and provides hints for mitigation strategies (Blake et al., 2014). Another very significant group of photochemical processes involves formation and loss of ozone (O_3), a significant atmospheric chemistry participant. Ozone is generated through complex photochemical reactions with the principal involvement of UV light interacting with oxygen molecules in the presence of VOCs and NO_x. Astonishingly enough, ozone distribution is extremely diverse across geographical locations and altitudes and exhibits a delicate equilibrium between formation and destruction. For example, in western North Pacific studies, it is stressed that ozone destruction tends to dominate its formation at altitudes lower than 6 km, while the opposite applies at higher heights (Anderson et al., 1996). Such findings highlight the importance of regional atmospheric dynamics and differential precursor emissions to influence trends in ozone and thereby focus on the varied implications of these reactions on the regional and global scales.

In addition, atmospheric state interactions can regulate photochemical reaction efficiency and outcomes. Deep convection and high-altitude transport are some of the processes that can amplify greatly photochemical processes. For instance, continental boundary layer high-altitude NO_x has been shown to greatly influence western Pacific region ozone production. The deposit of NO_x at high altitudes, where it is more likely to persist, can lead to spectacular enhancement of photochemical ozone and thus to air quality over long distances. Seasonal westerly flow extended this effect even further, demonstrating the impact of transport mechanisms on regional photochemical landscapes. Thus, knowledge of the types of photochemical reactions not only illuminates the complexity of atmospheric chemistry but also highlights the need for initiating comprehensive modeling in order to predict and regulate air quality and environmental health (Anderson et al., 1996) (Figure 1.6).

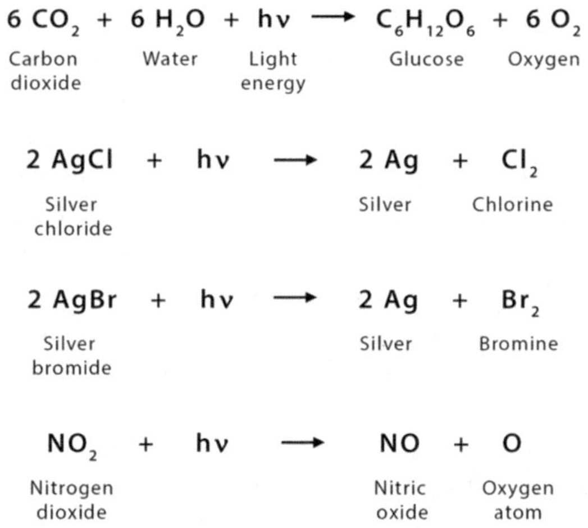

Figure 1.6. *Types of photochemical reactions.*

Source: https://www.chemistrylearner.com/wp-content/uploads/2020/09/Photo-chemical-Reaction-Examples.jpg.

1.2.1. Photosynthesis and Its Significance

Photosynthesis is a fundamental biological process that sustains life on our earth by the energy from light being converted into chemical energy. Through this intricate process, plants, algae, and some bacteria harness the energy of sunlight to synthesize glucose from carbon dioxide and water, yielding oxygen as a byproduct. This oxygen is crucial to the survival of aerobic organisms, rendering photosynthesis an indispensable contributor to maintaining Earth's atmosphere. In addition, the carbohydrates produced in photosynthesis not only provide the energy for plants but are also the basis for complex food webs that support a wide range of terrestrial as well as aquatic species. Therefore, it is essential to understand the dynamics of photosynthesis in order to realize the complex interactions that mark ecological stability and productivity. Photosynthesis's significance extends beyond mere biological process; it has strong links with environmental health and climate balance. In removing carbon dioxide from the atmosphere—one of the primary greenhouse gases—plants counteract climate change and help maintain the balance of the atmosphere. In addition, photosynthesis is subject to many environmental factors, including intensity of light, temperature, and the presence of water. The interplay of these factors can influence not only plant growth and production but even larger ecosystem processes. For instance, the alteration of photosynthetic rates of leading vegetation species during fluctuating climatic conditions can trigger alterations in habitat availability for different fauna, ultimately affecting biodiversity and ecosystem resilience. Understanding the implications of photosynthesis within an environmental framework is therefore important to enabling efficient conservation and resource management practices. Recent research emphasizes the interconnectedness of photosynthesis and broader ecological and environmental concerns. Studies have examined the effects of climate change on the photosynthetic performance of plants, with multiple feedback mechanisms that can exacerbate global warming (Etter et al., 2020). For instance, water scarcity due to drought results in alterations in the availability of water, hence influencing non-flow frequency in freshwater ecosystems, contributing to autotrophic activity. Furthermore, experimental analysis of leaf functional traits provides finicky insights into photosynthetic response variability between cohorts of plants across different climatic conditions (Puglielli et al., 2018). Such observations reiterate the importance of sustained research to disintegrate photosynthetic complexity through different environmental contexts. In relation, such observations not only elucidate the significance of photosynthesis but also highlight its central significance in ensuring ecological balance in relation to anthropogenic changes.

1.2.2. Photodegradation of Pollutants

Photodegradation of the pollutants represents a key field in environmental pollution removal, particularly in industrial and agro-waste. The application of photocatalytic materials, e.g., titanium dioxide (TiO_2), has proven efficient in the degradation of some organic pollutants under ultraviolet or visible light irradiation. Research indicates that the structural features of TiO_2 can be engineered by employing processes including sol-gel synthesis and hydrothermal treatment for enhancing its photocatalytic activity. The treatments produce improved surface characteristics and catalytic activity, which are necessary for the photodegradation of harmful substances like 2,6-dichlorindophenol.

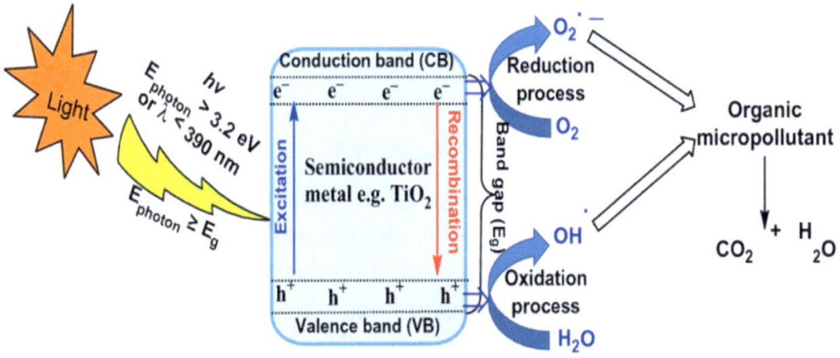

Figure 1.7. *Photodegradation of pollutants.*

Source: https://www.researchgate.net/publication/361828603/figure/fig4/AS:1143 1281154071974@1682647849564/Mechanism-of-photodegradation-of-organic-pollutants-Srikanth-et-al-2017.png.

Through the improvement of parameters such as TiO_2 loading and synthesis conditions, researchers can enhance the performance of photocatalytic systems in application, to promote cleaner air and reduced toxicity from pollutants (Královà et al., 2012). The photodegradation activity is not restricted to industrial contaminants but includes pharmaceutical contaminants, which are difficult due to being persistent and environmentally active. Quite intriguingly, antibiotics such as Ceftriaxone and Cefotaxime have been shown to remain in water bodies, and this is of concern to their influence on microbial communities. Studies indicate that photoirradiation of these antibiotics in the presence of photosensitizers such as riboflavin causes their breakdown by various reactive oxygen species mechanisms. These processes not only diminish the availability of these compounds in the environment but also lower their bactericidal activity against pathogens. This represents an important interface of photochemistry and environmental protection, as knowledge of the kinetics and mechanisms of such reactions can be applied to designing strategies for the containment of antibiotic pollution in water systems (Biasutti et al., 2014). Essentially, advancing knowledge on mechanisms of photodegradation can provide critical contributions to designing sustainable wastewater treatment methods. The addition of photocatalysts to treatment processes has the capability to reduce the expenses and energy requirements associated with traditional treatments, allowing remediation of complex waste streams to be more practical. Furthermore, further studies on the effectiveness of various photocatalysts under diverse light sources will enhance the creation of effective applications for environmental remediation. This larger vision places the ecological importance of photochemical processes in detoxification of pollutants, supporting an active response to ecosystem health, and offering cleaner water resources for the future (Králová et al., 2012) (Figure 1.7).

1.2.3. Photopolymerization Processes

Photopolymerization processes are one of the most significant photochemistry breakthroughs, primarily due to their capacity to employ light as an instrument for initiating polymerization. Unlike traditional thermal polymerization techniques that are likely to include high amounts of heat, photopolymerization speeds and effectively creates polymers with less energy levels. This change also enhances energy efficiency and is consistent with the principles of sustainability and is, thus, more relevant in the context of green chemistry. Photoinitiators enable the triggering of polymerization reactions such that it is simple to control the polymers properties, its molecular weight and percentage cross-linking. This regulation enables one to design tailored materials for a specific application, and therefore extend their application in coatings, adhesives, and biomedical devices, and thus demonstrate the far-reaching influence of photopolymerization on contemporary material science. Of additional significance is that photopolymerization has an influence not only on industrial application but also on the environment. Light-driven processes not only reduce energy consumption but also in many cases eliminate the use of volatile organic compounds (VOCs) that facilitate atmospheric pollution. In this sense, photopolymerization processes align with global attempts to minimize environmental harm associated with chemical manufacture and product formulation. The union of photocontrolled living radical and cationic polymerizations is particularly noteworthy, since these novel processes not only permit the generation of novel materials but the recycling and reutilization of existing ones, hence a closed-loop process for material use. Furthermore, such innovations reflect the necessity of developing ecologically friendly processes in light of growing environmental regulation and heightened public sensitivity towards sustainability. In the field of biomedicine, photopolymerization-based light-sensitive materials are being used as breakthrough solutions for noninvasive medical interventions. With the inherent features of photopolymerizable biomaterials, researchers can create devices that respond to light stimulation, enhancing their performance in drug delivery and tissue engineering. Light-sensitive systems offer precise control of drug release, a contrast with the ongoing trend towards individualized medicine. Additionally, as reported in current literature, the multifunctionality of materials such as carbon-based nanomaterials and gold nanoparticles demonstrates the extent of photopolymerization in designing next-generation biomedical devices (Lee et al., 2020). Not only do these advances address actual medical issues but also advance the role of photopolymerization in the design of environmentally friendly and sustainable health technology, proving the interconnection between photochemistry and environmental factors.

1.3. PHOTOCHEMISTRY IN THE ATMOSPHERE

The interaction between photochemistry and atmospheric processes is central to the determination of the Earth's atmosphere composition and dynamics. Photochemical reactions initiated by solar radiation-stimulated chemistry cause alterations of atmospheric constituents, particularly nitrogen oxides such as nitric oxide (NO) and nitrogen dioxide (NO_2). These species take part in complex photochemical cycles, which modulate air quality and climate. Recent studies have highlighted the significance of these signals in regions like the Antarctic plateau, where measurements from the air detected previously unknown levels of NO far beyond South Pole's immediate environment. These findings showcase the potential of photochemical processes to change atmospheric content and suggest that these interactions are more complex than previously thought, therefore reorganizing scientists' perception of polar chemical dynamics (Arimoto et al., 2008). Further, knowledge of the contribution made by photochemistry in atmospheric processes is crucial for studying the implications on environmental systems and the origin of life. Specifically, dissolved nitrogen oxide anions such as nitrate and nitrite have been identified as key players in the chemical pathways that were responsible for the emergence of life on early Earth. Current models have proposed the presence of these compounds in prebiotic oceans, representing a steady state of NO concentrations which are conducive to early biological reactions. However, mounting evidence indicates that photolytic destruction of these molecules, combined with reaction with reduced iron, could very likely cut their estimated concentrations significantly. The recognition of these precipitates a re-examination of the conditions conducive to prebiotic chemistry and leads to the necessity to further study NO dynamics in ultra-viable environments (Babbin et al., 2019). In the case of modern environmental challenges, photochemical implications extend far beyond to air pollution and climate change. The majority of nitrogen oxides result from combustion reactions and are responsible for the formation of ground-level ozone, a constituent of smog. Photochemical cycles of such gases can produce adverse health effects and environmental degradation. Understanding these photochemical processes not only helps resolve current environmental problems but also sheds light on potential feedback loops affecting climate regimes. In addition, through a study of the photochemical interactions occurring in different environments, such as the unique cases of the Antarctic plateau, scientists can develop improved means of mitigating the adverse effects of air pollution and ensuring atmospheric stability for future generations (Arimoto et al., 2008).

1.3.1. Role of Sunlight in Atmospheric Chemistry

Sunlight is an integral part of atmospheric chemistry, initiating most photochemical reactions responsible for determining the air quality and composition. Upon penetration of the Earth's atmosphere, solar radiation interacts with other atmospheric chemicals like gases, aerosols, and water vapor, inducing complex chemical transformations. Among the primary effects of such interactions is the formation of secondary pollutants such as ozone and photochemical smog, which has adverse effects on human health as well as the environment. The photochemical process is strongest in urban areas where the concentration of pollutants is high. Because sunlight powers these reactions, it highlights the intrinsic connection between solar radiation and atmospheric photochemical reactions and the necessity for understanding how changes in sunlight exposure, like those caused by climate change, might add to atmospheric chemistry. A key characteristic of photochemistry is the production of reactive oxygen species, like singlet oxygen, which is a product of sunlight-powered reactions in the atmosphere. Singlet oxygen is more energetic than in its ground state and, therefore, it is extremely reactive with other pollutants. Its reactivity will be responsible for forming secondary pollutants that are responsible for air quality degradation as well as respiratory disease. The role of sunlight in singlet oxygen generation has been the subject of significant interest in recent atmospheric chemical research, and emphasis has been placed on how it may be involved in smog formation and interact with other atmospheric chemicals. As Pitts illustrates, an understanding of the behavior of this reactive species is required to properly assess the impact of sunlight on atmospheric chemistry and the health implications of exposure to urban air pollution (Gross, 1969). Besides, sunlight's role extends to reactive species; it regulates the photolytic breakdown of contaminants. Ultraviolet light, for example, facilitates the decomposition of many organic substances, including volatile organic compounds (VOCs) and nitrogen oxides, which are reactants for ozone formation. The efficacy of the photolytic processes has the potential to impact the atmospheric makeup and the persistence of contaminants considerably. Additionally, the interaction between sunlight and aerosols in the atmosphere could be accountable for climate-forcing effects such as altering the Earth's radiation balance. In the context of worldwide concern regarding climate change and air quality, it is key to determine the central role of sunlight in atmospheric chemistry in finding effective environmental policies that would regulate pollution and achieve the world implications of atmospheric reactions on climate systems (García-Orenes, 2021) (Figure 1.8).

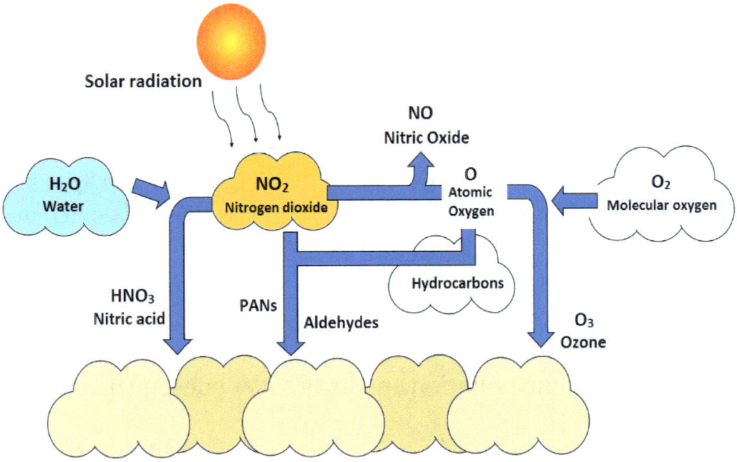

Figure 1.8. *Role of sunlight in atmospheric chemistry.*

Source: https://energyeducation.ca/wiki/images/b/b6/Photochemsmog.png.

1.3.2. Formation of Ozone and Its Implications

Tropospheric formation of ozone, especially in cities, is a secondary product of complex photochemical transformations of precursor pollutants like volatile organic compounds (VOCs) and nitrogen oxides (NO_x). When exposed to solar radiation, the compounds react in a sequence that results in the formation of ozone, which is a prominent air quality indicator. While ozone is important in the stratosphere to protect the earth from harmful ultraviolet radiation, its presence at ground level is a health risk to the environment and people.

Increased surface ozone concentrations, particularly in cities, are likely to instigate respiratory issues, exacerbate asthma, and lower crop yields, hence emphasizing the two-faced role of ozone as both a good and an evil in the atmosphere. The effects of increasing concentrations of tropospheric ozone are now evident in most areas, especially in the newly industrialized countries.

Recent studies have revealed a foreboding trend: even with emissions controls, the city air is filled with high levels of ozone. As noted in China studies, the increase in surface ozone concentration noticed is not only anthropogenically caused, but also by the changing meteorological conditions such as temperature, humidity, and wind patterns that have the ability to significantly enhance the formation of ozone (Liu et al., 2020). Thus, effective management interventions are to have a tacit understanding of emission control and meteorological impact because the latter can influence the effectiveness of intervention measures intended to mitigate urban air pollution and maintain public health. Under the conditions of global environmental change, there is a need to improve the predictability in tropospheric ozone concentrations to generate effective policy. Model experiments, such as those in the Tropospheric Ozone Assessment Report (TOAR), emphasize global atmospheric chemistry models' necessity to use in assessing ozone's distribution and variability (Paul et al., 2018). The models, apart from synthesizing observational data, allow scientists to understand the long-term impacts of ozone on climate change and air quality controls. A discussion of the differences observed between model predictions and real data will help to put into perspective the necessity for continued improvement in atmospheric models, which will ultimately allow more effective reaction to challenges posed by ozone pollution and its broader consequences (Figure 1.9).

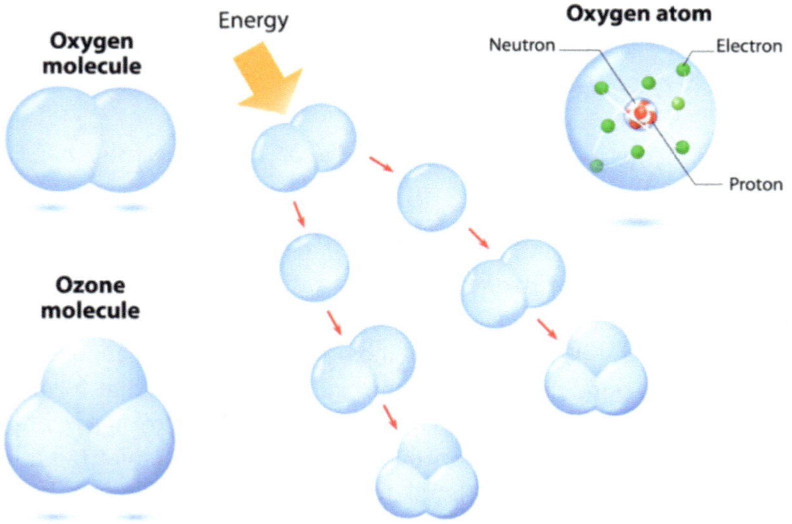

Figure 1.9. *Formation of ozone and its implications.*

Source: https://ozotech.com/wp-content/uploads/2022/01/Formation-of-Ozone-1024x724.jpg.

1.3.3. Impact of Photochemical Smog on Air Quality

The photochemical smog event significantly accelerates air quality deterioration, particularly in urban and industrial regions. Photochemical smog occurs when sunlight interacts with emissions, primarily volatile organic compounds (VOCs) and nitrogen oxides (NO_x), to create ground-level ozone and secondary pollutants. The composition of harmful chemicals created can have catastrophic impacts on human health, aggravating respiratory diseases such as asthma and bronchitis.

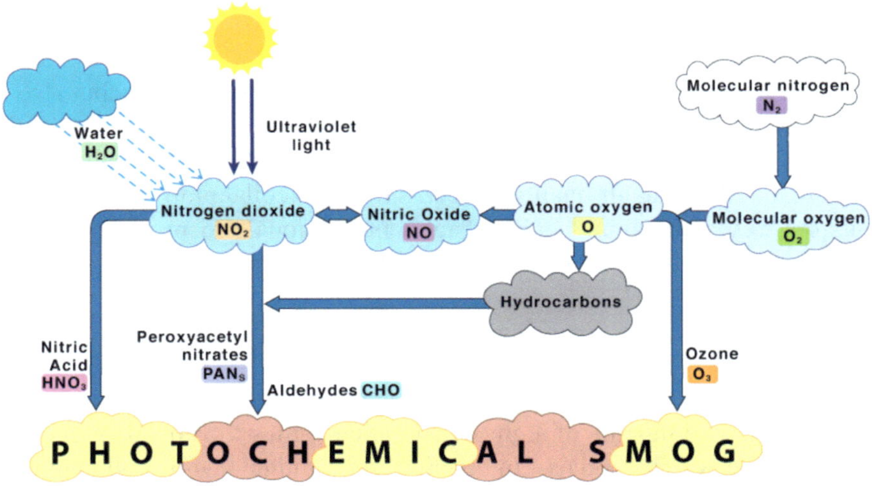

Figure 1.10. *Impact of photochemical smog on air quality.*

Source: https://iasgoogle.com/admin/common_uploads/blobid1686290215403.jpg.

In regions like northern China, severe haze events due to photochemical processes highlight the complexity of air quality management where emission sources and meteorological conditions interact to create hazardous conditions (An et al., 2019). It is thus required that an understanding of these interactions be incorporated in formulating effective regulatory policies as well as raising public awareness of the severity and implications of air pollution. Moreover, photochemical smog is closely associated with biomass burning, a process through which a large quantity of particulate matter and gas pollutants is released into the atmosphere. The emissions from biomass burning impact local and regional air quality concerns and range to the wide-ranging impacts on public health and climate. China itself has specifically recognized the significance of biomass as an air pollution source, and extensive research has been done on its effects (Chen et al., 2016). This points toward the need for systematic reviews that compile results of various studies to create an overall picture regarding the impacts of biomass burning. By addressing both anthropogenic and natural processes of photochemical smog generation, policymakers can create targeted strategies to reduce air pollution and protect the environment. Ultimately, photochemical smog's impact on air quality underscores the importance of multi-disciplinary approaches to environmental management. Integrating evidence from atmospheric chemistry studies and air quality measurements results in findings beneficial to designing better regulations and public policy programs. Understanding the connections among emissions, atmospheric processes, and weather can lead to improved forecast of the smog occurrence. Information of this nature is essential not only for the sake of improving air quality but also for improving public health as well as environmental sustainability. As scientific societies continue to clarify the details of photochemical smog, it only becomes clearer that addressing air quality concerns requires collaborative effort among regulators, researchers, and public health professionals to provide a cleaner environment for future generations (Figure 1.10).

1.4. PHOTOCHEMISTRY AND WATER QUALITY

Photochemistry plays a fundamental role in managing water quality, considering that it is the determinant of light-reaction interactions with other chemical compounds in water bodies. Interaction of sunlight with pollutants can transform them, either enhancing their toxicity or triggering their breakdown into innocuous compounds. For instance, photochemical processes in surface waters involve the degradation of organic pollutants such as pesticides and drugs beginning through photolysis. This is critical in maintaining the ecological balance of freshwater environments, particularly in the context of the continuous introduction of diversified pollutants through anthropogenic activities. This understanding of photochemical processes not only becomes imperative for water quality monitoring but also for the achievement of beneficial remediation processes involving natural sunlight to enhance pollution reduction in aquatic environments towards achieving goals associated with sustainable water resource management. In addition, photochemical influence is highly enhanced in temporary water streams, which would be negligible in broader freshwater ecology. Temporary water bodies, with their capacity to experience severe biogeochemical changes with respect to varied flow conditions, are important for biodiversity and ecological processes. Research indicates that alterations in flow regimes—driven by climate change—can potentially disrupt temporal dynamics of photochemical reactions, leading to adverse effects on aquatic organisms.

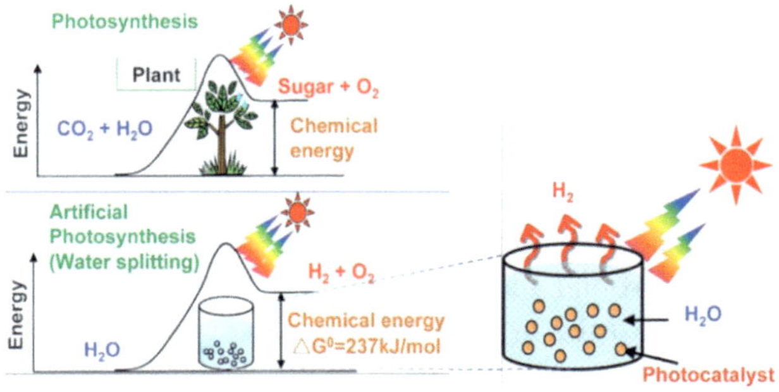

Figure 1.11. *Photochemistry and water quality.*

Source: https://www.researchgate.net/profile/Ioannis-Vamvasakis/publication/357152295/figure/fig7/AS:1102322478391306@1639825567781/Photosynthesis-by-green-plants-and-photocatalytic-water-splitting-as-artificial.png.

For example, it has been shown that variability in the frequency and duration of non-flow periods influences the growth and development of biofilms, significant components of freshwater ecosystems. The dynamics of these have critical importance for being studied, as the conservation strategies must consider the influence of photochemical processes on such vulnerable environments to render them more resistant to anthropogenic stress and climate variability (Etter et al., 2020). In cities, the situation is also complicated by the fact that photochemistry interacts with water quality in conjunction with added contaminants from various sources, including industrial pollution and vehicle emissions. For instance, experiments in the Seoul Metropolitan Area have illustrated the manner in which local photochemical activities generate dangerous ozone levels that have direct effects on water quality as well. Increased levels of ozone and its precursors have the potential to create altered chemical processes in water bodies, degrading the aquatic ecosystems. The findings suggest that an understanding of the sensitivities of ozone formation to various volatile organic compounds (VOCs), particularly in urban settings, is crucial for successful water quality management. The policymakers will have to consider such photochemical interactions while aiming to reduce air and water pollution, reflecting the interfaced nature of environmental systems and integrated solutions required in ecological conservation efforts (Ahn et al., 2020) (Figure 1.11).

1.4.1. Photochemical Processes in Aquatic Environments

Photochemical reactions are vital in aquatic communities; both the chemical composition of water bodies and the functioning of various ecosystems being influenced by them. Most essential among such reactions is that of sunlight interaction with dissolved organic matter (DOM), which gets converted in a way that can enhance or degrade water quality. Specifically, photolytic reactions are accountable for the degradation of volatile organic compounds and other xenobiotics, thereby detoxifying aquatic systems. For instance, scientific studies have established that sunlight has an influence on DOM structure and function, and thus microbial activity and nutrient cycling in water ecosystems. Such processes point to the critical role of photochemistry in regulating the ecological balance and overall health of water bodies, representing a critical component of environmental chemistry in aquatic water bodies (Parham et al., 2012). Besides, the understanding of such photochemical reactions is critical towards the solution of environmental issues such as eutrophication and the breakdown of pollutants.

Eutrophication, which often results from nutrient runoff, can lead to harmful algal blooms that destroy aquatic life and water quality. Photochemical reactions can mitigate the impacts of such blooms through the breakdown of algal toxins as well as enhancing the mineralization of organic matter. By enhancing the natural degradation processes by photolysis, we can promote healthier aquatic ecosystems. The research in various watersheds provides evidence that the temporal variations, regulated by seasonality and sunlight availability, exert a stronger control over the structure of dissolved organic carbon (DOC) than had been recognized. These observations point out that deployment of the photochemical capacities of aquatic communities can provide a feasible alternative for the improvement of water quality management (Parham et al., 2012).

Moreover, the significance of photochemical processes extends beyond mere ecological processes and encompasses more advanced environmental analysis. Studying these processes with techniques such as Life Cycle Assessment (LCA) can provide insight into the environmental impact of various agriculture practice and land use changes affecting aquatic systems. In regions such as the Waikato, for example, understanding what role photochemical processes play can inform the development of sustainable agriculture practices minimizing adverse environmental impacts with maximized productivity.

By employing both attributional as well as consequential LCA modeling approaches, researchers are able to adequately evaluate how intensification strategies in agriculture influence photochemical activity in water bodies. Both ends of raising productivity as well as preventing damage to the environment demonstrate the need for a comprehensive approach towards preserving and understanding photochemical activity in water bodies.

1.4.2. Effects of UV Radiation on Waterborne Pathogens

The ultraviolet (UV) radiation kinetics in inactivating waterborne pathogens is a rich interaction between environmental factors and microbial susceptibility. UV radiation, at the UVB wavelength band (280–320 nm), is pivotal for effective pathogen disinfection by viruses that are inherently resistant to other methods of death. Various studies describe how the effectiveness of sunlight in inactivating microorganisms is greatly reliant on the organism itself and the surrounding environment, such as natural organic matter (NOM) concentration. NOM may both absorb UV radiation and act as a sensitizer that generates reactive intermediates, and these intermediates are what cause microbial damage. Models for estimating the inactivation rates of pathogens under variable conditions emphasize the need for accurate knowledge of these processes to enhance water treatment technology (Boehm et al., 2018). Therefore, investigating the subtleties of UV radiations impact serves to accentuate its critical function in aquatic security. The public health connotations of the impacts of UV radiation on waterborne pathogens are especially urgent due to the increasing global water shortage and related illnesses. The United Nations has defined access to drinking water as a human right, yet millions lack access to safe water, which leads to the prevalence of pathogens like *E. coli* and *Cryptosporidium* even in developed nations (Byrne et al., 2014). Appropriate treatment systems, for example, the combination of solar disinfection (SODIS) with photocatalysis, have demonstrated great promise in the challenge. Photocatalyst technologies, such as titanium dioxide (TiO_2), can also be applied to enhance the efficiency of UV treatment through enhanced disinfection. This is particularly important where waterborne pathogens are extremely dangerous, and routine methods are unable to meet the challenge.

Thus, the application of photocatalytic technologies into UV disinfection processes is a serious option for public health protection and improved water quality. The future of water treatment technologies demands a global vigilance regarding UV radiations role in the environmental situation. Resolution of solar disinfection methods problems, i.e., optimization of reaction conditions and efficiency enhancement of the catalyst, holds the secret to commercialization of UV-based technologies. Recapitulating the new developments, researchers have been focusing on reactor configuration control and defining optimum catalyst loading conditions to achieve maximum disinfection efficiency. Parameters such as volume treated/total volume affect the potential to drive up rates of disinfection, reducing the time for effective pathogen inactivation by significant amounts. Their mass production and widespread implementation still remain challenging, however. Continuous research is needed in developing standard procedures and improving photocatalytic materials, thereby making the use of UV radiation as a means for waterborne pathogen management even more powerful (Boehm et al., 2018; Byrne et al., 2014).

1.4.3. Use of Photochemical Methods for Water Purification

The urgent need for effective techniques for water purification has spurred the research into photochemical technologies, specifically those based on semiconductor materials like titanium dioxide (TiO_2) in photocatalytic processes. Photocatalysis uses the light energy to activate these semiconductors to break down organic pollutants and microorganisms in water. TiO_2 is so effective because it generates reactive oxygen species when exposed to ultraviolet light that are able to break down obnoxious microorganisms and pollutants. The possibility of harnessing solar power to clean water not only provides a greener alternative in lieu of traditional techniques but also comes to the aid of the fundamental problem of access to clean drinking water globally. With advances in understanding of the principles of photocatalysis and catalyst properties, this technology has its promising potential for large-scale applications in the developed as well as developing world with water scarcity and water contamination issues. Moreover, the synergy between SODIS processes and photocatalytic devices presents a new frontier in water purification. While SODIS can destroy pathogens through the use of UV light, its effectiveness can be significantly boosted with the incorporation of photocatalysis. From research, it has been indicated that the window of mortality as far as SODIS is concerned can be significantly minimized with the utilization of semiconductor-based photocatalysts. These technologies present a two-way approach, not only using solar energy to induce disinfection but also stimulating the degradation of toxic chemical pollutants. The pair can be most effective where there are both waterborne diseases and chemical contamination, emphasizing the potential for photochemical treatments to address multidimensional challenges in water security. Bringing these technologies together then could revolutionize the manner in which clean water may be made available to all populations. Though promising advances have been achieved in photochemical water treatment, problems still face the real application and marketing of such technologies. The operation of photocatalysis is subject to numerous parameters, including photoreactor configuration, materials, and reaction conditions. Efforts have focused on enhancing parameters such as catalyst loading, illuminated volume ratios, and the position of immobilized catalysts relative to powdered catalysts. Besides, controversy regarding standardization in testing protocols and further developing TiO_2 photocatalysts' ability to capture visible light persists. Solving these issues is central to fully tapping the potential of photochemical techniques in water treatment purposes. Continued advances in reactor design and material processing could significantly enhance the efficiency and practicality of integrating solar photocatalysis systems into urban and rural settings, ultimately towards global water safety initiatives (Assadi et al., 2012; Byrne et al., 2014) (Figure 1.12).

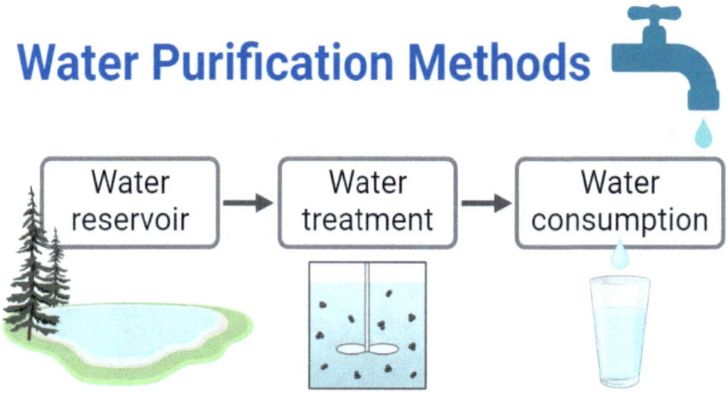

Figure 1.12. *Use of photochemical methods for water purification.*

Source: https://microbenotes.com/wp-content/uploads/2024/03/Water-Purification.jpeg.

1.5. PHOTOCHEMISTRY IN RENEWABLE ENERGY

The integration of photochemistry in the domain of renewable energy is a landmark in addressing environmental concerns. Photocatalytic reactions, which mimic natural photosynthesis, have high prospects for sustainability by transforming sunlight into chemical energy that can be harnessed. This transformation proves to be useful particularly in the production of clean fuels, such as hydrogen, from water via photocatalytic splitting of water. With the application of nanomaterials that absorb light and are more efficient, researchers are developing systems that convert light into chemical energy with higher efficiency. Such innovations not only reduce fossil fuel use but also ensure lower greenhouse gas emissions, thus mitigating climate change impacts. Therefore, the application of photochemical principles in green energy systems is a natural component of present environmental policy due to the global trend towards pollution-free sources of energy with negligible ecological effects. Photocatalysis also has implications beyond the production of energy to encompass the cleanup of the environment. Semiconducting-based photocatalysis is becoming economically viable and environmentally sound, as it is applied as a wastewater treatment process while breaking down pollutants using photochemical mechanisms. This technique adheres to the concept of zero waste discharge, pointing out the importance of sustainable operation in industrial effluent management.

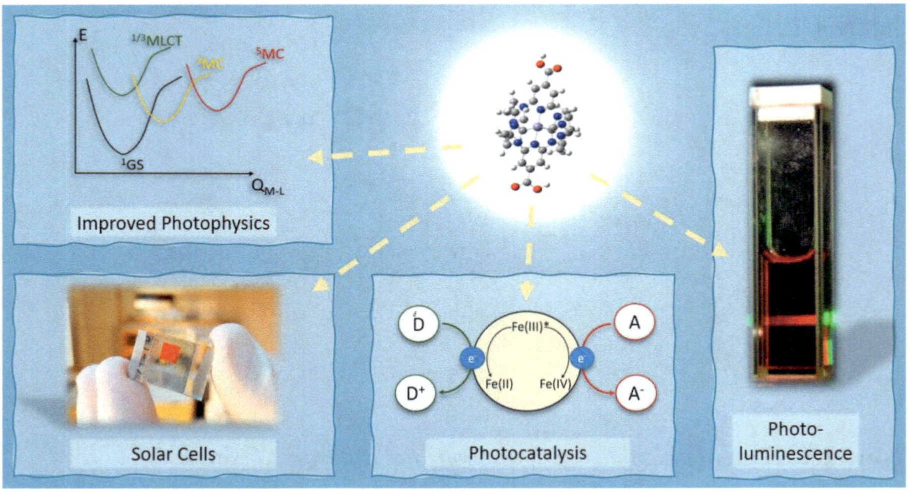

Figure 1.13. *Photochemistry in renewable energy.*

Source: https://pub.mdpi-res.com/catalysts/catalysts-10-00315/article_deploy/html/images/catalysts-10-00315-ag.png?1583847040.

The maximization of structural and compositional characteristics of nanomaterials enhances photocatalytic reaction efficiency, allowing organic pollutants to be degraded without the residue of harmful byproducts. This green synthesis process, which shuns traditional chemical methods with greener ones, demarcates the potential of photocatalysis toward designing sustainable technologies that prioritize ecological health. By stimulating the assembly of new photocatalysts, a method like this puts at the center stage the intersection of innovation and environmental science in solving pressing ecological challenges, hence paving the way to a greener future. Aside from applications in energy and environmental remediation, photochemistry also holds a lot of promise under the sweep of circular economy. With our initiatives in waste minimization and valuable material reharvesting, the prospect of photochemical reaction to synthesize useful chemicals from waste resources such as carbon dioxide resonates well within the sustainability paradigm. The integration of photochemical processes with biotechnological methods has the potential to advance in the production of bulk chemicals and plastics from renewable resources, thus diminishing our dependence on fossil fuel-based materials. Integration is necessary for the establishment of a sustainable chemical sector that comes into harmony with diverse biological feedstocks and promotes recycling of matter. With increasing focus on creative approaches that leverage the integration of photochemistry as well as biotechnology, potential in manufacturing eco-friendly materials and minimizing environmental impact is an integral aspect of contemporary environmental policy. The extensive integration emphasizes the importance of harnessing photochemistry for realizing a more sustainable and resilient future (Figure 1.13).

1.5.1. Solar Energy Conversion Technologies

The rapid advancement of solar energy conversion technologies has become a critical area of research in solving the environmental issues and providing sustainable energy. The photocatalytic materials, in particular titanium dioxide (TiO_2), have attracted widespread attention for their high solar energy conversion efficiency. Various preparation techniques, such as sol-gel and hydrothermal synthesis, have been studied to fabricate TiO_2 thin films with desirable characteristics. Research has established that the morphology and crystallinity of the layers have a direct influence on their photocatalytic activity, including applications in degradation of pollutants and purification of water. Printing of material and chemical vapor deposition (CVD) are new fabrication processes that provide controlled structures of TiO_2, which enhance their utility in a range of environmental processes (Králová et al., 2012). The use of these materials to solar energy systems is indicative of the growing intersection of photochemistry and ecological sustainability in contemporary energy. Solar energy conversion technology has the secondary purpose beyond traditional electricity generation, such as new uses for the efficient harnessing of solar power. Photovoltaic cells, for example, convert sunlight into electricity directly, while solar thermal systems use the sun's heat to produce energy. These technologies are supplemented by still more advanced materials to achieve greater efficiency and decreased costs. Nanostructured materials capable of maximizing light absorption and minimizing energy losses in conversion stages are under active investigation. Furthermore, it is essential to have scalable techniques for production to facilitate increased utilization of these technologies. As part of a comprehensive environmental policy, solar energy conversion technologies utilized directly address overarching problems such as fossil fuel dependency and greenhouse gas emissions and thus form necessary building blocks toward a sustainable future of energy (Devincenzi et al., 2025). Furthermore, the environmental implications of optimizing solar energy conversion technologies cannot be overestimated. Their ability to mitigate climate change is enormous, as they provide a cleaner way of harnessing conventional energy sources. Increasing numbers of research highlight the connection between solar power systems and reduced air pollution, pointing towards their population health benefits.

The development of good photocatalytic systems, which degrade harmful pollutants in sunlight, is a double blessing of generating clean energy while purifying contaminated places. By studying various materials and synthesis paths, researchers are able to maximize photocatalytic performance and therefore facilitate new applications that cater to world sustainability demands. Ultimately, advances in solar-to-energy conversion technologies represent a fundamental step toward a cleaner and sustainable future on our planet (Králová et al., 2012) (Figure 1.14).

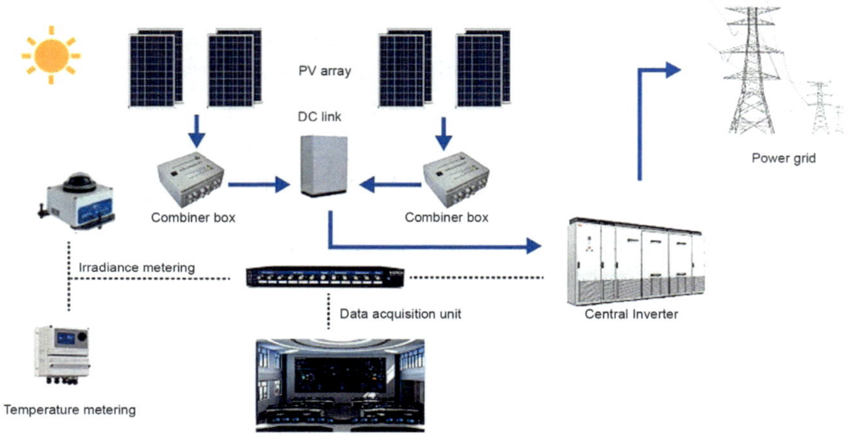

Figure 1.14. *Solar energy conversion technologies.*

Source: https://eepower.com/uploads/articles/magnetics-applications-in-solar-energys-power-conversion-fig3.jpg.

1.5.2. Photochemical Cells and Their Applications

The role of photochemical cells in the overall context of environmental sustainability cannot be overstated. Photochemical cells harvest light energy to drive chemical reactions, thereby converting solar energy into usable forms, reducing the demand on fossil fuels. By using processes such as photosynthesis-like processes, the photochemical cells produce chemical energy from sunlight that can be stored and harnessed in power generation and hydrogen gas production. Recent breakthroughs have focused attention on the development of improved and cost-effective materials that further enhance the performance of such cells. Metal complexes, for instance, have a tendency to function as vital building blocks with specific emissions that facilitate imaging of cell interactions. These developments hold the promise of a new era in energy solutions that augment the world's fight against climate change, disclosing the enormous potential of photochemical technology. Applications of photochemical cells also extend beyond energy generation to medicine and biotechnology. The advent of light-controlled release of bioactive molecules using photoactivatable metal complexes allows for the targeted therapy, which is very promising in diseases such as cancer. Furthermore, these complexes can be made to respond to light, enabling spatial and temporal control over therapeutic deliveries, enhancing their efficacy while limiting side effects. The new advancements in theoretical models also enter the scene, allowing the scientists to characterize the mechanisms of these photochemical reactions with greater accuracy and to create better systems for function applications. Hence, the merge of photochemical cells with biotherapeutic practices unveils an interdisciplinarity cutting-edge methodology, providing an integrative approach toward solving health and environmental problems through the regulation of light. Yet, the construction of photochemical cell technologies is not without its own issues. Low penetration of ultraviolet and visible light, necessary for most photochemical reactions, is a limiting factor in applications with real-world potential. Methods such as the use of caged molecules, which are activated by light to release bioactive agents, have been explored for their capability to overcome these limitations. With the integration of nanomaterials into these systems, researchers are working to find ways to elevate excitation wavelengths required for uncaging and therefore potentially enhance photochemical therapy to be more effective and efficient. This variety of approach not only showcases the use of photochemical systems in natural environments but also illustrates the great demand for continued research in this highly potential field. These advancements illustrate the maturity and interdisciplinary nature of photochemistry, opening the door to synergistic technological advances that counteract both health and environmental sustainability challenges.

1.5.3. Role of Photochemistry in Hydrogen Production

The role of photochemistry in hydrogen production increases in significance as our pursuit of clean energy becomes more and more relentless. Photochemical reactions harness the energy of sunlight to drive chemical reactions, and hydrogen production through water splitting is a good example. When semiconductor materials absorb light from the sun, charge carriers are formed that drive the oxidative and reductive processes needed for hydrogen generation. The reaction, or photocatalysis, splits water into hydrogen and oxygen and is a promising technology that leverages inexpensive solar energy. By optimizing the catalyst materials and enhancing light absorptance, researchers aim to maximize the photochemical efficiency of hydrogen production to enable a robust hydrogen economy. Thus, understanding and formulating this intricate interaction between light, water, and catalysts lies at the center of renewable energy solution development.

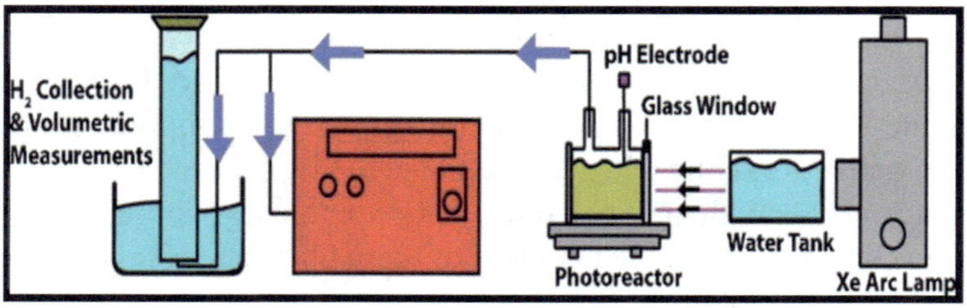

Figure 1.15. *Role of photochemistry in hydrogen production.*

Source: https://www.researchgate.net/profile/Ali-T-Raissi/publication/245367675/figure/fig4/AS:667082702217227@1536056318503/Experimental-setup-for-the-production-of-hydrogen-via-photochemical-oxidation-of-ammonium.jpg.

The technological advancement in photochemical hydrogen production is also directly related to several environmental concerns and sustainability. Electrochemical devices, based on both renewable electricity and photochemical reactions, provide an alternate approach. Such devices have the capability to utilize renewable energy resources, like solar power, to split water into hydrogen effectively. In addition, the ecological benefits of the production of hydrogen through photochemistry are notable. Hydrogen is a green fuel that, upon burning, produces only water vapor as a residue and reduces greenhouse gas emissions from fossil fuels. Advances in model and experimental approaches, as identified in recent reviews, include the importance of understanding effects like solubility, temperature, and redox conditions on efficiencies of hydrogen production. By treating such variables in an integrated way, researchers can enhance the effectiveness of hydrogen production strategies and make them consistent with global sustainability objectives. Photochemistry integration in environmental systems enhances the overall relevance of hydrogen production technology. As stressed through research on atmospheric chemistry, photochemistry is a central force behind diverse natural processes, some of which affect pollutant degradation and greenhouse gas cycles. The link between photochemical reactions and environmental factors brings out the importance of planning an ecosystem where hydrogen production will maintain ecological balance in place. Knowing that local atmospheric conditions may also be influenced by processes of hydrogen production, it is clear now that cautious modeling and assessment are necessary. This interconnectivity reflects the intrinsic union of the modes of generating energy, environmental health, and atmospheric chemistry, paving the way towards innovative solutions for climate change mitigation and energy sustainability through the pursuit of hydrogen generation through photochemistry (Figure 1.15).

1.6. ENVIRONMENTAL IMPACTS OF PHOTOCHEMICAL PROCESSES

The role of photochemical processes in the calculation of environmental dynamics cannot be overstated. Central to such processes is tropospheric ozone generation, a powerful greenhouse gas with a significant radiative forcing influence. Of special significance is the oxidation of volatile organic compounds such as isoprene to ozone generation in tropical regions. By complex processes involving nitrogen oxides, they bring about spectacular fluctuations in the concentrations of ozone, hence influencing air quality both locally and globally. Research indicates that isoprene nitrates are capable of providing a very substantial fraction of local nitrogen oxides, changing the ratio between the production and destruction of ozone. This modulation is of utmost significance, especially with anthropogenic tropical emissions being forecast to grow and thus further worsen the air quality and climate change problems. Understanding these dynamics is crucial for the formulation of efficient environmental policy interventions in mitigation of the harmful effects of photochemical processes on the atmosphere. Moreover, studies into photochemical behavior under different geographic conditions have demonstrated drastic differences in ozone production rates. A study analysis of extratropical western North Pacific found that two distinct photochemical environments exist above and below latitude 30°N. Surprisingly, the regions north of 30°N had considerable increase in total column density of ozone and exhibited different ozone formation rates, leading to net production even at lower altitudes, which is unusual for marine systems. This finding emphasizes how continental ozone precursor emissions are able to drive their influence deep into marine atmospheric environments, often creating higher concentrations of ozone as far as 2,000 kilometers from their sources. These findings reveal the complex relationship between regional emissions and their broader environmental impacts, emphasizing the importance of localized research in dictating global atmospheric conditions. These findings emphasize the need for integrated environmental management policy that considers the interdependence of photochemical reactions. With ozone concentrations still fluctuating under natural and anthropogenic stresses, implications for human health, ecosystems, and climate systems are extensive. The coupling of air quality with photochemistry presents intricate behaviors that must be investigated to facilitate coherent mitigation strategies. The ongoing research on the dynamic relationship between emissions, photochemistry, and ozone distribution presents the need for a holistic approach to environmental science. A qualitative understanding of these processes, particularly on growing anthropogenic stresses, is a vital foundation for policy formulation on dealing with these environmental concerns. The demand for action given these findings is clear; without coordinated action, the impacts of photochemical processes will likely grow more extreme in the future.

1.6.1. Contribution to Climate Change

The interface between photochemistry and global change is of extreme importance, as increased greenhouse emissions create radical modifications of atmospheric structure and energy distribution. Photochemistry, driven primarily by solar radiation, causes the breakdown of a number of pollutants and the formation of secondary pollutants that further exacerbate climate change effects. For example, while having the capacity to purify the air, photochemical smog also contributes to global warming due to the release of ozone, a highly effective greenhouse gas at lower altitude levels. Furthermore, interference with these natural photochemical processes by human interventions—fossil fuel burning, for example—attests to the urgent requirement of using sustainable methods. These contributions to world climate change can be mitigated by using photochemistry to fight back against pollutant tactics, thus emphasizing the double role of sunlight as both the causative force in chemical reactions and a pathway towards the solution of environmental crises related to atmospheric quality. By an examination of individual ecosystems, the impact of climate change on keystone species like the Himalayan cedar (*Cedrus deodara*) illustrates the wide-ranging impacts of human pressures. The alteration of the forest habitats under climate change causes a cascade of environmental effects, leading to reduced biodiversity and degraded ecosystem services. It has been established through research that Himalayan cedar populations are under threat of reducing regeneration rates due to increased human-induced pressure, such as logging and grazing (Gillani et al., 2025). These anthropogenic activities not only pose a threat to the existence of the cedars but also constitute landscape-level alterations that have the effect of enhancing the impacts of climate change. The interaction between species, environmental conditions, and human activities is a complex interaction with climate processes, and there is a need for informed conservation strategies that consider both ecological balance and climate resilience. Moreover, the evolution of ecosystems in stable as opposed to shifting environments—e.g., those in the East African Rift System—gives valuable information on the impact of climate change. The studies pertinent to this region investigate how biological and climatic development influences tectonic and consequently human development. For example, analyzing sedimentation processes through core sampling can clarify the ways in which previous climate changes had affected species adaptations, including hominin evolution (Beck et al., 2024). These findings are testaments to the complexities of climatic interactions, functioning not only currently but throughout life on Earth history. With the speeding up of climate change, the integration of learning from past and present environmental data is now critical to inform adaptive approaches and responses for mitigating future effects on ecosystems and human communities (Figure 1.16).

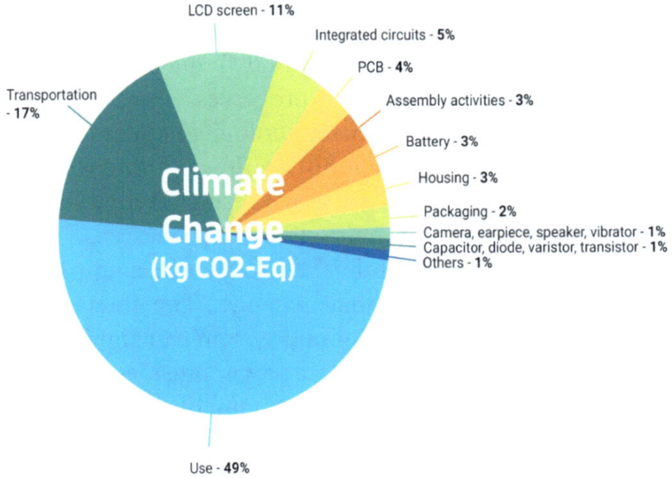

Figure 1.16. *Contribution to climate change.*

Source: https://www.fairphone.com/wp-content/uploads/2015/01/03-ClimateChange-1024x702.png.

1.6.2. Effects on Biodiversity and Ecosystems

The intricate interaction between photochemistry and biodiversity is often underestimated but plays a vital role in organizing communities. Photochemical processes, including sunlight absorption by other organisms, contribute to primary production and nutrient cycling in the environment. Chemical energy produced via sunlight through photosynthesis by plants and algae forms the foundation of most food webs. This early productivity provides directly support to a diverse range of species, pointing towards the importance of photochemistry to biodiversity. Alterations in the photochemical processes, be they human-induced in nature such as pollution and climate change, have significant impacts on the ability of ecosystems to provide for a broad range of life forms, thus threatening to overall biodiversity. Furthermore, the implications of changed hydrological regimes, which follow from climatic change, provide a better explanation of the photochemistry-ecosystem health relationship. Freshwater ecosystems that are subjected to regimes of varying flow and undergo transitions from permanent to temporary bodies of water undergo dramatic changes in biodiversity as has been demonstrated in recent studies. These changes can have cataclysmic effects on both autotrophic and heterotrophic communities within the ecosystems. For example, prolonged durations of non-flowing conditions may stress stream biofilms, which are crucial for nutrient cycling and energy transfer in freshwater ecosystems. Experiments at the Experimental Stream Facility indicate that the duration and frequency of non-flow events control biofilm recovery and ecosystem processes. Temporary water bodies, generally underrepresented, are dominant ecological agents and require more attention to exploring their response to shifting flow regimes (Etter et al., 2020). Biodiversity loss due to damaged ecosystems results in a pervasive ecological crisis, one driven by natural and anthropogenic processes. Ongoing species and habitat diminution sustains diminished resilience against environmental change, increased sensitivity to climate change. The interconnectedness of ecosystems illustrates how the health of one system can significantly influence others as species migrate or change due to new circumstances. Thus, preserving ecosystems through the implementation of sustainable practices and mitigation of photochemical and environmental impacts is essential in biodiversity conservation. With the globe facing more environmental issues, conserving the most impacted ecosystems is necessary to combat the preservation to maintain the conservation of our relied-upon biodiversity (Bussotti et al., 2016).

1.6.3. Mitigation Strategies for Photochemical Pollution

It is crucial to reduce photochemical pollution, especially in urban areas where motor vehicles are the primary source. One of the most effective methods is the implementation of strict transport emissions controls. Through the use of electric and hybrid cars, urban areas can readily put an end to VOCs, which consist of photochemical smog. Evidence from New Delhi research demonstrates the ways in which geographically localized transport and industrial emissions exacerbate air quality issues, necessitating the development of policies that incentivize clean technology adoption. Moreover, augmenting the public transport network aids in alleviating traffic congestion, thus lowering the concentration of VOCs in the urban atmosphere, which is essential in suppressing ozone formation and other photochemical pollution impacts (Tripathi et al., 2022). Apart from traffic restructuring, green area promotion within the urban landscape can serve as a natural method of limiting photochemical pollution. Urban vegetation plays a vital role in the removal of air pollutants by absorbing excessive carbon dioxide and releasing oxygen via photosynthesis. Furthermore, as quoted in recent research, plants have the ability to neutralize photochemical pollution by helping to regulate temperature and humidity, and thereby modifying the local meteorological factors that govern the formation of ozone. Through the incorporation of community gardens, urban forests, and green roofs, city planners can create sustainable ecosystems that reduce air pollution. These interventions not only address the issue of direct pollution but also enhance biodiversity and quality of life for human beings (Faqir et al., 2021). Lastly, public awareness campaigns and education are vital in reducing photochemical pollution. Engaging the public in education on air pollution sources and their health impacts can give the momentum towards greater grassroots support for environmental campaigns. Campaigns to lower private car usage, recycle, and conserve energy can have a dramatic cumulative impact on VOC emission savings. Educational programs can enlighten individuals regarding cleaner habits, such as the use of environmentally friendly products and low air pollution technologies. Educated citizens, through concerted efforts, can drive local governments toward more sustainable policies and practices that ultimately lower photochemical pollution levels (Faqir et al., 2021; Tripathi et al., 2022).

SUMMARY

In short, the intricate interactions between photochemistry and environmental processes emphasize the essential role that sunlight plays to sustain life on Earth. As the primary source of energy, sunlight stimulates photosynthesis to enable the conversion of light to chemical energy. This process not only sustains plant life but also influences larger ecological systems through complex redox reactions. According to recent studies, reactive oxygen species (ROS) generated during photosynthesis have a central role in signaling pathways that regulate energy and metabolic fluxes in chloroplasts. These mechanisms highlight the delicate equilibrium between energy generation and the activation of potential oxidative stress in photochemical systems. Understanding these dynamics is crucial to counteract environmental challenges and improve photosynthetic efficiency in applications from agriculture to renewable energy. Besides, the development from our photochemistry will contribute a lot to solving the issue of sustainable energy. A good example is the creation of photoelectrochemical (PEC) devices where solar light is employed to provide clean, carbon-free fuels through artificial photosynthesis. As outlined in recent research work, the advancement of semiconductor materials and co-catalysts plays a central role in increasing PEC process efficiency, which are fundamental in addressing the global energy crisis. Mechanisms like optimizing charge transfer mechanisms and stabilizing photoelectrodes are reflective of the need to overcome existing technological limitations. With the innovations in these areas, scientists are able to create stable systems that effectively store solar energy, thereby providing a clean alternative to fossil fuels. In essence, the use of photochemistry in our agendas to cope with climate change and environmental degradation is both timely and necessary. The application of solar energy not only assists in global sustainability goals but also accentuates the ability to harness natural phenomena for the enrichment of human society. With continued progress in the field of photochemistry, its implications on ecological health and utilization of resources will be profound. The future of studies should ensure that both the biochemical nature and technological applications of photochemistry are studied and a greater understanding of its importance in the environment is established. Addressing these opportunities and challenges will pave the way for innovations that will be able to make positive contributions towards global energy policy and custodianship of the environment, showcasing the interconnectedness of nature, science, and technology. An investigation of photochemistry demonstrates that it is a vital component in numerous branches of science, particularly for environmental studies. Electronically excited states are the central element of photochemical processes, as they are the place where photochemistry and photophysics intersect with material science. Theoretical account of such states involves highly accurate quantum chemical calculations, which, even though they are required, may become computationally intensive. The newest advancements in machine learning are now proving to have vast potential in enhancing the efficiency of such calculations and consequently enabling more intense and quicker simulations of excited-state dynamics. The application of artificial intelligence in this situation not only makes research quicker but also hints at possibilities for new uses in other areas, emphasizing the need to use excited-state behavior proficiently (Westermayr et al., 2020). The environmental significance of photochemistry cannot be overstated, particularly in the context of sustainable processes. Green chemistry, in its focus on minimizing toxic byproducts and energy use, is increasingly converging with photochemical methodology to develop more eco-friendly processes. One such example is the transition from batch manufacturing systems to continuous manufacturing systems that capture some of the most significant Green Chemistry principles in maximizing reaction efficiency and minimizing waste. Flow chemistry, one of the salient aspects of this transformation, allows for more regulated reactions, minimizing the creation of hazardous byproducts. Adding to the dream of sustainable chemical processes is the development of new separation techniques, which enable industries to adopt more eco-friendly methods. The synergy between photochemistry and Green Chemistry not only enables green objectives but also promotes economically viable production methods that are able to adapt to current ecological standards (Rogers et al., 2019). With ongoing development of photochemistry, impacts in environmental science and industry continue to be important. The union of sophisticated computational methods, including machine learning, with traditional quantum chemical descriptions is creating a more advanced understanding of photochemical reaction and their relevance to environmental systems. This progression in research design not only accelerates the exploration of excited states but enables the optimization of effective photochemical processes for environmental cleanup and renewable resource utilization. With ongoing advancements in analytical methods and strong determination to include environmental parameters in chemical protocols, the potential of photochemistry to be of benefit in preserving the environment is elevated. Therefore, understanding key issues in photochemistry highlights its significance in devising eco-friendly technologies and practices for the future.

REFERENCES

1. Ahn, J. Y., Blake, D. R., Blake, N. J., Chang, Y. S., et al. (2020). Observation-based modeling of ozone chemistry in the Seoul metropolitan area during the Korea-United States Air Quality Study (KORUS-AQ). https://core.ac.uk/download/287622411.pdf (accessed on 28th September 2025).
2. Al-Abadleh, H. A. (2015). A review on the bulk and surface chemistry of iron in atmospherically relevant systems containing humic like substances (HULIS). https://core.ac.uk/download/143689511.pdf (accessed on 28th September 2025).
3. An, Z., Huang, R. J., Zhang, R., Tie, X., Li, G., Cao, J., Zhou, W., et al. (2019). Severe haze in northern China: A synergy of anthropogenic emissions and atmospheric processes. *Proceedings of the National Academy of Sciences, 116*, 8657–8666. https://doi.org/10.1073/pnas.1900125116 (accessed on 28th September 2025).
4. Anderson, B. E., Avery, M. A., Barrick, J. D., Blake, D., et al. (2003). An assessment of western North Pacific ozone photochemistry based on springtime observations from NASA's PEM-West B (1994) and TRACE-P (2001) field studies. https://core.ac.uk/download/286728796.pdf (accessed on 28th September 2025).
5. Anderson, B., Bachmeier, A., Barrick, J., Blake, D., et al. (1996). Assessment of ozone photochemistry in the western North Pacific as inferred from PEM-West A observations during the fall 1991. https://core.ac.uk/download/287622907.pdf (accessed on 28th September 2025).
6. Anderson, B., Barrick, J., Blake, D., Bradshaw, J., et al. (1997). An assessment of ozone photochemistry in the extratropical western North Pacific: Impact of continental outflow during the late winter/early spring. https://core.ac.uk/download/287622908.pdf (accessed on 28th September 2025).
7. Arimoto, R., Blake, D., Buhr, M., Chen, G., et al. (2008). A reassessment of Antarctic plateau reactive nitrogen based on ANTCI 2003 airborne and ground-based measurements. https://core.ac.uk/download/286361411.pdf (accessed on 28th September 2025).
8. Assadi, M. H. N., Hanaor, D. A. H., Li, S., Sorrell, C. C., et al. (2012). *Ab initio* study of phase stability in doped TiO_2. http://arxiv.org/abs/1210.7555 (accessed on 28th September 2025).
9. Aydoğan, C., Yılmaz, G., Shegiwal, A., Haddleton, D. M., & Yağcı, Y. (2022). Photoinduced controlled/living polymerizations. *Angewandte Chemie International Edition, 61*. https://doi.org/10.1002/anie.202117377 (accessed on 28th September 2025).

10. Babbin, A. R., Ranjan, S., Rimmer, P. B., Sasselov, D. D., et al. (2019). Nitrogen oxide concentrations in natural waters on early Earth. http://arxiv.org/abs/1902.07740 (accessed on 28th September 2025).
11. Bandy, A., Blake, D., Chen, G., Davis, D., et al. (1999). A mass-balance/photochemical assessment of DMS sea-to-air flux as inferred from NASA GTE PEM-West A and B observations. https://core.ac.uk/download/287622774.pdf (accessed on 28th September 2025).
12. Beck, C., Berke, M. A., Feibel, C., Foerster, V., Olaka, L., Roberts, H. M., Scholz, C. A., et al. (2024). ICDP workshop on the Deep Drilling in the Turkana Basin project: Exploring the link between environmental factors and hominin evolution over the past 4 Myr. *Scientific Drilling*. https://www.semanticscholar.org/paper/e5a146890890ca333db288a178eee37dc4c14bca (accessed on 28th September 2025).
13. Biasutti, M. A., Criado, S. N., Garcia, N. A., Reynoso, M. M., et al. (2014). Riboflavin-sensitized photooxidation of Ceftriaxone and Cefotaxime: Kinetic study and effect on *Staphylococcus aureus*. https://core.ac.uk/download/159289952.pdf (accessed on 28th September 2025).
14. Blake, D. R., Chai, F., Chen, Y., Ding, A., et al. (2014). On the use of an explicit chemical mechanism to dissect peroxy acetyl nitrate formation. https://core.ac.uk/download/286361364.pdf (accessed on 28th September 2025).
15. Boehm, A., Davies-Colley, R., Dodd, M., Kohn, T., et al. (2018). Sunlight-mediated inactivation of health-relevant microorganisms in water: A review of mechanisms and modeling approaches. https://core.ac.uk/download/323080860.pdf (accessed on 28th September 2025).
16. Bussotti, F., Manes, F., Pollastrini, M., Salvatori, E., et al. (2016). Editorial to: "A challenge for the analysis of environmental stress. The measurement of chlorophyll a fluorescence (ChlF) from plants to ecosystems." https://core.ac.uk/download/54537122.pdf (accessed on 28th September 2025).
17. Byrne, A. J., Dionysiou, D. D., Dunlop, P. S. M., Ibanez, M., et al. (2014). Solar photocatalysis for water disinfection: Materials and reactor design. https://core.ac.uk/download/pdf/60775675.pdf (accessed on 28th September 2025).
18. Chen, J., Li, C., Ristovski, Z., Milić, A., Gu, Y., Islam, M. S., Wang, S., et al. (2016). A review of biomass burning: Emissions and impacts on air quality, health and climate in China. *The Science of The Total Environment, 579*, 1000–1034. https://doi.org/10.1016/j.scitotenv.2016.11.025 (accessed on 28th September 2025).

19. Chobtang, J. (2016). *Appraisal of the Environmental Sustainability of Milk Production Systems in New Zealand: A Thesis Presented in Partial Fulfilment of the Requirements for the Degree of Doctor of Philosophy in Science in Life Cycle Management at Massey University, Manawatū, New Zealand*. https://core.ac.uk/download/148644006.pdf (accessed on 28th September 2025).
20. Devincenzi, D. L., & Pleasant, L. G. (2025). The 1st Symposium on Chemical Evolution and the Origin and Evolution of Life. https://core.ac.uk/download/pdf/42852549.pdf (accessed on 28th September 2025).
21. Dibb, J. E., Huey, L. G., Lefer, B., Liao, J., et al. (2012). Modeling chemistry in and above snow at Summit, Greenland–Part 2: Impact of snowpack chemistry on the oxidation capacity of the boundary layer. https://core.ac.uk/download/72054666.pdf (accessed on 28th September 2025).
22. Etter, A., & Fischer, S. (2020). Stream biofilm response to an increasing number of non-flow periods. https://core.ac.uk/download/289121486.pdf (accessed on 28th September 2025).
23. Ezell, M. J., Finlayson-Pitts, B. J., Gerber, R. B., Ni, R., et al. (2019). Experimental and theoretical studies of the environmental sensitivity of the absorption spectra and photochemistry of Nitenpyram and analogs. https://core.ac.uk/download/287622486.pdf (accessed on 28th September 2025).
24. Faqir, Y., Ma, J., & Chai, Y. (2021). Chitosan in modern agriculture production. *Plant Soil and Environment, 67*, 679–699. https://doi.org/10.17221/332/2021-pse (accessed on 28th September 2025).
25. Ford, P. C. (2013). Photochemical delivery of nitric oxide. https://core.ac.uk/download/286028122.pdf (accessed on 28th September 2025).
26. Foyer, C. H. (2018). Reactive oxygen species, oxidative signaling and the regulation of photosynthesis. *Environmental and Experimental Botany, 154*, 134–142. https://doi.org/10.1016/j.envexpbot.2018.05.003 (accessed on 28th September 2025).
27. García-Orenes, F. (2021). Comment on soil-2021-17. https://www.semanticscholar.org/paper/4296b3d77ebf997eb40a25589810992e6a77d061 (accessed on 28th September 2025).
28. Gillani, S. W., Ahmad, M., Manzoor, M., Waheed, M., Tribsch, A., Shaheen, H., Mehmood, A. B., et al. (2025). Synergizing population structure, habitat preferences, and ecological drivers for conservation of *Cedrus deodara*. *BMC Plant Biology, 25*. https://www.semanticscholar.org/paper/cc3736a7cba91075d6dad56159bd3b7999048f7e (accessed on 28th September 2025).

29. Gozem, S., Luk, H. L., Schapiro, I., & Olivucci, M. (2017). Theory and simulation of the ultrafast double-bond isomerization of biological chromophores. *Chemical Reviews, 117*, 13502–13565. https://doi.org/10.1021/acs.chemrev.7b00177 (accessed on 28th September 2025).
30. Hassaan, M. A., El-Nemr, M. A., Elkatory, M. R., Ragab, S., Niculescu, V. C., & El Nemr, A. (2023). Principles of photocatalysts and their different applications: A review. *Topics in Current Chemistry, 381*. https://doi.org/10.1007/s41061-023-00444-7 (accessed on 28th September 2025).
31. Henze, D. K., Paulot, F., & Wennberg, P. O. (2012). Impact of the isoprene photochemical cascade on tropical ozone. https://core.ac.uk/download/4894284.pdf (accessed on 28th September 2025).
32. Jiang, C., Moniz, S. J. A., Wang, A., Zhang, T., & Tang, J. (2017). Photoelectrochemical devices for solar water splitting–materials and challenges. *Chemical Society Reviews, 46*, 4645–4660. https://doi.org/10.1039/c6cs00306k (accessed on 28th September 2025).
33. Králová, M. (2012). Self-cleaning properties of thin printed layers of titanium dioxide. https://core.ac.uk/download/30310816.pdf (accessed on 28th September 2025).
34. Law, C. S., Mopper, K., Robinson, C., Stubbins, A., et al. (2006). Open ocean carbon monoxide photo-production. https://core.ac.uk/download/2772700.pdf (accessed on 28th September 2025).
35. Lee, H. P., & Gaharwar, A. K. (2020). Light-responsive inorganic biomaterials for biomedical applications. *Advanced Science, 7*. https://doi.org/10.1002/advs.202000863 (accessed on 28th September 2025).
36. Lefohn, A. S., Malley, C. S., Smith, L., Wells, B. B., Hazucha, M. J., Simon, H., Naïk, V., et al. (2018). Tropospheric ozone assessment report: Global ozone metrics for climate change, human health, and crop/ecosystem research. *Elementa: Science of the Anthropocene, 6*. https://doi.org/10.1525/elementa.279 (accessed on 28th September 2025).
37. Li, L., Scheiger, J. M., & Levkin, P. A. (2019). Design and applications of photo responsive hydrogels. *Advanced Materials, 31*. https://doi.org/10.1002/adma.201807333 (accessed on 28th September 2025).
38. Liu, Y., & Wang, T. (2020). Worsening urban ozone pollution in China from 2013 to 2017–Part 1: The complex and varying roles of meteorology. *Atmospheric Chemistry and Physics, 20*, 6305–6321. https://doi.org/10.5194/acp-20-6305-2020 (accessed on 28th September 2025).
39. Parham, L. M. (2012). *Spatial and Temporal Variation in Degradation of Dissolved Organic Carbon on the Main Stem of the Lamprey River.*

https://core.ac.uk/download/215516333.pdf (accessed on 28th September 2025).

40. Pechen, A., Agrachev, A. A., Billingsley, P., Brif, C., Craig, N. C., Rabitz, H., Isidori, A., et al. (2010). Unified analysis of terminal-time control in classical and quantum systems. http://arxiv.org/abs/1011.1269 (accessed on 28th September 2025).

41. Puglielli, G., & Varone, L. (2018). Inherent variation of functional traits in winter and summer leaves of Mediterranean seasonal dimorphic species: Evidence of a 'within leaf cohort' spectrum. https://core.ac.uk/download/158822376.pdf (accessed on 28th September 2025).

42. Rogers, L., & Jensen, K. F. (2019). Continuous manufacturing–the green chemistry promise? *Green Chemistry, 21*, 3481–3498. https://doi.org/10.1039/c9gc00773c (accessed on 28th September 2025).

43. Sadler, P. J., & Smith, N. A. (2013). *Photoactivatable Metal Complexes: From Theory to Applications in Biotechnology and Medicine*. https://core.ac.uk/download/16663845.pdf (accessed on 28th September 2025).

44. Schwartz, T., & Hutchison, J. (2024). On the importance of experimental details: A comment on "Non-Polaritonic Effects in Cavity-Modified Photochemistry." https://www.semanticscholar.org/paper/6c78237217df243e8ef00186df6c1c458f0f4c88 (accessed on 28th September 2025).

45. Sheldon, R. A., & Brady, D. (2022). Green chemistry, biocatalysis, and the chemical industry of the future. *ChemSusChem, 15*. https://doi.org/10.1002/cssc.202102628 (accessed on 28th September 2025).

46. Tee, S., Kong, J., Koh, J. J., Teng, C., Wang, X., Wang, X., Teo, S. L., et al. (2024). Structurally and surficially activated TiO_2 nanomaterials for photochemical reactions. *Nanoscale*. https://www.semanticscholar.org/paper/fcc13e5be33da9c5665141f104b98be6af318eb7 (accessed on 28th September 2025).

47. Tripathi, N., Sahu, L. K., Wang, L., Vats, P., Soni, M., Kumar, P., Satish, R., et al. (2022). Characteristics of VOC composition at urban and suburban sites of New Delhi, India in winter. *Journal of Geophysical Research: Atmospheres, 127*. https://doi.org/10.1029/2021jd035342 (accessed on 28th September 2025).

48. Westermayr, J., & Marquetand, P. (2020). Machine learning for electronically excited states of molecules. *Chemical Reviews, 121*, 9873–9926. https://doi.org/10.1021/acs.chemrev.0c00749 (accessed on 28th September 2025).

CHAPTER 2

PRINCIPLES OF PHOTOPHYSICAL AND PHOTOCHEMICAL PROCESSES

Contents

Introduction 45
2.1. Basic Concepts of Light 50
2.2. Photophysical Processes 56
2.3. Photochemical Processes 61
2.4. Spectroscopy Techniques 66
2.5. Applications in Material Science 71
2.6. Environmental Implications 76
Summary 81
References 83

INTRODUCTION

Photophysical and photochemical reaction studies occupy the top spot in today's chemistry because such processes form the foundation of a broad spectrum of applications: solar energy conversion, photochemistry in the environment, organic synthesis, etc. Photophysical processes determine the matter-light interactions, focusing on the excited state of the molecules after absorption of the photon. These underlying processes define the efficiency of energy transfer, fluorescence, and phosphorescence, and power technology in lighting, sensors, and display. An understanding of these processes also allows one to design photochemical reactions that are more efficient, as electronic-to-vibrational energy level transition dictates the viability and outcome of chemical transformations. With the growing significance of light as a catalyst for chemical reactions, a thorough understanding of such concepts is essential in order to make the best out of them. In organic synthesis, photoredox catalysis has evolved as a pioneering method employing light as an engine for chemical reactions, generally under very mild conditions not possible by conventional thermal processes.

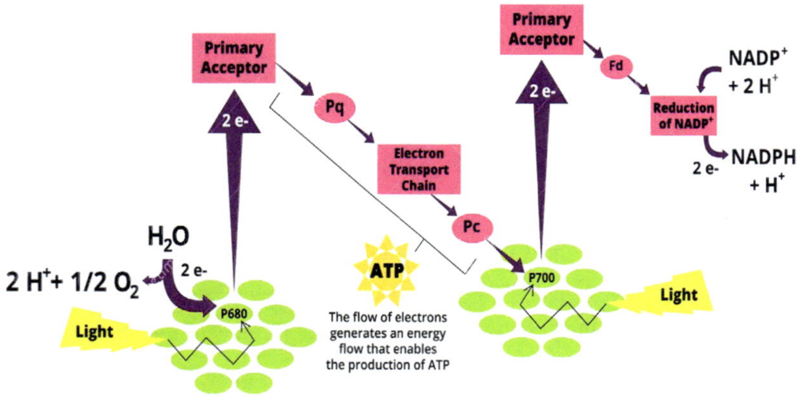

Figure 2.1. *Principles of photophysical and photochemical processes.*

Source: https://thumbs.dreamstime.com/z/photochemical-phase-photosynthesis-biological-vector-illustration-diagram-scheme-conversion-light-water-carbon-dioxide-oxygen-248161835.jpg.

Acridinium dyes, organic photocatalysts, have been highlighted for their unique electronic properties, especially their high excited-state reduction potentials. The properties enable acridinium dyes to catalyze a variety of synthetic transformations efficiently, thereby making them pillars in the evolution of sustainable chemistry. Recent improvements such as the discovery of new synthetic pathways for functionalizing acridinium cores illustrate the promise of fine-tuning their photocatalytic properties. These improvements are paving the way for transforming intricate syntheses of chemicals that otherwise need stringent conditions, enhancing the efficiency and accessibility of organic reactions (Yan et al., 2021). Despite the promising application of photochemical processes, there are still a multitude of practical challenges, not least the issues of solvent compatibility and organic photoswitch solubility. Photoswitches, as central components in many photochemical systems, are not water-soluble due to them possessing a very lipophilic nature. This renders them unavailable for applications in biological systems and requires innovative solutions to solubilization. Research on the effect of water on the photochemical activity of these molecules gives opportunities for their solubility enhancement without the need to compromise their active functions. As the demand for photochemical application in biologically pertinent environments increases, addressing these solubility issues becomes ever more important. The further development of science in this area raises the fine-tuned balance between photophysical theory and their applied reality further, even more so underlining the significance of this area in building basic science as well as technology (Volari et al., 2021) (Figure 2.1).

Definition of Photophysical Processes

An understanding of the definition of photophysical processes is important with regard to understanding the general principles of photophysical and photochemical phenomena. Photophysical processes are concerned with interactions and light conversions in materials, primarily with the absorption, emission, and scattering of electromagnetic radiation. Photophysical processes are directly linked with the electronic levels of molecules, and the absorption of light leads to electrons ascending to a higher level of energy, resulting in temporary states. For instance, on excitation, a molecule might enter the singlet state and relax through non-radiative processes or fluoresce by light emission. Reversible switching between ground and excited states is pivotal in numerous applications like photonic devices, sensors, and solar energy conversion devices. The kinetic aspects of such processes have significant roles to play in terms of determining efficiency and efficacy of photophysical systems for useful applications. Furthermore, the intricacy of photophysical processes is also revealed through their sensitivity to external parameters such as temperature, polarity of the solvent, and molecular structure. The influence of these parameters becomes highly interesting when applied to sophisticated materials such as covalent organic frameworks (COFs) with unique electronic and structural features. The understanding gained from molecular organic photochemistry guides the synthesis and application of COFs as efficient visible-light photocatalysts with an address to the crossroads of photophysical theory and application. For instance, COFs tunability in structure and porosity significantly enhances their photocatalytic activity with a focus on the importance of managing photophysical processes in material science development. This inherent relationship helps to demonstrate that the more advanced our comprehension of photophysical processes, the more we are capable of innovating in photochemistry and in material applications (Alberto et al., 2022). Aside from traditional definitions and interaction, water's effect on photophysical processes helps to bring a new direction to this field. Several organic compounds, especially molecular photoswitches, are characterized by lipophilicity, leading to insolubility and distribution challenges within aqueous solution. These water interactions can change the photochemical nature of these molecules, and therefore attempts are made to solubilize them without losing their function. Ultimately, understanding the definition and implications of photophysical processes extends beyond theoretical paradigms; it plays a direct role in real-world applications where materials interact with light. Therefore, as researchers explore photophysical processes and their interaction with various environments, they provide a window of opportunity for innovations in areas such as biochemistry and renewable energy technologies (Volari et al., 2021).

Definition of Photochemical Processes

Photochemical processes refer to the various chemical reactions that are caused or influenced by the absorption of light. Such a contact of light with matter triggers energy transfer that can lead to extensive modification of molecular structure and properties. In essence, these processes represent a wide variety of phenomena including isomerization, bond cleavage, and the formation of new molecules upon irradiation. With their knowledge of the mechanisms that operate these processes, scientists can access light to produce new products, such as solar energy technologies and advanced photonic materials. A profound comprehension of photochemical mechanisms is therefore crucial to unlocking new avenues in both experimental research and practical applications in the field of chemistry. In order to adequately describe photochemical processes and their implications, it is essential to consider the role of excited states in the course of these reactions. When a photon is absorbed, a molecule may occasionally transition from the ground state into a higher energy excited state. Excitation alters the molecular potential energy surface, which, in turn, modifies reaction paths and products. The excited state may be short-lived, facilitating instantaneous reactions, or may be quite prolonged, allowing complex dynamics to occur. The study of these excited states is significant to the understanding of processes such as photostability, energy transfer, and light-harvesting material efficiency. Therefore, the electron excitation and relaxation dynamics are essential to the explanation of photochemical reaction complexities. In addition, photochemical processes research is increasingly being combined with computational methods modeling molecular reactions upon irradiation. Computational chemistry developments today, primarily through platforms like OpenMolcas, enable researchers to investigate sophisticated reaction mechanisms and predict the outcome of photochemical reactions more accurately. These computational tools elucidate such features as conical intersections, which are key in the understanding of nonradiative processes and dissipation pathways for excited states. Through this emerging body of photochemistry, more stable applications between fields ranging from materials science to biochemistry become possible as theoretical models converge with experimental data. Lastly, the ability to simulate photochemical processes computationally is central to the stimulation of innovation within various technological fields and an increase in our understanding of light-matter interactions (Fdez et al., 2019; Volari et al., 2021) (Figure 2.2).

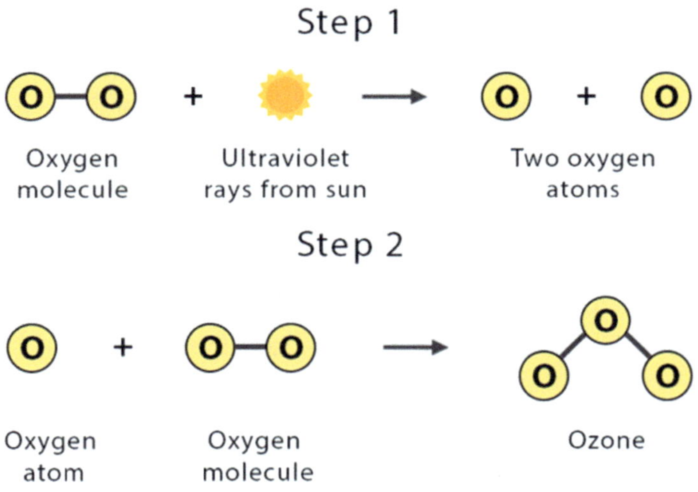

Figure 2.2. *Photochemical processes.*

Source: https://www.chemistrylearner.com/wp-content/uploads/2020/09/Ozone-Photochemical-Reaction.jpg.

Importance of Studying These Processes in Science

Photophysics and photochemistry have been a crucial field of research within the context of the mechanisms of light-matter interactions for many years. The knowledge of these processes has elucidated how energy is distributed and reorganized at molecular levels with profound implications in all fields of science. For example, what has been understood from molecular organic photochemistry has actually guided the assembly of covalent organic frameworks (COFs) as potential photocatalysts for organic conversions. This understanding allows scientists to design and engineer materials that are able to harness and store light energy. Consequently, this area of learning propels research development in fields like materials science, biochemistry, and environmental science towards the development of technologies with the ability to address major global challenges like energy conversion and sustainable chemical transformation. Furthermore, studying such interactions allows for the training of future scientists who can tackle complex problems in interdisciplinarity. Progress in smart nanoparticles proves most effectively the significance of exploring photophysical and photochemical processes in scientific research. Smart nanoparticles have the unique ability to respond to biological cues and are thus more effective drug delivery systems for treating cancer. Information regarding photophysical properties that govern light absorption and mechanisms of energy transfer can be used to enable researchers to design better nanoparticles for particular therapeutic requirements. By functionalizing these nanoparticles with tumor-specific ligands, as referenced in recent literature, researchers can improve targeting and lower off-target effects. This technology is not only optimal for treatment effect, but it is also a direction towards personalized medicine, where the treatments can be customized for patient-specific profiles. These developments underscore the drug development potential of these scientific concepts to give rise to safer and more effective treatments, directly affecting public health outcomes and quality of life for the patient (Sun et al., 2023). The research into photochemical processes also holds tremendous potential for future research and technological developments. As in COF research, the intricate structure and properties of these materials can be directly controlled through understanding excited electronic states and energy transfer processes. The flexibility of the COFs' properties, including high surface area and porous framework, allows for a wide range of applications, most prominently in photocatalysis. Investigating how such materials can be utilized to facilitate efficient energy conversion and chemical reactions is necessary to design sustainable solutions for energy. While scientists investigate the challenges and opportunities available in this field, they pave the way for innovative applications that can revolutionize the arenas of renewable energy and atmospheric clean-up, paving the way toward a brighter future (Alberto et al., 2022).

2.1. BASIC CONCEPTS OF LIGHT

An understanding of the principles of light gives a platform for understanding the principles of photophysical and photochemical processes. Light can be conceived to be both particle and wave, a principle which is a core of quantum mechanics. The wave-particle nature renders light to be able to exhibit wave behavior like diffraction and interference, as well as photoelectric phenomena that exhibit the energy of photons in the form of packets. As light travels through matter, its energy is absorbed, transmitted, or reflected, producing various outcomes depending on material properties and wavelength. This phenomenon is at work in applications ranging from plant photosynthesis to the operation of photovoltaic cells in solar-powered renewable technologies. Through interactions, the fundamental principles of light extend down to applied applications that expand our understanding of photochemical reactions. In the study of lights role in photophysical processes, it is feasible to examine how energy transfer occurs at a molecular level.

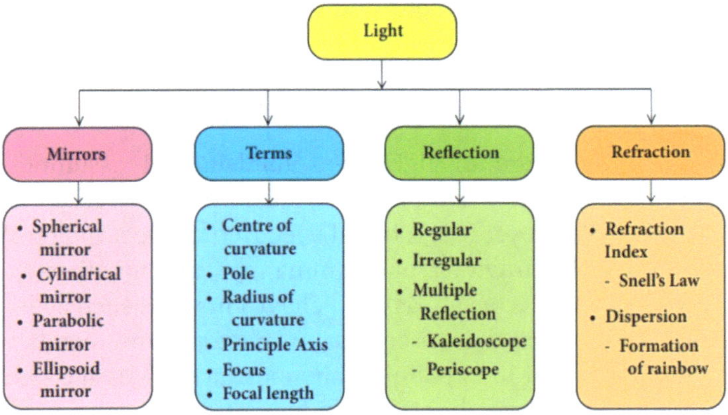

Figure 2.3. *Basic concepts of light.*

Source: https://img.brainkart.com/imagebk44/AFNmjM5.jpg.

Excitation of electrons to higher energy levels by light is something that is critical in natural and artificial systems. For instance, when molecules are excited with light, they are able to undergo the transition that leads to excited states, which then disappear via various channels, including fluorescence or non-radiative relaxation. Understanding the mechanisms of these energy transfer processes is vital in photochemical research areas where chemical changes occur due to the absorption of light. Furthermore, as in the synthesis of complex organic architecture through radicals based on nitrogen, advanced understanding of lights interaction with matter optimizes researchers abilities to conceive creative reactions. The interaction of excitation and relaxation processes indicates the importance of light in triggering dynamic chemical transformations in technology and nature (Pratley et al., 2022). With the development of technology, the application of light to generate energy has emerged as a major sector of research. Photocatalysis has been gaining prominence as a powerful method to utilize solar energy to trigger chemical reactions, transforming energy production and storage. Such substances like hybrids of graphene have been observed to be capable as photocatalysts, their potential having been demonstrated to improve the efficiency in solar fuel production (Kommula et al., 2021). Such novel photocatalysts utilize the light-absorbing characteristics that are part of their compositions, thus allowing various chemical conversions under solar irradiation. As photocatalysis has evolved, a deeper understanding of the nature of light is required to push the boundaries of solar energy application, both challenging and promising technology for sustainable development. The nature of light not only forms the basis of elementary scientific research but also guides the evolution of practical solutions to the current energy challenges (Figure 2.3).

2.1.1. Nature of light as electromagnetic radiation

The electromagnetic nature of light is a fundamental assumption in the examination of photophysical and photochemical processes. Light itself is not only visible to the naked eye; it is a broad spectrum of electromagnetic waves encompassing infrared, ultraviolet, and gamma rays, among others. This range highlights light's two facets: as a wave, it behaves in ways such as interference and diffraction, and as a particle, it consists of photons with quantized energy. This dual nature is crucial in explaining phenomena such as the photoelectric effect, which illustrates how light can induce energetic transformations in material. Through proper examination of lights various forms, we can appreciate its application in sparking chemical reactions, making way for enhanced areas like photodynamic therapy and environmental monitoring (José et al., 2021). Upon interaction of light with matter, there are different photophysical effects that occur, such as absorption, emission, and scattering. When light collides with material, photons get absorbed by the material, moving electrons to higher levels of energy. It is this absorption process that is essential to photochemical reactions, as it controls the available energy to produce changes in chemical compounds. The efficiency of absorption by light often depends on specific wavelengths of the electromagnetic spectrum to provide selective reactivity in various substances. For example, varying absorption characteristics in chlorophyll in vegetation enable photosynthesis, wherein light energy is transferred to chemical energy. This direct correlation of light absorption and energetic change also underscores the significance of understanding electromagnetic radiation in biological and chemical systems (Sharma et al., 2021). Aside from this, theories about light as electromagnetic radiation do not only extend to its physical interactions but also to its social and technological impacts. Advances from our understanding of light have revolutionized different disciplines, including communications, energy production, and medicine. For instance, application of lasers to surgery shows the beneficial application of lights properties in an advanced medical application, while photonics technologies ensure efficient data transmission in telecommunication systems. Photodynamic therapy discovery, as noted in recent studies, shows a specific application where light is used for targeting and killing cancerous cells, exhibiting lights applicability in modern therapeutic strategies (José et al., 2021). As society evolves, the multifaceted understanding of light as electromagnetic radiation continues to guide advancements in new technologies and approaches that tackle critical global issues.

2.1.2. Interaction of Light with Matter

The interaction of light with matter is a general concept in photophysics and photochemistry alike, on which there are a vast array of processes and applications. Upon incidence on a material, the light can be absorbed, transmitted, reflected, or scattered, leading to any one of a range of phenomena depending on the material's characteristics. Photophysical processes primarily include photon absorption that may lead to electrons being promoted to higher energy levels. This excitement can bring about different outcomes, such as phosphorescence or fluorescence upon relaxation of electrons to the ground state. These interactions at the microscopic level are of utmost significance in technologies such as solar cells and organic light-emitting diodes (OLEDs), in which efficient light emission and absorption is important for maximum performance and functioning. Therefore, a thorough understanding of how light behaves in different materials provides one with an insight into how to create better devices and materials that exploit these effects in an improved form. Beyond the photophysical aspects, the photochemical aspect of light-matter interaction is what throws the window open towards understanding the transformatory potential of photons when they interact with molecules. Photochemistry is all about chemical transformations brought about or catalyzed by light, which can create enormous changes in molecular structure and properties. For instance, certain molecular systems are defined by isomerization reactions upon photo-exposure to specific wavelengths of light. In these aspects, the reversible photochemical reactions, as in the case of quadricyclane/norbornadiene, demonstrate how light can induce complex chemical reactions with the potential to be stored or released under particular conditions (Wang et al., 2021). This interaction between light and matter not only offers possibilities for novel solar energy storage methods but also highlights the promise of photochemistry towards building responsive materials that change over time based on their environment. Another promising sector of investigation within the interaction of light and matter is the development of stimuli-responsive materials, particularly aggregation-induced emission (AIE) forms of materials. These compounds show typical photophysical responses to various kinds of stimuli and, therefore, find widespread applications from biology to environmental monitoring. Upon stimulation with varying types of stimuli, such as light, temperature, and pH, AIEgens show interesting changes in their luminescent properties and find applications as multifunctional probes for sensing and signaling (Zhang et al., 2021). This versatility not just enhances their uses in scientific studies but also drives developments in material science, where the design principles behind these interactions are pivotal for the design of advanced luminescent materials. Understanding how light-activated transitions can tune material properties offers a basis to develop next-generation responsive systems that can revolutionize various fields of technology (Figure 2.4).

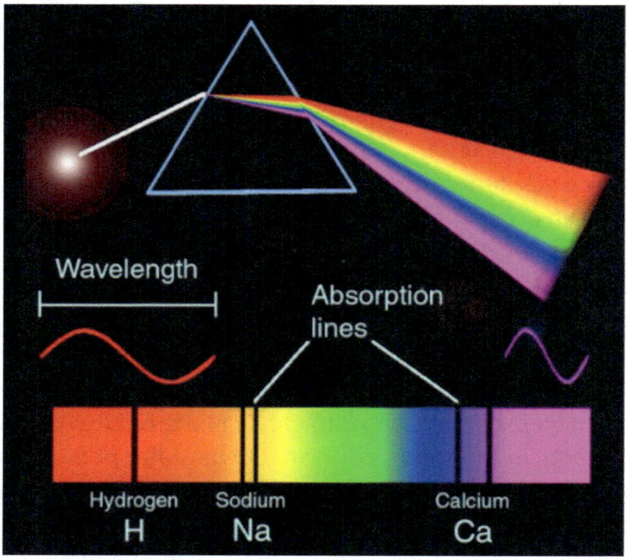

Figure 2.4. *Interaction of light with matter.*

Source: https://image1.slideserve.com/2621731/slide1-l.jpg.

2.1.3. Quantum Theory of Light

Quantum theory of light transforms the fundamental understanding of how photons interact with matter and forms the basis of photophysical and photochemical phenomena. In the traditional methodology, light had been conceptualized as a wave, but the realization of quantum mechanics brought in the concept of light as discrete packages called photons. It is this duality that allows one to explain such a complex phenomenon as absorption, emission, and light scattering by atomic and molecular systems. These processes comply with the probabilistic mechanism of quantum mechanics, in which entangled photonic states can lead to sophisticated energy transfer processes. Consequently, quantum light interactions are relevant to use in applications ranging from solar energy conversion to medical therapy and thus even more so in the application toward furthering both theoretical research and applied innovations in all fields of science.

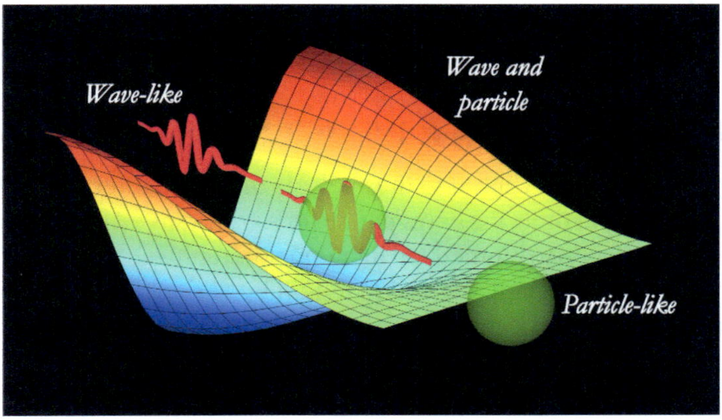

Figure 2.5. *Quantum theory of light.*

Source: https://cdn.mos.cms.futurecdn.net/gyuT23ZwvfNoevmxW6XMte.jpg.

The incorporation of quantum theory in the design of photoactive molecular systems is a significant advancement in photophysical and photochemical applications. As is described in more recent research, understanding the associated quantum mechanical behavior is essential to unlocking the potential of such systems in various technologies such as photovoltaics and photocatalysis. Development of the OMNI-P2x, a general neural network potential, is this union of quantum mechanics for practical applications. It is capable of efficient simulation of excited states of molecules at much lower computational expense than traditional quantum approaches but with precision equivalent to that of established quantum approaches. This breakthrough ushers in high-throughput investigations of photochemical reactions, thus paving the way for the rational development of new materials that can efficiently harvest light, highlighting the revolutionary impact of quantum theory in modern science (Mikołaj et al., 2025). The shift from precious limited elements to low-cost first-row transition metals in photophysics is a reflection of the new face of quantum-based materials science. Recent advances have also expanded the range of metallic complexes that are available, which display rich photoluminescent properties and reactive phases. The advances underscore the diversity and promise of low-cost solutions in luminescent technologies. By taking advantage of transition metals' unique electronic features, researchers are now finding new uses for the materials in applications such as photodynamic therapy and organic light-emitting diodes. This method enhances sustainability in material selection at the cost of previous paradigms owing to scarcity. With scientists exploring the quantum nature of these elements in greater depth, they will hopefully uncover additional dynamical photosensitizers and usher in a new material design frontier founded on quantum theory (Wegeberg et al., 2021) (Figure 2.5).

2.2. PHOTOPHYSICAL PROCESSES

Photophysical effects are central to the analysis of light-matter interactions, and these interactions constitute the cornerstone for a wide range of applications from solar energy conversion to biomedical imaging. Photophysics is essentially about the absorption, emission, and scattering of light by various substances, above all electronic molecular transitions. When it absorbs a photon, a molecule typically undergoes an electronic transition from a ground state to an excited state. Short-lived as this energetic excitation is, it has a significant influence on the spectral and reactivity properties of the molecules. The levels of such electronic states are controlled by molecular structure and surroundings and, therefore, not only influence the efficiency of processes like photosynthesis but also determine the design of cutting-edge materials like smart nanoparticles (Sun et al., 2023). These principles need to be understood for nanotechnology innovations as well as photonics innovations. The interplay between photochemical and photophysical processes elicits a sensitive interchange between absorbed energy and chemical reaction. While photophysics takes responsibility for the initial processes of electronic excitation, photochemistry extends this relationship into molecular transformation. The excited states of a molecule, once established, may go on to have a range of fates: they can return to the ground state by giving out light, become involved in reactive processes that produce new products, or take part in energy transfer reactions. In particular, in terms of photothermal therapy and drug delivery, as with the example of intelligent nanoparticles, such understanding allows for optimization of conditions required for effective targeted therapeutic action (Sun et al., 2023). In addition, the pathways that these excited states may take are of primary importance in applications such as solar energy harvesting, where the efficiency of energy conversion mechanisms significantly relies on such principles. Recent technological advances in remote sensing have unveiled new avenues of photophysical processes as expressed in solar-induced chlorophyll fluorescence (SIF). While scientists explore the dynamics of SIF, they get information regarding the intricate interconnections between light absorption and photosynthesis efficiency in ecosystems. Not only does it bring together photophysical principles with environmental and agricultural use, but it also sheds light on the broader ramifications of the interaction of light in the natural world. By studying terrestrial vegetation structure response to light stimulation, researchers can extrapolate significant ecological processes and a better appreciation of how light, vegetation, and climate change interact. This new field demonstrates the universal relevance of photophysical effects throughout the sciences and points to the promise of the integration of this knowledge into new applications, such as advances in agricultural production through more effective light utilization (Figure 2.6).

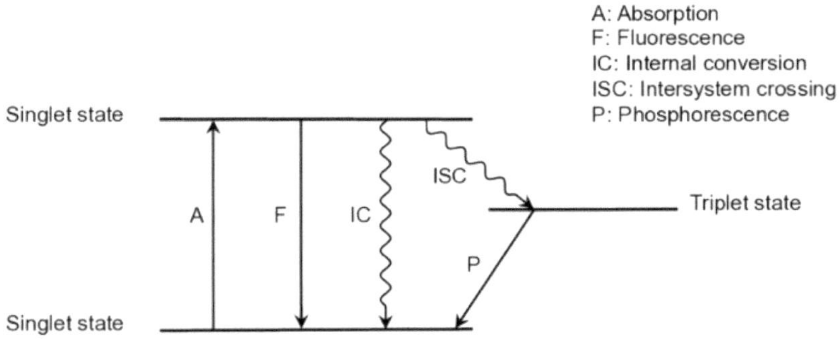

Figure 2.6. *Photophysical processes.*

Source: https://encrypted-tbn0.gstatic.com/images?q=tbn:ANd9GcT42LUv1dDFmMmgEV5d1vHo2mmemWuCCEBf7Q&s.

2.2.1. Absorption of Light and Electronic Transitions

Light absorption is the key to the understanding of electronic transition in molecules, as it varies directly with the energy states of electrons in different states. When a photon possessing sufficient energy hits a molecule, it can excite an electron from a lower energy ground state to an excited high energy state. This process is the foundation of many photophysical and photochemical reactions and enables multifarious applications in solar energy conversion, phototherapy, and photosynthesis. The light wavelengths absorbed are specific to the molecular architecture and electronic environment influencing the energy gap between the ground and excited states. Therefore, different compounds have unique absorption spectra, which serve as signatures for their identification, thereby becoming valuable resources in material characterization and spectroscopic analysis. With research unfolding, the emergence of new, cheaper, and more accessible materials has changed the understanding of emissive complexes.

Traditionally, precious metals have been the king of photophysical applications, but recent studies have proven that first-row transition metals V, Cr, and Ni are capable of displaying interesting electronic properties upon incorporation into coordination complexes as well. These complexes have shown unusually high photoluminescence properties and enhanced reactivity in electronically excited states, which suggest new directions for the design of efficient luminophores and photosensitizers. Through the study of differing oxidation levels and structural variations, researchers have uncovered new pathways that increase the distance photochemical reactions reach, into new possibilities in lighting technologies and solar energy harvesting. The breakthrough helps emphasize the central relationship between light absorption and electronic transitions that govern molecular operation with light (Wegeberg et al., 2021). Apart from all these developments, aqueous solubility of organic photoswitches has gigantic challenges and possibilities for their practical use in the world. Molecules designed to be photoswitched tend to be lipophilic, thus the problem of dissolving them in water-based systems. However, an understanding of how water interacts with such molecules can bring innovative means of solubilization, hence diversifying their functional applications. Changing the chemical composition of such photoswitches or employing surfactants allows one to stabilize and enhance their efficiency in aqueous solutions. Such a modification is relevant to fields such as biochemistry, where one has to regulate molecular behavior within biological systems. Research on such interactions demonstrates the importance of considering solvation effects in investigations into the light absorption and related electronic transitions (Volari et al., 2021).

2.2.2. Emission Processes: Fluorescence and Phosphorescence

It is crucial in photophysics research to comprehend emission mechanisms, i.e., fluorescence and phosphorescence, which occur when a molecule excites by absorbing energy and subsequently releases it as light, but which are quite distinct in mechanism and timescales. Fluorescence is virtually instant, typically in nanoseconds, when the excited electron falls back to ground state directly, emitting a photon of light in the process. This fast emission renders fluorescence a critical tool in applications like fluorescence microscopy and biosensing, enabling real-time visualizations of biological functions at the molecular level. Knowledge of these fast emission dynamics underpins the development of numerous imaging methods throughout biology and medicine, in which fluorescence labeling facilitates the observation of cellular behavior. Phosphorescence, however, is a slow emission process by intersystem crossing, a process whereby an excited electron crossover between two spin states, from a singlet state to a triplet state. This crossing leads to a long lifetime of the excited state, typically from microseconds to hours. The light emission occurs when the electron returns to the ground state from the triplet state, which is hindered by the spin-forbidden nature of the triplet state. Phosphorescent materials, therefore, have far-reaching applications in areas requiring prolonged emission of light, such as glow-in-the-dark materials and display devices. Phosphorescent material research is moving rapidly, with ongoing efforts to expand its efficiency and convert it into use in consumer products, an arena where understanding the underlying processes is essential. Recent advances in materials science have stimulated interest in stimulating phosphorescence by introducing certain heteroatoms into organic molecules, which can facilitate intersystem crossing and extend the emission lifetime. For instance, the incorporation of sulfur and phosphorus into molecular structure has been found to increase spin-orbit coupling, thus the phosphorescent characteristics of most chromophores. These are advances that are central given that they not only add to the palette of available phosphorescent materials but also inform the improvement of more efficient optoelectronic systems. As discussed in organic room-temperature phosphorescence research, understanding the connection between molecular structure and emission kinetics is required for designing next-generation lighting and display technology, highlighting the role of photophysical principles in utilitarian applications (Mazumder et al., 2024; Wang et al., 2022).

2.2.3. Energy Transfer Mechanisms

In photophysical and photochemical reactions, the processes of energy transfer play a fundamental role in explaining how systems interact with electromagnetic radiation. Energy transfer is by various different mechanisms, and some of them include radiative and non-radiative mechanisms. Radiative transfer typically involves emission of light, in which excited electronic states revert back to their ground state by emitting a photon. Non-radiative transitions, such as internal conversion and intersystem crossing, are transitions in which there is no emission of photons, but electronic energy is transferred to vibrational energy or energy moves from one state to another. These mechanisms are crucial to explain the behavior of phenomena in organic light-emitting diodes (OLEDs) and other luminescent materials as they establish energy utilization efficiency and light emission in these mechanisms, referring to their significance in technology advancement (Zhi-Cui et al., 2020). Between different energy transfer mechanisms, Förster resonance energy transfer (FRET) and Dexter energy transfer are two of the primary routes for energy transfer in molecular systems. FRET is a dipole-dipole interaction mechanism for the transfer of energy between chromophores over a range of up to 10 nanometers, making it highly applicable in biological systems and nanotechnology. FRET efficiency relies on the spectral overlap of the donor emission and acceptor absorption, and on their relative orientation. In contrast, Dexter energy transfer is a quantum-mechanical mechanism involving electron exchange at closer distances, generally less than 1 nanometer. Such a process entails direct overlap of molecular wave functions and is applicable in systems where short-range interactions are predominant.

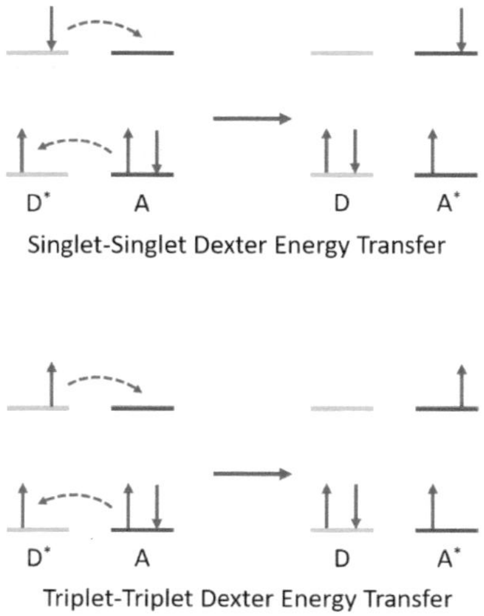

Figure 2.7. *Energy transfer mechanisms.*

Source: https://www.edinst.com/wp-content/uploads/2024/05/Quenching_Dexter_ Mechanism.png.

Understanding of these processes enables the creation of materials for photonic devices as well as improved fundamental knowledge of energy migration within complex molecular aggregates (Fdez et al., 2019). Recent progress in luminescent materials, specifically radical-based OLEDs, highlights the importance of energy transfer processes in optimizing device performance. Research in stable luminescent radicals has led the way in developing very efficient OLEDs that capitalize on the uniqueness of open-shell systems. Doublet emission processes from the novel spin structures in radicals circumvent traditional limitations observed within singlet and triplet states to enhance substantially external quantum efficiency (EQE). With researchers continuing to develop new structures and modifications of luminescent radicals, understanding mechanisms of energy transfer becomes pivotal to optimizing device structures toward advances in light-emitting technologies. The ability to manipulate such mechanisms holds promise to enable new applications in solid-state lighting and future display technology, highlighting the importance of energy transfer in modern photophysical science (Zhi-Cui et al., 2020) (Figure 2.7).

2.3. PHOTOCHEMICAL PROCESSES

Photochemical reactions are significant in most scientific disciplines, from physical chemistry to materials science and biology. Photochemical reactions are referred to as photochemical processes in which matter interacts with light to form any amount of chemical changes or reactions. The basis of photochemistry is from the understanding that light has the ability to provide energy that can promote electrons from a ground state to an excited state at a higher energy level, leading to a range of photochemical conversions. The ability of optical energy to initiate chemical reactions is the basis of the advances in technologies such as solar energy conversion and photoredox catalysis. In particular, the development of photochemical principle-based stimuli-responsive materials reflects the considerable potential for innovation across many areas, as indicated by new developments in photochemical application and the emerging research field focusing on these elaborate processes (Zhang et al., 2021). Among the myriad uses of photochemical processes, their use of acridinium photocatalysts is an example of the advancement of this area. Acridinium dyes, which have been reported to exhibit excellent photophysical properties, have gained significant interest due to their robust capabilities in photoredox catalysis. The organic photocatalysts enable different synthetic processes to be carried out under mild conditions efficiently, greatly enhancing the efficiency and sustainability of chemical conversions. The fine adjustment of functionalized acridinium photocatalysts through new methods, such as site-selective C–H alkylation, has shown the potential of their fine-tuning in respect to performance and reactivity (Yan et al., 2021). The malleability of these photocatalysts not only highlights the future of organic synthesis but also signifies a new trend toward greener chemistry strategies involving light as the energy source for driving chemical reactions. With increased research development, the fusion of photochemistry and photophysics offers ample opportunity for more in-depth study and design of new materials and applications. For instance, the development of light-intensity- or wavelength-sensitive materials encourages close integration of photonic with chemical change. These materials can be engineered to best utilize them for drug delivery and diagnostics. With deeper knowledge of the underlying photochemical processes, scientists can improve the properties of such materials, and this offers new opportunities for energy storage, cleanup of the environment, and very advanced analysis techniques. Lastly, the underlying principles of photochemical processes are at the forefront of science advancement, serving as the foundation of future innovation and technological advancements in a large range of fields (Zhang et al., 2021) (Figure 2.8).

$$H_2 + Cl_2 \xrightarrow{\text{Dark}} \text{No Reaction}$$

$$H_2 + Cl_2 \xrightarrow{\text{Light}} 2HCl$$

Figure 2.8. *Photochemical processes.*

Source: https://chemistnotes.com/wp-content/uploads/2022/03/photochemical-reaction.png.

2.3.1. Mechanisms of Light-Induced Chemical Reactions

The mechanisms of photochemical reactions are intrinsically connected with the light-matter interaction, primarily by electron excitation. Upon absorption of photons by molecules, the latter can excite electrons to higher energy states to establish excited electronic states essential for subsequent chemical change. Intra- and intermolecular energy transfers with multiple paths are significant for the generation of radicals due to initial excitation. Such photogenerated radicals in organic photochemistry are part of highly reactive processes, making distinctive synthetic routes unavailable to conventional approaches. As evidenced in the review of advances in organic photochemistry, researchers have used the understanding to construct covalent organic frameworks (COFs) as photocatalysts to illustrate how structural tunability affects the photocatalytic reaction efficiency in organic conversions (Alberto et al., 2022). Application of light in chemical reaction promotion is not only restricted to the activation of processes but also has a crucial impact on reaction dynamics and mechanisms themselves. Of particular importance is photoredox catalysis, which has been a game-changing tool in mediating selective C–H functionalization and alkenamine amination, exhibiting the power of light-driven reactions. Nitrogen-centered radicals (NCRs) are one of the key intermediates in these approaches, offering new avenues for building complex organic molecules. The creation of NCRs is typically accomplished using photolytic processes, as irradiation with light facilitates breaking bonds in precursor molecules. This advance not only streamlines synthesis pathways but facilitates selective functionalization of diverse moieties without prefunctionalization schemes. Thus, the research of these extreme organisms pushes the scope of synthetic opportunities as well as accessing the unique properties of light to achieve effective chemical conversion (Pratley et al., 2022). Even the dimensionality and crystallinity of photocatalytic materials such as COFs contribute significantly to regulating their activity in light-mediated reactions. The porosity of materials facilitates higher light harvesting and ensures the presence of available reaction sites for substrates. These parameters dictate parameters dictate the photocatalysts to be assessed in detail to optimize reaction rates and selectivity for photo-driven processes. Furthermore, the relationship between molecular structure and photophysical properties such as quantum yield and excited state lifetimes defines the efficiency of light-induced conversions. As science in this field continues to advance, the nuance of these interactions will be crucial to the creation of new materials and expanding the range of application for photochemical processes in such areas as environmental remediation and green synthesis. Ultimately, the photochemical mechanisms of light-induced chemical reaction not only illuminated basic principles of photochemistry but also laid the groundwork for revolutionary developments in modern synthetic chemistry (Figure 2.9).

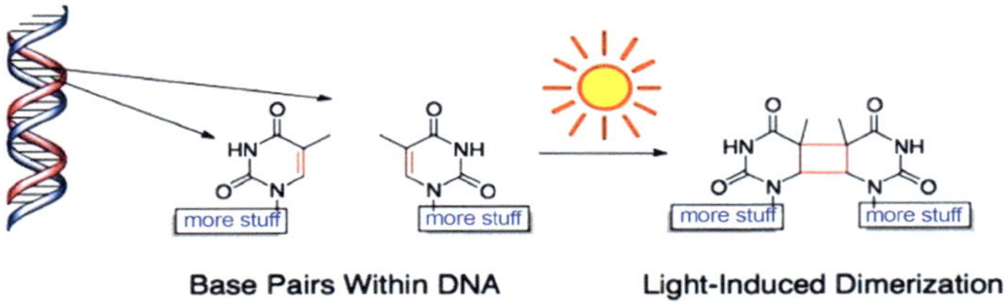

Figure 2.9. *Mechanisms of light-induced chemical reactions.*

Source: https://icanhasscience.com/wp-content/uploads/2011/03/Photodimerization-of-Thymines-e1301279663215.jpg.

2.3.2. Role of Catalysts in Photochemical Reactions

The selectivity and efficiency of photochemical reactions are greatly improved through the use of catalysts. Practically, such chemicals permit the conversion of reactants to products without permanent alteration. As a characteristic of photochemistry, catalysts permit specific pathways through which reactants absorb light energy and become excited states. The excited state is required in initiating the reaction, particularly in light-driven systems. The employment of catalysts has the tendency of reducing the activation energy and inducing reactions with less stringent conditions with less environmental impact.

With the advent of photocatalysis and photoredox chemistry, the knowledge of the mechanism of these catalysts gains more significance. Investigating the parameters governing these processes not only aids in the foundation of new innovative synthetic methods but also solidifies the foundation for future research breakthroughs in the arena of photochemical transformations (Buzzetti et al., 2018). The latest advances in photoredox catalysis emphasize the pivotal role played by catalysts to facilitate light-driven transformations. Through mechanisms such as single-electron transfer and energy transfer channels, catalysts enable reactions that are otherwise challenging under visible light. These responses prove particularly useful in green chemistry since they minimize reliance on hazardous reagents and traditional energy sources. Besides, investigation of visible-light-induced energy-transfer reactions is increasingly gaining attention, providing unique opportunities for the application in synthesis (Quan-Zhou et al., 2018). As mature approaches to the employment of catalysts in photochemical reactions are developed, it is increasingly crucial to uncover the subtlety of their mechanism and reactivity. This understanding allows chemists to design more efficient catalysts, tune the conditions of reaction, and raise yields for products, further expanding the horizon of organic synthesis applications. The combination of photochemical reaction and catalytic process opens up new windows to a more sustainable chemical synthesis. By leveraging catalysts, scientists can explore multifunctional and complex transformations that reduce waste and energy consumption in applications from pharmaceuticals to materials science. Catalysts do more than accelerate reaction rate and selectivity; they assist with the rational design of sustainable reaction pathways. With researchers continuing to uncover mechanistic specifics of these processes, opportunities for novel catalysts and reaction pathways increase significantly. Thus, such continued development underscores a necessity for experimental and theoretical research to drive future innovation in photocatalysis. Last but not least, understanding the role of catalysts in photochemical systems will serve to increase the efficiency and sustainability of chemical production globally (Buzzetti et al., 2018; Quan-Zhou et al., 2018).

2.3.3. Applications in Organic Synthesis

The application of photophysical and photochemical principles to organic synthesis has revealed new methodology that enhances the efficiency and selectivity of the reaction. Photochemical conversion, where light energy is employed, allows chemists to directly manipulate molecular entities to facilitate complex organic transformation with minimal waste. For instance, antitumor drugs and natural product synthesis have been aided by photochemical processes that make bond formation selective at mild conditions. This aligns with growing demand for the principles of green chemistry, where the reduction of hazardous solvents and byproducts is of highest priority. By employing light to catalyze organic transformations, scientists are able to tap into reactions that traditional methods—often reliant on vigorous reagents and multistep protocols with numerous steps—cannot provide. This kind of transformation is not only a reflection of environmental benefits but also of the development of sustainable synthetic processes that are increasingly being used in academic and industrial settings. In addition, advances in stimuli-responsive materials have opened up new fields of application for organic synthesis. The use of aggregation-induced emission (AIE) is only one of the ways in which light-activatable properties enhance material performance.

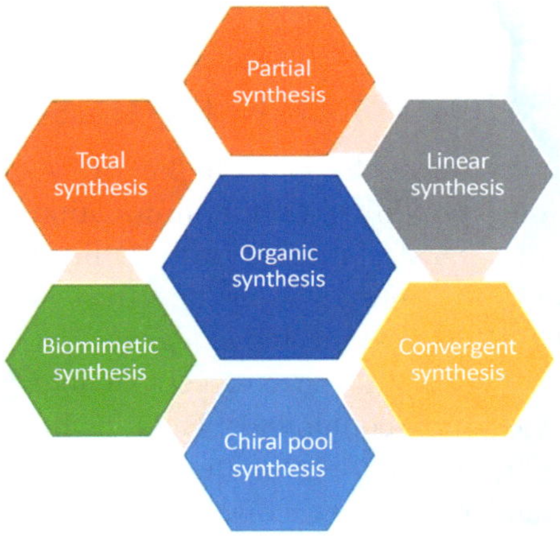

Figure 2.10. *Applications in organic synthesis.*

Source: https://scienceinfo.com/wp-content/uploads/2023/06/Organic-synthesis-1024x536.jpg.

The promising potential of AIEgens shows that they can respond to various external excitations, such as temperature and pH, thereby providing tailored performance for syntheses. This is of far-reaching significance in areas ranging from materials science to biology, where photophysical properties of these materials are utilized for detection and diagnostics. By turning the principles of AIE on the tools of synthetic protocols, chemists are unlocking for progressively more sophisticated applications that push the boundaries of traditional organic synthesis. The concept that light stimulations can control chemical outcomes positions such protocols at the bleeding edge of modern synthetic strategies (Zhang et al., 2021). Another investigation in the area of organic synthesis typically involves sensor design with enhanced performance and sensitivity.

Silicon-based organic crystals are a prime example where the photochemical principles enable ultralong RTP at room temperature that is very important in oxygen sensing. High-level oxygen measurement sensitivity is of gigantic importance in industrial processes and biological systems. The characteristic structural properties of these crystals, which allow for effective oxygen permeability while maintaining luminescence characteristics, are characteristic of the interplay between chemical structure and photophysical behavior. Although scientists continue to further enhance the creation of materials taking advantage of AIE and phosphorescence concepts, the field is going to witness its sensors and responsive systems revolutionally enhanced, thus sustaining the importance of photophysics knowledge in working towards organic synthesis equipment (Wu-Guo et al., 2021) (Figure 2.10).

2.4. SPECTROSCOPY TECHNIQUES

Spectroscopy techniques are at the core of the study of photophysical and photochemical reactions, providing useful information regarding molecular interactions and energy transfer mechanisms. Utilizing light to study the electronic structures of molecules, various spectroscopy techniques such as ultraviolet-visible (UV-Vis) spectroscopy, fluorescence spectroscopy, and nuclear magnetic resonance (NMR) spectroscopy enable scientists to study the properties of a wide variety of compounds. By allowing the quantification of light emission and absorption, these techniques make it possible to investigate molecular behavior under different conditions. Additionally, advances in computational technology have led to improved simulation and interpretation of spectroscopic data, whereby the possibility of modeling molecular dynamics better has been made available. Fluorescence spectroscopy, specifically, has drawn considerable interest due to its ability to probe transient states and track the dynamic movement of molecules in real time. This technique is based on the unique properties of fluorophores—molecules that are capable of re-emitting light they have absorbed—giving information regarding energy transfer efficiencies, conformational changes, and biomolecular interactions. Adding advanced techniques such as time-resolved fluorescence and resonance energy transfer (FRET) has further enhanced intricate studies of molecular assemblies and interactions in biological systems. These advances have driven applications into realms as varied as materials science and biochemistry, to illustrate the strength and overarching importance of fluorescence as a method in unraveling complex molecular processes. In addition, the creation of spectroscopy computational engines, such as OpenMolcas, has revolutionized how researchers approach spectroscopic phenomena. This open-source software not only combines thorough methods of modeling molecular interactions but also aids the analysis of spectroscopic data by molecular dynamics and photochemical reaction simulations. The unique capabilities offered in OpenMolcas, including adiabatic and nonadiabatic dynamics simulation and conical intersection investigation, provide a rich platform for understanding multifaceted photophysical processes. Because molecular photoswitches also have some solubility problems, e.g., they are naturally lipophilic and poorly soluble in water, such computational capabilities can help overcome these limitations to ultimately expand the scope of spectroscopy applications in photochemical sciences and other fields (Fdez et al., 2019; Volari et al., 2021) (Figure 2.11).

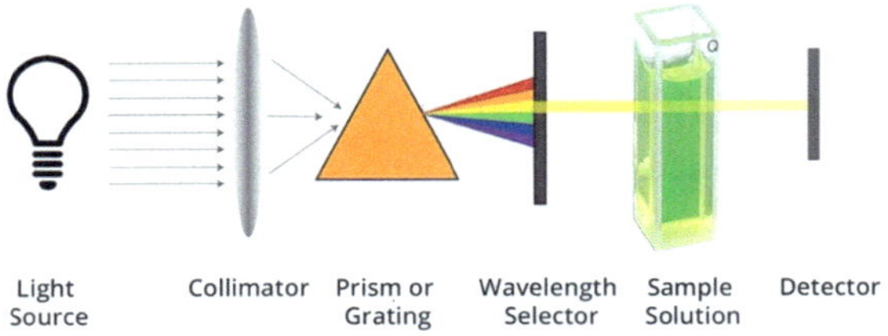

Figure 2.11. *Spectroscopy techniques.*

Source: https://microbiologynotes.org/wp-content/uploads/2020/12/Spectroscopy-Introduction-Principles-Types-and-Applications.jpg.

2.4.1. UV-Vis Spectroscopy in Studying Photophysical Processes

The application of UV-Vis spectroscopy in the study of photophysical processes is a fundamental paradigm for comprehension of excited state dynamics in various molecular systems. The technique exploits the absorption of ultraviolet or visible radiation by molecules, leading to electronic transitions that are informative for the elucidation of their photophysical behavior. From the absorption spectra, the scientists not only find the energy levels and electronic structures of the molecules but also the interactions with the surrounding world. This is especially crucial in photoredox catalysis, in which excited-state redox potential of photosensitizers becomes a strong factor in initiating target chemical reactions. Since recent studies highlight, the potential to interpret such dynamics opens the door for designing more efficient photochemical systems and thereby expanding the chances for applications from solar energy conversion to organic synthesis (Kudisch, 2025). Moreover, transient absorption spectroscopy, a member of UV-Vis techniques, has turned out to be a precious tool for comprehending ultrafast processes in excited states. This technique makes it possible for scientists to capture transient species and their processes in real time after exciting them. Being able to generate and observe radicals, such as radical ions, has enabled researchers to explore photophysical pathways that govern elaborate reactions. For instance, studies aimed at radical anions indicate that excited-state properties can influence diffusion-limited electron transfer reactions significantly, by directly impacting photochemical efficiency of applications. As researchers delve further into the excited-state dynamics of these radical species, they are uncovering mechanisms for such super-redox processes and learning new methods of tailoring photoactive materials (Kudisch, 2025). UV-Vis spectroscopy is also a flexible platform for examining solvation influences and the effect on photophysical phenomena. Spectral fingerprints obtained from different solvents can yield information on solvation dynamics and excited-state stabilization. By identifying characteristic solvate-specific absorption bands, researchers are able to make quantitative photogeneration efficiency measurements in real time. This is particularly true in research on solvated electrons, which are relevant intermediates in numerous photochemical reactions. Being able to correlate photophysical properties to environmental conditions opens up larger knowledge on electron transfer processes in solution towards creating personalized approaches in designing new materials with selected photophysical properties. These works not only contribute to our knowledge of photochemical reactions but also guide future developments in supramolecular chemistry.

2.4.2. Fluorescence Spectroscopy for Analyzing Photochemical Reactions

Fluorescence spectroscopy has been a most useful technique in the analysis of photochemical reactions owing to its sensitivity and ability to provide instantaneous information. By measuring the fluorescence emitted from a sample under excitation, scientists can gain valuable information on molecular interactions, kinetic processes of reactions, and the environment surrounding fluorescent species. The technique finds particular application in studying complex processes involving transient states and intermediate species. For instance, for photocatalytic reactions, fluorescence spectroscopy makes it possible to follow changes in excited-state lifetime and emission spectra, and they can be connected to reaction pathways and efficiency of energy transfer. Thus, the introduction of fluorescence spectroscopy to the study of photochemical processes clarifies mechanisms otherwise lost in traditional analysis. Besides, fluorescence spectroscopy can serve for the investigation of the energy harvesting systems' dynamics, namely nanostructure-based ones like quantum dots (QDs). These materials exhibit unique photochemical features enhancing light absorption and electron-hole pair formation and hence are suitable for application in energy conversion, e.g., photocatalytic hydrogen evolution. As noted by recent publications, the ability to monitor photochemical processes with QDs using fluorescence techniques allows for better insights into factors affecting hydrogen generation efficiencies, such as synthesis methods and excited-state stability (Su et al., 2023). Subtle information that can be obtained from fluorescence spectra can be employed to guide the rational design of better QD-based catalysts by determining how structural modifications affect energy transfer efficiencies and reaction products. Apart from this, the utility of fluorescence spectroscopy can be extended towards environmental applications such as monitoring pollutants in agricultural ecosystems. Considering the complexity of pesticide residues and heavy metals, fluorescence-based sensing technologies can potentially provide rapid and sensitive determination of toxic substances in food and agro-products. The ability is crucial for food safety guarantee and management of public health concerns associated with contaminants that cause a number of health disorders (Sharma et al., 2021). Utilizing fluorescence to analyze the interaction between pollutants and fluorescent probes, scientists are able to develop efficient analysis methods which simplify the process of detection without sacrificing accuracy. Essentially, fluorescence spectroscopy not only provides an even better view of photochemical reactions but also bridges a significant gap in environmental monitoring by allowing more sustainable agriculture (Figure 2.12).

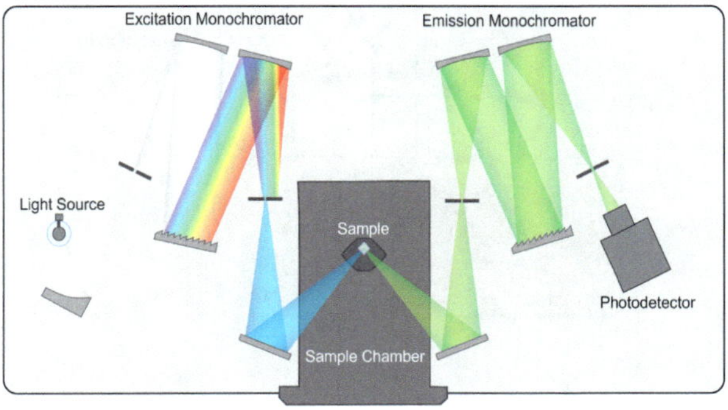

Figure 2.12. *Fluorescence spectroscopy for analyzing photochemical reactions.*

Source: https://edinst.com/wp-content/uploads/2024/05/Fluorescence-Spectrometer.png.

2.4.3. Time-Resolved Spectroscopy for Dynamic Studies

Time-resolved spectroscopy is now an essential technique for the examination of photophysical and photochemical processes, enabling scientists to observe dynamic events occurring at ultrafast timescales. By using ultra-short light pulses, it is feasible to resolve the transient states of molecular systems during the instant when they absorb photons, providing information on energy transfer, electronic relaxation, and reaction mechanisms. The ability to freeze these fleeting moments allows the in-depth understanding of the behavior of molecules under varying conditions, including alterations in solvent dynamics and interaction with the local environment.

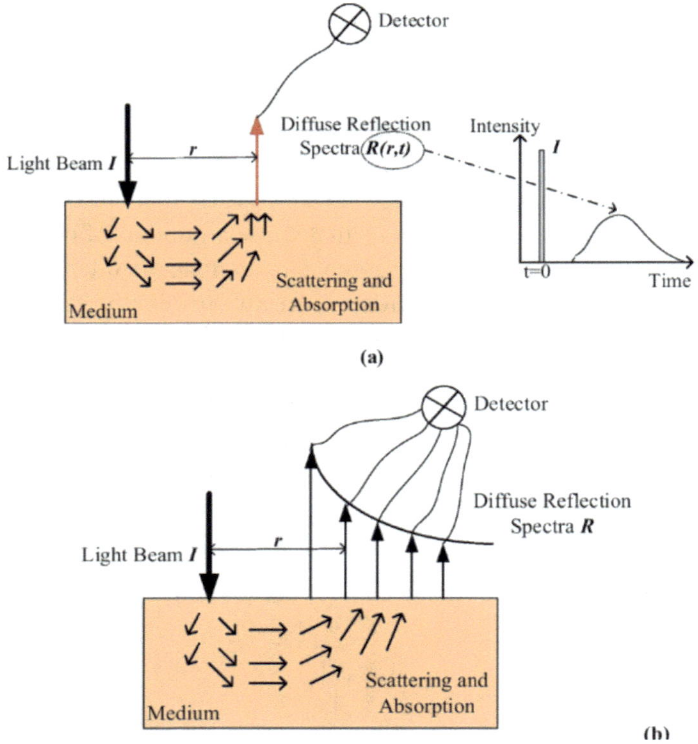

Figure 2.13. *Time-resolved spectroscopy for dynamic studies.*

Source: https://www.researchgate.net/publication/357659917/figure/fig1/AS:11431281084470176@1663208449786/Schematic-diagram-of-measurement-principle-of-a-time-resolved-spectroscopy-and-b.png.

Above all, time-resolved spectroscopy facilitates the understanding of mechanisms governing processes such as photoprotection in photosynthesis and excited-state dynamics in photochemical reactions, making possible advances in fields ranging from material science to biochemistry. Of special interest is time-resolved spectroscopy's application to the understanding of energy-dependent quenching (qE) in photosystem II (PSII), particularly within the framework of plant reaction to shifting light environments. Dynamic pH changes across the thylakoid membrane are directly linked to shifts in energy transfer pathways between chlorophyll molecules, enabling the system to transition from an unquenched to a quenched state upon environmental stimulation. Time-resolved techniques enable the simultaneous monitoring of these photophysical changes and the physiological state of the thylakoid membrane, which is of critical significance for an integrated understanding of qE mechanisms. As is highlighted in recent publications, advanced experimental methods are necessary to explore these processes on various timescales and molecular interactions, as the interplay between light absorption and molecular quenching remains a complex area yet to be explored in full (Amarnath et al., 2013). Additionally, advances in time-resolved spectroscopy will also assist us in enhancing our understanding of nanomaterial interaction dynamics, for example, through studies involving organic polymer-doped nanoparticles. Experiments have proven that, for instance, time-resolved techniques can accurately explain singlet oxygen formation process initiated by triplet excited states in such nanoparticles, demonstrating how molecular structure changes influence photophysical phenomena. Conjugated polymer nanoparticle studies illustrate how the introduction of specific dopants can lead to exceedingly high quantum yields of singlet oxygen, thereby offering insight into technological applications, including photodynamic therapy. With constant strict research that employed stationary and time-resolved spectroscopy, the potential for further understanding selective energy transfer reactions and mechanisms exists, while with regard to emphasizing time-resolved methods towards the development of new materials and therapeutic methods (Chesta et al., 2018) (Figure 2.13).

2.5. APPLICATIONS IN MATERIAL SCIENCE

The photochemical and photophysical principles have far-reaching implications in material science, especially as researchers shift focus to employing more common metals in the form of luminescent materials. The field has traditionally been biased towards precious and rare elements but is in the middle of a massive upheaval through the use of first-row transition metals such as V, Cr, Mn, and Fe. These metals have exhibited interesting new behaviors under electronically excited states previously unknown, and this has made it possible to form novel coordination compounds with excellent photoluminescence. Thus, the discovery and design of 3d metal-based luminophores and photosensitizers are extremely important in terms of expanding the scope of applications, particularly in environments that may be less optimal for traditional materials. This evolution not only underscores the importance of low-cost resources but also introduces novel challenges and opportunities into the materials science discipline, ultimately resulting in the creation of multifunctional materials across multiple uses. Within the domain of new materials, the creation of Aggregation-Induced Emission (AIE) luminogens offers a particularly exciting future prospect for photochemical application. These molecules possess unique photophysical properties through their fluorescing ability upon being in an aggregated state, which is quite dissimilar from the common fluorescent dyes that are prone to be quenched when aggregated.

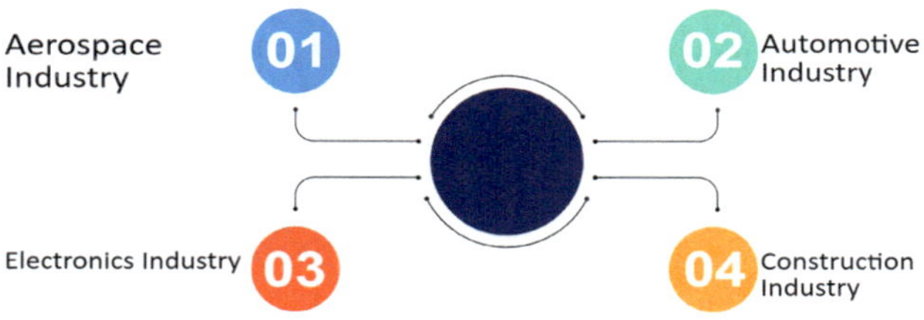

Figure 2.14. *Applications in material science.*

Source: https://fastercapital.co/i/Building-Tomorrow--The-Role-of-Materials-Science-in-Manufacturing--Applications-of-Materials-Science-in-Different-Industries.webp.

The potential of stimuli-responsive AIEgens has inspired massive research endeavors with expected breakthroughs in numerous fields including physical chemistry, biology, and medicine. By responding to varied stimuli such as light, temperature, and pH, such materials are designed precisely for sole applications ranging from biosensing to drug delivery systems. The progress seen in the design and use of stimuli-responsive AIEgens summarizes the nature of innovation powering material science in positioning such materials at the cutting edge of material technology. With material science more and more embracing photochemistry and photophysics principles, it is becoming increasingly apparent that the understanding of how luminescent materials function is key to future developments. For instance, knowledge of how different stimuli affect material performance enables researchers to tailor their properties for a specific application. By activating their reactive properties, scientists are able to design next-generation photosensitizers that work efficiently in a variety of applications such as solar energy conversion and environmental monitoring. This data also forms the basis for the next generation of luminescent materials with enhanced uses without being unsustainable by using prevalent metals. The dynamics of such sustained growth emphasize the revolutionizing impact of integrating advanced photochemical techniques in materials science to ultimately facilitate a more adaptive and ecological trajectory of technological development (Figure 2.14).

2.5.1. Photonic Materials and Their Properties

In photonic materials, synthesis and integration of light-sensitive inorganic biomaterials are transforming the universe of medical technology. The materials possess the unique property of developing highly controllable and noninvasive medical devices that can operate under exacting light stimuli. Carbon nanomaterials and gold nanoparticles, for instance, have played a key role in generating new therapies such as photothermal and photodynamic therapies. These techniques utilize the unique interactions of matter with light to target diseases specifically and not cause harm to the nearby healthy tissues. Modulation of cellular activity and regulation of extracellular microenvironments also exemplify the strengths of these biomaterials towards enhanced therapeutic response in regenerative medicine and drug delivery systems (Lee et al., 2020). As technology continues to advance, the use of these materials in the clinic becomes greater, with precision medicine potentially being guided by light alone in the future. The addition of molecular photoswitches to a variety of photonic materials brings a fascinating element to the manipulation of material properties. These systems enable the control of properties such as conductivity, magnetism, and even luminescence with unprecedented levels of spatial and temporal resolution. By employing the geometrical and electronic transitions at the molecular level, scientists are now beginning to see the potential of photoresponsive mechanisms to alter the properties of bulk material. In particular, the work has indicated the potential for the use of these systems across the range of materials from amorphous to crystalline state, with this work leading on to an intriguing field full of promise. The challenges of converting microscale manipulation into macroscale behavior are daunting, but the advance in photonic materials holds promise that it can overcome hurdles to provide revolutionary applications in optoelectronics and smart materials (Goulet et al., 2020). The future is full of promising applications with the interaction of photonic materials with their diverse properties. Ongoing research into light-sensitive systems has posed intriguing questions about their potential in a wide range of scientific fields, from materials science to nanotechnology. With each innovation by scientists in material design and makeup, the possibility of optically programmable systems becomes more tangible. Future advances would include specially engineered light-sensitive biomaterials that not only produce desired therapeutic effects but adapt to dynamic biological conditions in real time. The application of these systems to real-world technologies has the potential to transform telecommunications and environmental monitoring, personal medicine, and more. The ongoing creation of photonic materials is thus the key to greatly advancing our understanding of photophysical and photochemical phenomena, the possibility of paradigm-shifting technologies and medicines.

2.5.2. Solar Energy Conversion Technologies

The harnessing of solar energy into forms that can be exploited is increasingly viewed as a key technology in solving the world's energy crisis. This involves the utilization of sunlight by a variety of processes, largely photophysical and photochemical ones. These photophysical processes involve the absorption of solar photons by matter, leading to electronic excitations that can cause charge separation. Advancements in materials science have been at the forefront of evolving the efficiency of such processes, particularly through the design of better photovoltaic cells. The use of materials with suitable bandgap energies has enhanced the levels of conversion, consequently providing maximum outputs of electric energy from solar radiation. Nevertheless, material stability problems and the cost-effectiveness of bulk deployment persist, necessitating concerted research aimed at the technology development as well as commodity availability to the general market. In addition to photovoltaic systems, photochemical technologies present a very interesting potential of sunlight transformation into energy via chemical processes. These processes would typically entail the exploitation of light as a catalyst for endothermic reactions, such as fuel generation from carbon dioxide. The MOST system concept illustrates the potential in this area, where molecular isomers have the ability to hold energy from sunlight in reversible light-induced processes (Wang et al., 2021). Some systems, for example, norbornadiene and quadricyclane, illustrate how one can increase energy storage capability through deliberate molecular engineering. By focusing on photochemical reactions that are oriented towards high-energy isomerization and subsequent thermal release, researchers are addressing the core questions of energy storage duration and efficiency. These advances help provide a sustainable energy cycle that balances energy production with use. In addition, recent advances on stimuli-responsive materials continue to push the potential applications of solar energy conversion technology. For instance, the development of aggregation-induced emission luminogens (AIEgens) showcases the potential of such materials to be rendered responsive to various stimuli, including light, in a dynamic manner. Such responsiveness opens up innovative applications across various fields, such as biomedicine and material science, through energy storage and controlled release as a function of environmental stimuli (Zhang et al., 2021). In understanding the fundamental principles governing AIEgen behavior, researchers can design future materials that can effectively tap solar energy and respond to external stimuli while enhancing function and efficiency. Overall, the merging of new material science and photophysical and photochemical processes is essential to the development of efficient and sustainable solar energy conversion technologies (Figure 2.15).

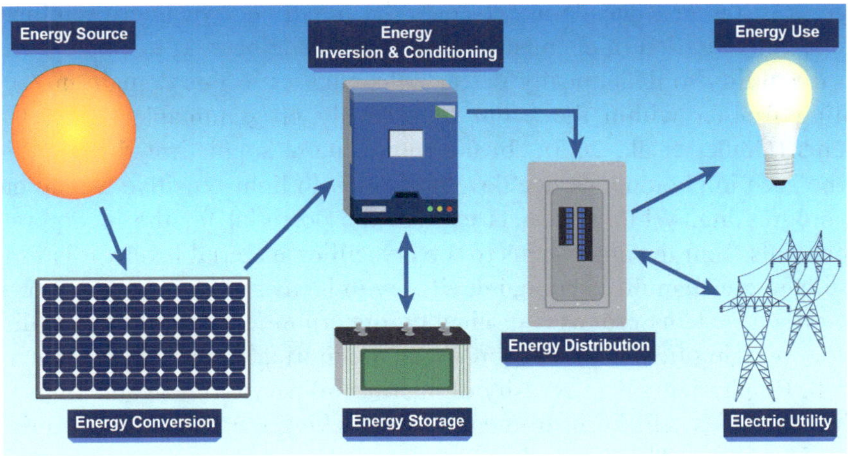

Figure 2.15. *Solar energy conversion technologies.*

Source: https://energyresearch.ucf.edu/wp-content/uploads/2017/01/pv_system.png.

2.5.3. Development of Photonic Devices

The development of photonic devices has significantly played a role due to the inclusion of new light-sensing materials. The devices apply the concepts of photophysics and photochemistry to govern processes with highly precise measurements in many applications, particularly in the biomedical field. Inorganic light-sensitive biomaterials have made possible noninvasive clinical therapies like photothermal therapy and drug delivery systems. Such materials, including nanoparticles and carbon nanomaterials, facilitate the creation of light-guided nanovehicles to modulate extracellular milieus and regulate cellular functions. This innovation points to the intersection of medical technology and materials science, emphasizing the need for rigorous examination of how photonic devices must be created to optimize therapeutic outcomes in clinical use (Lee et al., 2020). In the last 10 years, the combination of molecular photoswitches with a broad spectrum of materials has revolutionized the operation of photonic devices. These photoswitches allow for large spatiotemporal resolution, and material properties can be fine-tuned by scientists with large precision using optical stimuli. While the capabilities of these systems are vast, problems exist in fully achieving their full potential at both the molecular and macroscopic scales. The idea of photoswitchable materials is complex, requiring an understanding of how electronic and geometric changes affect bulk properties. Progress in the last decade underscores the importance of interdisciplinarity and the collaboration of chemistry, physics, and engineering to create devices with the ability to dynamically detect light. Studies in this domain unveil an exciting frontier within the realm of optically programmable material and systems (Goulet et al., 2020). In the future, more sophisticated applications will be seen in photonic device development with light-sensitive technologies still progressing. While scientists explore the potential for the incorporation of materials from the amorphous to the crystalline material levels, it is further possible to create multifunctional devices with increased control. The future of light-driven systems promises fascinating opportunities not only for medicine but for emerging fields like environmental monitoring and intelligent materials as well. Emphasizing the necessity of interdisciplinary research and innovative design strategies will be instrumental in unlocking the complete potential of photonic devices. The future developments anticipated can bring innovations that transform how we utilize light in technology and advance various industries to make them efficient and responsive to environmental factors (Goulet et al., 2020).

2.6. ENVIRONMENTAL IMPLICATIONS

The interplay between photophysical and photochemical processes is dynamic with significant environmental consequences, particularly in the formation and transformation of various pollutants. Phenolic compounds (PhCs) are global environmental pollutants that are derived from industrial effluent discharge, combustion, and the oxidation of volatile organic compounds (VOCs). Phenolic compounds have aromatic rings with hydroxyl substituents, whose presence causes them to be susceptible to photochemical processes when exposed to sunlight. When PhCs undergo transformation, they can be accountable for secondary organic aerosol (SOA) formation. SOAs not only impact regional air quality but also potentially have broad-reaching effects on human health and climate regimes since they can modulate the absorption and scattering of solar radiation. Therefore, understanding PhC behavior during photophysical processes is critical for the development of efficient environmental monitoring techniques to counterbalance their harmful effects on ecosystems and human health (Yang et al., 2024). The organic and inorganic fractions of soil interact significantly with each other, particularly the fate of organic pollutants in the environment. Soil quality and its capacity to act as a contaminant buffer is closely related with the complex interactions among soil minerals, organic matter, and microbes. These interactions can modify the transformation process of organic pollutants, which can affect their mobility, persistence, and potential toxicity in the environment. For example, the sorption pattern between organic contaminants and soil components tightly regulates retention or discharge of contaminants like PhCs in soils. Therefore, the effectiveness of natural attenuation or biodegradation processes depends on the fine equilibrium of these interactions. Understanding how the dynamics of soil moisture levels, soil structure, and microbial activity influence these processes can further define our approaches to the use of soil as an effective medium in the fight against environmental pollution (Huang, 2023). The effects of photophysical and photochemical processes are not only confined to atmospheric chemistry but also encompass essential functionalities in soil systems, influencing nutrient cycling and sustainable agriculture. Because photochemical reactions can alter the molecular makeup of organic matter in soil, they can change the availability of nutrients for plant and microbial life. Breakdown of agrochemicals through light-catalyzed processes also enhances the process. This agrochemical breakdown, if not controlled, could lead to chronic soil pollution and decreased soil health. The recognition of how such processes control the transformation and stability of organic pollutants in different environmental environments is ever more critical in the context of growing concern regarding climate change and soil degradation. In this regard, a collective perception of photophysical and photochemical fundamentals may optimize more effective environmental regulation and development of sustainable solutions, improving ecosystem resilience under the context of anthropogenic stress (Huang, 2023; Yang et al., 2024).

2.6.1. Role of Photochemical Processes in Atmospheric Chemistry

Photochemical reactions account for a large fraction of atmospheric chemistry, usually by virtue of interactions between sunlight and atmospheric substances. These interactions take up energy, which is commonly followed by the formation of reactive radicals and intermediate species with the capacity to significantly alter chemical balance and composition in the atmosphere. These reactions are required for the breakdown of pollutants and greenhouse gases and contribute to the formation of secondary pollutants, such as ozone in the troposphere. The impacts of these processes are not only in air quality but also have implications for climate systems and biogeochemical cycles. For instance, photochemical formation of organic aerosols suggests the sensitive interface between atmospheric chemistry and climate modeling, highlighting the importance of adopting a comprehensive approach towards understanding such processes.

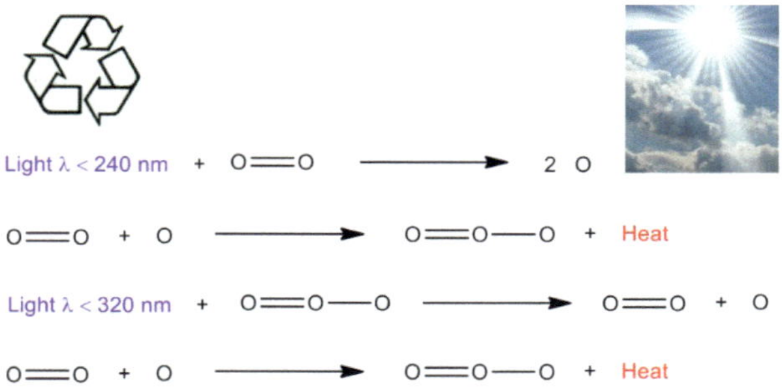

Figure 2.16. *Role of photochemical processes in atmospheric chemistry.*

Source: https://image1.slideserve.com/1901455/mechanism-of-ozone-formation-in-the-upper-atmosphere-l.jpg.

In addition, atmospheric photochemical reaction dynamics suggest their vulnerability to physical and environmental factors, which can result in varied outputs. Temperature, humidity, and the availability of other atmospheric constituents dictate the efficiency of these reactions. Surprisingly, photolysis of large compounds such as methane and halocarbons yields products with significant global warming effects. Research on nonadiabatic molecular dynamics simulations has reported varying results based on the initial conditions of the molecular system, as is the case with methyl hydroperoxide (MHP) (Prlj et al., 2023). These findings highlight the need for precise modeling in atmospheric chemistry, demonstrating that changes in reaction pathways have a profound effect on atmospheric constituents and their interactions with one another over long periods of time. Beyond effects on air quality and climate, photochemical processes lead to a wide range of applications in environmental science. For instance, the production of advanced materials that can exploit photochemical reactions is being researched more and more, particularly in air purification and pollution management uses. Recent studies have focused on inorganic photochromic materials as being highly promising in photochemical devices to counteract environmental problems (Du et al., 2024). These advances not only represent a milestone in the science of materials but also underscore the interdisciplinary nature of atmospheric chemistry, which blends the principles of physical chemistry with environmental technology applications. Lastly, understanding and control of photochemical reactions are crucial in solving present-day environmental problems and continuing towards sustainable approaches (Figure 2.16).

2.6.2. Impact of Photophysical Processes on Climate Change

The connection between photophysical processes and worldwide climate change is a crucial field of study that requires extensive research. Included in the critical elements is the role played by solar-induced chlorophyll fluorescence (SIF), a critical parameter used in signifying vegetation growth and photosynthetic activity. SIF, which is emitted during the light reactions of photosynthesis, has acquired considerable prominence in following vegetation trends and ecosystem responses to climatic fluctuations. Over the past two decades, advancements in SIF remote sensing have enabled scientists to capture high-resolution data, facilitating the analysis of terrestrial ecosystems under the pressure of climate change. This technological innovation also emphasizes the complex interaction between photosynthetic processes and environmental factors, suggesting that SIF can not only improve our comprehension of climate changes but also in designing strategies for mitigating the adverse impacts of climate fluctuation on food security and biodiversity at the global level (Sun et al., 2023). Moreover, photochemical processes, particularly hole-transporting layer (HTL) processes in solar energy technology, play a pivotal role in the changing trend of climate change mitigation. Tuning efficient solar cell architectures like perovskite solar cells (PSCs) relies heavily on HTL optimization to improve energy conversion efficiencies. Organic HTL choices are highly favorable, as they can provide the advantages of tunable bandgaps and low-cost synthesis. Through enhanced performance and stability of PSCs, photochemical process improvements can be a part of the transition of the world to renewable energy, which will contribute significantly towards curbing greenhouse gas emissions. As the world experiences a rise in climate issues, incorporating photochemical knowledge into energy is a double-edged sword that can promote sustainable energy while also reversing existing climate changes (Yao et al., 2022). Ultimately, the applicability of photophysical and photochemical phenomena moves beyond theoretical importance to practical uses that have direct bearings on climate change. Understanding how these phenomena influence environmental parameters enables more profound insight into ecosystem response and energy technologies. From employing SIF data to inform conservation policy to the use of sophisticated photochemical methods for strengthening renewable energy infrastructure, the capacity for the deployment of these scientific principles to power and facilitate climate action is vast. Development of an integrated framework to cross-link these disciplines is essential to the endeavor of climate research and policy. With growing climate change challenges, photophysical and photochemical processes knowledge will be vital in formulating efficient, evidence-based approaches to enhancing resilience in human and natural systems (Sun et al., 2023; Yao et al., 2022).

2.6.3. Photodegradation of Pollutants

Photodegradation of pollutants is a milestone for environmental photochemistry and photophysics in which sunlight acts as a catalyst for the destruction of the toxicants. It largely utilizes photocatalysis in which semiconductor materials such as titanium dioxide promote the destruction of organic pollutants upon exposure to light. The process involves the formation of electron-hole pairs on light absorption to yield reactive species that enhance degradation of complex organic molecules into less harmful by-products. Deep insight into such photophysical interactions is decisive since it allows scientists to calibrate photocatalytic systems for the mitigation of environmental problems as a result of industrial effluents and dyes, which are extremely common in wastewater effluents (Alberto et al., 2022).

The functionalization and design of covalent organic frameworks (COFs) have been major breakthroughs in the photodegradation of contaminants. The very tunable structural and electronic properties of the porous materials can be engineered to maximize photocatalytic activity under different light conditions. The unique characteristics of COFs, i.e., stability and high surface area, enable them to effectively adsorb the pollutants, thereby having a greater opportunity to be photodegraded when exposed to light. Literature survey indicates the ever-growing possibilities with COFs, which reveal their versatility in various applications from the treatment of wastewater to green chemistry synthesis. This move towards the employment of advanced materials for the degradation of pollutants presents a necessity to fill knowledge gaps from molecular organic photochemistry in an effort to innovate strategy in environmental remediation (Alberto et al., 2022). As photodegradation continued to develop, it is noteworthy that the dynamics of mixed pollutants in aquatic ecosystems should be taken into account.

Existing literature indicates that competitive photocatalytic degradation of dyes in mixed solutions presents challenges and opportunities for optimization. The intermolecular interactions between different dye molecules also have a significant influence on their degradation rates, where molecular conformation and the availability of functional groups play key roles. These interactions provide valuable information not only for the improvement of more efficient photocatalyst designs but also for enhancing photocatalytic system predictability. This ongoing investigation of pollutant behavior in mixtures emphasizes the intricacy of real-world environmental conditions, which demand a multi-dimensional approach in order to achieve holistic solutions for contamination issues and to make photocatalytic technology sustainable (Amanina et al., 2022).

SUMMARY

In wrapping up an overview of the principles of photochemical and photophysical processes, it is evident that these fields are at the forefront of advancing many areas of science. Understanding phenomena such as fluorescence, phosphorescence, and energy transfer not only avails the doorway to the intricacies of molecular interactions but also to new avenues of applications in physical chemistry, materials science, and biology. With new studies uncovering the precise mechanisms behind such spectacular processes day after day, the potential for new materials and technologies grows more radiant with each passing day. The application of special luminescent materials in the form of stimulus-responsive aggregates showcases this potential most dramatically, showing just how principles of photophysics can be harnessed to create sophisticated systems with engineered functions (Zhang et al., 2021). Further, the interfacing of photophysics with biological systems holds immense potential for medical advancement, notably in treatments like cancer therapy. Photodynamic therapy (PDT) illustrates how light manipulation leads to immunogenic cell death (ICD), offering a valuable option for combating malignancies. By exploiting the action of photosensitizers and their response to light, PDT not only enhances tumor selectivity but also elicits an adaptive immune response, thus setting the stage for sustained tumor control (Alzeibak et al., 2021). Through ongoing advances in PDT protocols and demonstration of their potential, scientists emphasize interdisciplinary research with the principles of photophysics dictating more effective therapeutic strategies. Photophysical and photochemical phenomena have a bright future in the coming years with ongoing innovation set to usher in a revolution in most sectors. Advances in nanotechnology and material science promise to improve the efficacy and selectivity of phototherapeutic techniques so that they can be made more suitable for application in the clinic. There are still problems, particularly concerning the hypoxic tumor microenvironment that can reduce the effectiveness of techniques such as PDT. Nevertheless, as researchers continue to uncover new solutions and refine existing techniques, the application of these concepts to real-world use will be refined, encouraging continuing study and research. The continued study of photophysical and photochemical processes will thus become a mainstay in shaping the limits of science and technology and laying the groundwork for advancements that improve human health and environmental well-being. The principles of photophysical and photochemical processes are the core of knowledge about the interaction of light with matter as well as the resulting chemical transformations. Highest priority amongst these principles is energy absorption, in which molecules are transformed from a ground state to an excited state by absorbing photons, leading to the formation of radical species, which is a key characteristic of most chemical reactions. For instance, NCR chemistry has been particularly useful in organic synthesis, in which it allows for complex molecules to be synthesized directly by functionalization of sp^2 units. The capability to synthesize and control such reactive intermediates says a lot about the intricate interdependence between photochemical energy conversion and efficiency of synthetic processes in modern chemistry, thereby opening up an avenue towards efficient streamlined synthesis (Pratley et al., 2022). Another key aspect of photophysical processes is the dynamics of excited-state returns to the ground state. This is achieved by a variety of different mechanisms, including non-radiative decay, phosphorescence, and fluorescence. Also, reversible isomerization is possible for some molecular systems upon illumination, thereby illustrating the interplay between photochemistry and thermodynamics. Molecular solar thermal systems are a case in point that illustrates how photoisomerization may facilitate heat storage through reversible processes. The research on isomeric systems such as norbornadiene/quadricyclane parallels the evolving methods of harvesting solar energy more effectively. Through the creation of materials that can perform these light-induced conversions effectively, scientists are fostering photochemical-based sustainable energy solutions (Wang et al., 2021). Moreover, application of these principles also transcends elementary chemical understanding to technology and material science application. The nitrogen-focused radicals catalyzed heterocycle formation is a sign of the way complexity in photochemical reactivity can translate to innovation in pharmaceutical and synthetic chemistry fields. Augmented reactivity through photoredox catalysis is a good example of the way excitation state behaviors and radical formation can give way to new modes of C–H amination and olefin functionalization. Convergence of photochemical and photophysical events is paramount since it highlights the requirement for interdisciplinary research in pushing novel ways of taking advantage of the power of light in chemical synthesis as well as energy utilization. This has a wider connotation where the mode of light-driven chemical reactions is utilized to address current matters in energy conversion as well as storage, paving the way for future enhancement in chemistry (Pratley et al., 2022).

REFERENCES

1. Algorri, J. F., Ochoa, M., Roldán-Varona, P., Rodríguez-Cobo, L., & López Higuera, J. M. (2021). Photodynamic therapy: A compendium of latest reviews. *Cancers, 13*, 4447–4447. https://doi.org/10.3390/cancers13174447 (accessed on 28th September 2025).
2. Alzeibak, R., Mishchenko, T. A., Shilyagina, N. Y., Balalaeva, I. V., Vedunova, M. V., & Krysko, D. V. (2021). Targeting immunogenic cancer cell death by photodynamic therapy: Past, present and future. *Journal for ImmunoTherapy of Cancer, 9*, e001926–e001926. https://doi.org/10.1136/jitc-2020-001926 (accessed on 28th September 2025).
3. Amarnath, K., Fleming, G. R., Sylak-Glassman, E., Zaks, J., et al. (2013). Models and measurements of energy-dependent quenching. https://core.ac.uk/download/323080709.pdf (accessed on 28th September 2025).
4. Buzzetti, L., Crisenza, G. E. M., & Melchiorre, P. (2018). Mechanistic studies in photocatalysis. *Angewandte Chemie International Edition, 58*, 3730–3747. https://doi.org/10.1002/anie.201809984 (accessed on 28th September 2025).
5. Chesta, C. A., Hernández, L., Ibarra, L. E., Lorente, M., et al. (2018). Amplified singlet oxygen generation in metallated-porphyrin doped conjugated polymer nanoparticles. https://core.ac.uk/download/288802218.pdf (accessed on 28th September 2025).
6. Cui, Z. Y., Abdurahman, A., Ai, X., & Li, F. (2020). Stable luminescent radicals and radical-based LEDs with doublet emission. *CCS Chemistry, 2*, 1129–1145. https://doi.org/10.31635/ccschem.020.202000210 (accessed on 28th September 2025).
7. Du, J., Yang, Z., Lin, H., & Poelman, D. (2024). Inorganic photochromic materials: Recent advances, mechanism, and emerging applications. *Deleted Journal, 2.* https://doi.org/10.1002/rpm.20240004 (accessed on 28th September 2025).
8. Fdez. Galván, I., Vacher, M., Alavi, A., Angeli, C., Aquilante, F., Autschbach, J., Bao, J. J., et al. (2019). OpenMolcas: From source code to insight. *Journal of Chemical Theory and Computation, 15*, 5925–5964. https://doi.org/10.1021/acs.jctc.9b00532 (accessed on 28th September 2025).
9. Goulet-Hanssens, A., Eisenreich, F., & Hecht, S. (2020). Enlightening materials with photo switches. *Advanced Materials, 32*. https://doi.org/10.1002/adma.201905966 (accessed on 28th September 2025).

10. Guo, W. J., Chen, Y. Z., Tung, C. H., & Wu, L. Z. (2021). Ultralong room-temperature phosphorescence of silicon-based pure organic crystal for oxygen sensing. *CCS Chemistry, 4*, 1007–1015. https://doi.org/10.31635/ccschem.021.202100932 (accessed on 28th September 2025).

11. Huang, P. (2023). Environmental impacts of soil component interactions. https://www.semanticscholar.org/paper/bd02c6a2a3fa48af96e1a79fc5caab1e3a3ab0e5 (accessed on 28th September 2025).

12. Kommula, B., & Bhattacharyya, S. (2021). Challenges and future prospects of graphene-based hybrids for solar fuel generation: Moving towards next generation photocatalysts. *Materials Advances, 3*, 142–172. https://doi.org/10.1039/d1ma00748c (accessed on 28th September 2025).

13. Kudisch, B. (2025). (Invited) Fundamental photochemistry of electrochemically-prepared radical ions. *ECS Meeting Abstracts*. https://www.semanticscholar.org/paper/aa39152a2df38cbf2de2592fd5bf22d09bf961a1 (accessed on 28th September 2025).

14. Lee, H. P., & Gaharwar, A. K. (2020). Light-responsive inorganic biomaterials for biomedical applications. *Advanced Science, 7*. https://doi.org/10.1002/advs.202000863 (accessed on 28th September 2025).

15. López-Magano, A., Daliran, S., Oveisi, A. R., Mas-Ballesté, R., Dhakshinamoorthy, A., Alemán, J., García, H., et al. (2022). Recent advances in the use of covalent organic frameworks as heterogeneous photocatalysts in organic synthesis. *Advanced Materials, 35*. https://doi.org/10.1002/adma.202209475 (accessed on 28th September 2025).

16. Martyka, M., Tong, X., Jankowska, J., & Dral, P. O. (2025). OMNI-P2x: A universal neural network potential for excited-state simulations. https://doi.org/10.26434/chemrxiv-2025-j207x (accessed on 28th September 2025).

17. Mazumder, A., Li, K., Liu, Z., Wang, Y., Pei, Y., Peteanu, L. A., & Jin, R. (2024). Isomeric effects of $Au_{28}(S\text{-}c\text{-}C_6H_{11})_{20}$ nanoclusters on photoluminescence: Roles of electron-vibration coupling and higher triplet state. *ACS Nano, 18*, 21534–21543. https://doi.org/10.1021/acsnano.4c06702 (accessed on 28th September 2025).

18. Pratley, C., Fenner, S., & Murphy, J. A. (2022). Nitrogen-centered radicals in functionalization of sp^2 systems: Generation, reactivity, and applications in synthesis. *Chemical Reviews, 122*, 8181–8260. https://doi.org/10.1021/acs.chemrev.1c00831 (accessed on 28th September 2025).

19. Prlj, A., Hollas, D., & Curchod, B. F. E. (2023). Deciphering the influence of ground-state distributions on the calculation of photolysis observables. *The Journal of Physical Chemistry A, 127*, 7400–7409. https://doi.org/10.1021/acs.jpca.3c02333 (accessed on 28th September 2025).
20. Schalley, C. (2006). *Analytical Methods in Supramolecular Chemistry*. https://www.semanticscholar.org/paper/7601a3562ccfc20255d3f3856c66f64d9b3d822b (accessed on 28th September 2025).
21. Sharma, P., Pandey, V., Sharma, M. M. M., Patra, A., Singh, B. P., Mehta, S., & Husen, A. (2021). A review on biosensors and nanosensors application in agroecosystems. *Nanoscale Research Letters, 16*. https://doi.org/10.1186/s11671-021-03593-0 (accessed on 28th September 2025).
22. Su, H., Wang, W., Shi, R., Tang, H., Sun, L., Wang, L., Liu, Q., et al. (2023). Recent advances in quantum dot catalysts for hydrogen evolution: Synthesis, characterization, and photocatalytic application. *Carbon Energy, 5*. https://doi.org/10.1002/cey2.280 (accessed on 28th September 2025).
23. Suhaimi, N. A. A., Kong, C. P. Y., Shahri, N. N. M., Nur, M., Hobley, J., & Usman, A. (2022). Dynamics of diffusion- and immobilization-limited photocatalytic degradation of dyes by metal oxide nanoparticles in binary or ternary solutions. *Catalysts, 12*, 1254–1254. https://doi.org/10.3390/catal12101254 (accessed on 28th September 2025).
24. Sun, L., Liu, H., Ye, Y., Yang, L., Islam, R., Tan, S., Tong, R., et al. (2023). Smart nanoparticles for cancer therapy. *Signal Transduction and Targeted Therapy, 8*. https://doi.org/10.1038/s41392-023-01642-x (accessed on 28th September 2025).
25. Sun, Y., Gu, L., Wen, J., van der Tol, C., Porcar-Castell, A., Joiner, J., Chang, C. Y., et al. (2023). From remotely sensed solar-induced chlorophyll fluorescence to ecosystem structure, function, and service: Part I—Harnessing theory. *Global Change Biology, 29*, 2926–2952. https://doi.org/10.1111/gcb.16634 (accessed on 28th September 2025).
26. Volarić, J., Szymański, W., Simeth, N. A., & Feringa, B. L. (2021). Molecular photo switches in aqueous environments. *Chemical Society Reviews, 50*, 12377–12449. https://doi.org/10.1039/d0cs00547a (accessed on 28th September 2025).
27. Wang, Z., Cheng, X., Xie, Y., Liu, S., Dong, M., Zhao, J., Liang, F., et al. (2022). Recent advances in organic room-temperature phosphorescence of heteroatom (B/S/P)-containing chromophores. *CCS Chemistry, 5*, 292–309. https://doi.org/10.31635/ccschem.022.202202242 (accessed on 28th September 2025).

28. Wang, Z., Erhart, P., Li, T., Zhang, Z., Sampedro, D., Hu, Z., Wegner, H. A., et al. (2021). Storing energy with molecular photoisomers. *Joule, 5*, 3116–3136. https://doi.org/10.1016/j.joule.2021.11.001 (accessed on 28th September 2025).
29. Wegeberg, C., & Wenger, O. S. (2021). Luminescent first-row transition metal complexes. *JACS Au, 1*, 1860–1876. https://doi.org/10.1021/jacsau.1c00353 (accessed on 28th September 2025).
30. Yan, H., Song, J., Zhu, S., & Xu, H. C. (2021). Synthesis of acridinium photocatalysts via site-selective C–H alkylation. *CCS Chemistry, 3*, 317–325. https://doi.org/10.31635/ccschem.021.202000743 (accessed on 28th September 2025).
31. Yang, Y., Li, X., Cai, M., & Chen, K. (2024). Review of the mechanisms of liquid-phase transformation of atmospheric phenolic compounds: Implications for air quality and environmental health. *Atmosphere.* https://www.semanticscholar.org/paper/c3398f9797c908afce2aea2e0f071629cca62957 (accessed on 28th September 2025).
32. Yao, Y., Cheng, C., Zhang, C., Hu, H., Wang, K., & De Wolf, S. (2022). Organic hole-transport layers for efficient, stable, and scalable inverted perovskite solar cells. *Advanced Materials, 34.* https://doi.org/10.1002/adma.202203794 (accessed on 28th September 2025).
33. Zhang, J., He, B., Hu, Y., Alam, P., Zhang, H., Lam, J. W. Y., & Tang, B. Z. (2021). Stimuli-responsive AIEgens. *Advanced Materials, 33.* https://doi.org/10.1002/adma.202008071 (accessed on 28th September 2025).
34. Zhou, Q. Q., Zou, Y. Q., Lu, L. Q., & Xiao, W. J. (2018). Visible-light-induced organic photochemical reactions through energy-transfer pathways. *Angewandte Chemie International Edition, 58*, 1586–1604. https://doi.org/10.1002/anie.201803102 (accessed on 28th September 2025).

CHAPTER 3

PHOTOCATALYSIS – MECHANISMS AND MATERIALS

Contents

Introduction 87
3.1. Fundamental Mechanisms of Photocatalysis 91
3.2. Types of Photocatalysts 96
3.3. Factors Influencing Photocatalytic Activity 102
3.4. Applications in Environmental Remediation 107
3.5. Applications in Energy Conversion 112
3.6. Challenges and Limitations 117
Summary 123
References 125

INTRODUCTION

The multifaceted phenomenon of photocatalysis has been of considerable interest in scientific and industrial communities, primarily due to the profound importance it holds for environmental sustainability and energy generation. Through the harnessing of light energy, photocatalysis has the capability to fuel chemical reactions that facilitate the degradation of pollutants, production of renewable energy, and other applications that attest to its diversity. At the heart of this process are photocatalysts, materials that absorb light and, in the process, catalyze chemical reactions. Among these chemicals is one that stands out, namely titanium dioxide (TiO_2), which boasts unparalleled stability, cost-effectiveness, and environmentally friendliness, virtues that qualify it for practical application in activities like waste processing and solar energy transformation (Nematov, 2024). With increased scope of research into photocatalysis, the contribution of the underlying mechanisms and interactions between materials becomes increasingly significant. Photocatalytic process mechanisms are founded upon the generation of charge carriers subsequent to light absorption. When photons are struck upon a photocatalyst, the electrons are lifted from the valence band to the conduction band, forming electron-hole pairs for further redox reactions.

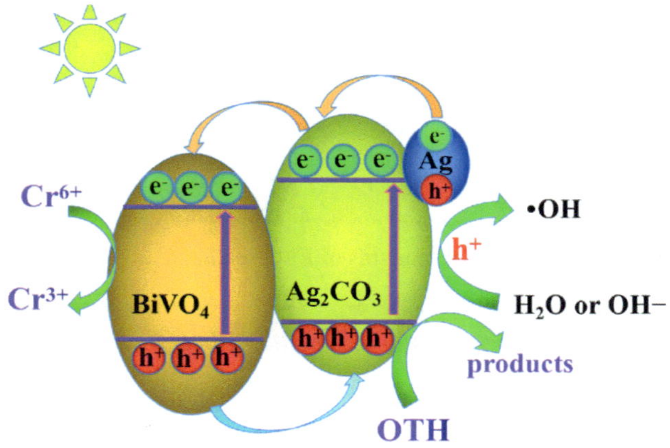

Figure 3.1. *Photocatalysis – mechanisms and materials.*

Source: https://www.researchgate.net/publication/365698237/figure/fig2/AS:1143 1281099723870@1669296675973/The-photocatalysis-mechanism-of-plasmonic-heterojunction-photocatalyst.png.

This electron transfer is necessary for enabling reactions such as the oxidation of organic contaminants and reduction of water for hydrogen evolution. Notably, the photocatalytic efficiency can frequently be regulated by factors such as crystallinity, morphology, and surface area of photocatalysts. Experiments have proven that structural properties of TiO_2, for instance, can be designed for peak photocatalytic performance and broader application in environmental remediation and clean renewable energy technologies (Nematov, 2024). Thus, the accurate study of these processes not only advances our surface-level understanding but also provides the key to innovative material design and application. As photocatalysis becomes more widespread in everyday use—through application in air purification systems and self-cleaning surfaces—

the demand for more advanced and novel photocatalytic materials has risen. This has initiated research into various materials and composite systems that produce improved photocatalytic performance. The most promising among these are hybrid materials that result from the combination of TiO_2 with other semiconductors and metals, which extend the range of light absorption and increase the efficiency of charge separation. Moreover, advances in synthesis methods are yielding materials with their properties engineered for specific applications. Criticisms of these materials reveal their inadequacies and the challenge to overcome, for example, photostability and recyclability. A combined view of how the properties of a material correlate with mechanisms and function will drive future advancements, eventually transforming the role of photocatalysis in the development of sustainable technology (Figure 3.1).

Definition of Photocatalysis

Photocatalysis in environmental chemistry and materials science is a process that accelerates a photoreaction using the assistance of a catalyst in light irradiation. The process is the application of semiconductor materials that provide for the adsorption of light to generate excited electronic states. These excited states enable the generation of reactive species which, in turn, can cause a series of chemical reactions such as pollutants degradation, hydrogen evolution, and greenhouse gases conversion into useful resources. Utilizing sunlight as energy, photocatalysis is one of the novel approaches for sustainable chemistry with the promise to come up with solutions to some of the most serious environmental challenges of our time, notably water purification and air treatment. Photocatalysis follows the process of excitation of electrons from the valence band of a semiconductor to the conduction band, hence generating electron-hole pairs. These carriers can subsequently perform redox reactions with surface-adsorbed species. For instance, the photogenerated holes would oxidize water molecules to form highly reactive hydroxyl radicals, which can efficiently degrade organic contaminants. The electrons, on the other hand, may be engaged in reduction reactions whereby carbon dioxide is reduced to form hydrocarbons or hydrogen gas is generated. This bimodal application is an illustration of the versatility of photocatalysts as, in addition to their capacity for degradation of toxic substances, they can also bring about desirable chemical conversions, hence making photocatalysis an integral part of green chemistry endeavors (Jiang et al., 2022). Nanomaterials have turned into a central focus of photocatalysis due to their unique features, some of which include increased surface area and available electronic structures. Use of nanostructured catalysts enhances light absorption and promotes charge carrier separation, thereby increasing the efficiency of photocatalytic reactions. For example, metal oxides such as titanium dioxide are among the most researched for their photocatalytic behaviors whose performances in contaminant degradation and antimicrobial applications are remarkable. However, such are the challenges as to efficiency and cost-effectiveness, particularly when using such processes at large scale for commercial application. Future progress in photocatalyst material engineering must solve these issues at a low environmental cost and best performance under practical conditions in order to enable future uses in sustainable technology (Yaqoob et al., 2020) (Figure 3.2).

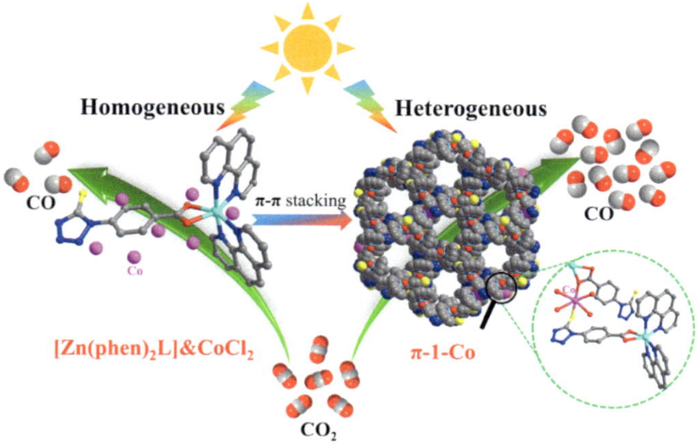

Figure 3.2. *Photocatalysis.*

Source: https://www.pnas.org/cms/10.1073/pnas.2118278119/asset/38ae147b-ce45-49a0-87d4-2c6b356590de/assets/images/large/pnas.2118278119fig06.jpg.

Importance in Environmental and Energy Applications

In the past decade, the significance of photocatalysis has amplified significantly, with more relevance placed in environmental remediation and energy. The process, which utilizes semiconductor materials, enables the transformation of contaminants from water and air into less harmful substances under the influence of light. This activity is priceless in addressing structural environmental issues such as water scarcity and air contamination, driven by industrial activity and urbanization. By degrading harmful organic contaminants to harmless products, photocatalysis is a green technology for environmental protection against degradation. The advancement of photocatalytic materials in the recent past, particularly metal oxides, has depicted enhanced efficiency in catalyzing the process of reactions with good stability and selectivity and hence are indispensable in environmental management practice. Notably, incorporation of photocatalysis in energy development has enabled the avenue towards the realization of sustainable energy production. Photocatalysis not only has a vital role to play in water purification but also can play a role in hydrogen production through the photocatalytic water splitting, thereby offering a cleaner energy source in comparison to traditional fossil fuels. This is subject to the global shift in the usage of renewable energy sources where hydrogen can be produced, which can serve as a common fuel between various industrial processes. The enhancement of photocatalytic activity from advancements in material science has led researchers to explore diverse catalysts, including ternary metal oxides, which have been discovered to improve light absorption and enable greater reaction rates. Photocatalysis is thus a symbol of innovation with ecological sustainability coupled with the necessity for clean energy sources. Further, the overall importance of photocatalysis is reflected through the capability for carbon dioxide utilization and prevention of global warming. Rising atmospheric levels of CO_2 as a result of fossil fuel combustion necessitate effective means of reduction of the greenhouse gas emissions. Photocatalytic reactions are extensively researched for their potential for conversion of CO_2 to valuable products, e.g., methane or methanol, by chemical conversion. Not only does this solve the urgent problem of carbon emissions but also results in the circular economy in that it provides a means of recycling waste CO_2 into useful chemicals. The most recent studies on metal oxide catalysis show their high potential for CO_2 capture and conversion, and they lead the way in the search for sustainable environmental processes and sources of energy (Bisht, 2024). Embedding photocatalytic technologies into wide-ranging environmental and energy applications illustrates their potential to transform engineering towards a more sustainable future.

3.1. FUNDAMENTAL MECHANISMS OF PHOTOCATALYSIS

The analogy of the fundamental steps in photocatalysis lies in the interaction between photons and the semiconductor material, leading to electron excitation. When light with the appropriate energy falls on a photocatalyst, one can get electrons from the valence band excited to the conduction band and form electron-hole pairs. Electron transfer is crucial as the freshly formed electrons can participate in reduction reactions and holes can participate in oxidation reactions. The efficiency of this process not only relies on the character of the photocatalyst, such as bandgap energy and surface area, but also on the wavelength of approaching light that must match the absorption tendency of the material. Furthermore, environmental conditions, including pH and temperature, have the potential to determine the overall efficiency of photocatalytic reactions, highlighting the importance of understanding these mechanisms in order to optimize photocatalytic systems. Apart from the intrinsic electron excitation, charge carrier stabilization—affecting electrons and holes—is important in enhancing photocatalytic activity. The charge carriers formed are susceptible to recombination that can reduce their effective lifetime and decrease reaction yields.

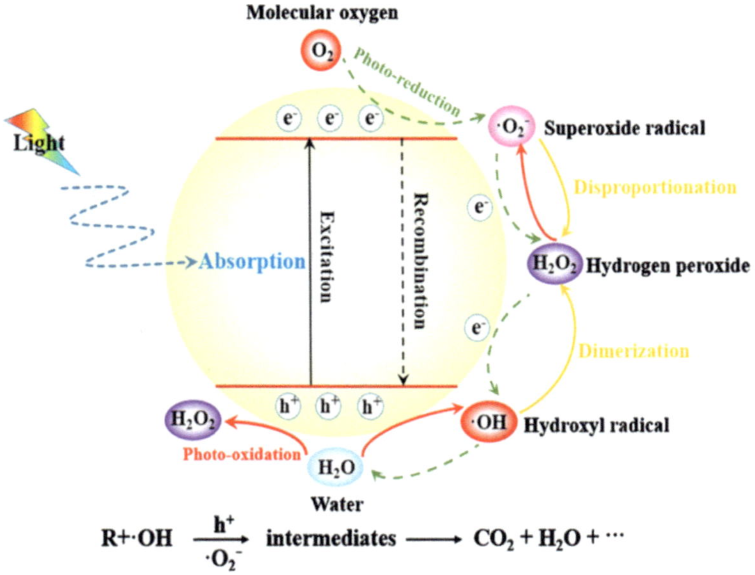

Figure 3.3. *Fundamental mechanisms of photocatalysis.*

Source: https://www.researchgate.net/publication/367368272/figure/fig1/AS:11431281114574377@1674567889669/Schematic-diagram-illustrating-the-mechanism-of-photocatalysis-Reproduced-with.ppm.

Avoidance of this has given rise to a wide range of approaches aimed at enhancing charge separation. For instance, the introduction of cocatalysts has been shown to increase electron transfer, basically decreasing recombination rates and increasing active life of charge carriers. Also, the engineering of photocatalytic structures at the nanoscale can enhance surface interactions, thereby enhancing the likelihood of reaction on the surface of the catalyst. These design rules follow from a profound understanding of the rate-determining steps of photocatalytic processes, as reflected in recent advances in the research on photocatalytic materials and functionalities, capturing the multi-dimensionality of such processes in a comprehensive manner. One of the more recently explored areas of research is the study of plasmon-excited hot carriers in photocatalysis, which has shown great promise for the catalysis of reaction under solar light. When light is illuminated on metal nanoparticles, it may form hot carriers—high-energy electrons—possessing sufficient energy to induce chemical changes regardless of traditional bandgap excitations. Such a mechanism extends photocatalytic materials' operation range by allowing them to utilize the energy of light efficiently. However, challenges arise, particularly regarding their utilization efficiency of the hot carriers. Researchers are engaged in an effort to describe the process of generation and transport of these carriers and develop new approaches towards enhancing their application in catalytic processes. Monitoring of hot carrier dynamics and their interaction with reactants must be facilitated in order to utilize their energy to enhance photocatalytic performance, a new frontier towards the quest for sustainable energy sources (Sun et al., 2025; Yang et al., 2024) (Figure 3.3).

3.1.1. Light Absorption and Excitation of Electrons

The intricate relationship between the absorption of light and excitation of electrons is behind the photocatalytic processes. When a material is exposed to light, it experiences an energy level rise that is crucial in initiating a number of chemical transformations. Specifically, by absorbing photons, valence band electrons in the photocatalyst can be promoted to the conduction band and create electron-hole pairs. This procedure is not only a result of energy transformation, but rather more the inherent capability of catalysts to effectively capture the light energy. The efficiency of the process largely relies on the electronic structure and band gap of materials, which dictate the manner and timing by which the electrons can transfer. As such, the finesse of light absorption holds the most significance since it has direct implications on photocatalysts efficacy in triggering light-induced reactions in different environments. Further research into light absorption emphasizes the role of specific materials in enhancing photocatalytic activity. Notably, plasmonic nanoparticles are prominent due to their remarkable properties that compensate for traditional photocatalysts. Upon absorption of light by these nanoparticles, they are able to generate "hot" charge carriers through plasmon damping, and this can subsequently power chemical transformations in the neighboring medium. The new hybridization of plasmonic and photonic resonances can efficiently promote hot electron generation at wavelengths, thereby maximizing the photocatalytic process (Huang et al., 2020). The ability to accurately engineer the electronic environment and propel the charge carrier dynamics is reflective of how material design can maximize photocatalysis performance and indicates that the future innovation will rest on these subtle interactions between light and nanomaterials. Besides, specific photocatalysts such as hydroxyapatite (HAp) exhibit the complexity of electron excitation processes in light. The research has proved that in the presence of ultraviolet (UV) radiation, oxygen vacancies in HAp are accountable for the surface radicals formation, which play a vital role in the photocatalysis process. Such a model of defect states allows electrons to be promoted from the valence band to donor states related to the aforementioned vacancies (Bystrov et al., 2019). The threshold excitation energy for these transitions has been calculated to be around 3.6–3.9 eV, in agreement with the onset absorptions observed in HAps photocatalytic performances under continuous UV irradiation. This reinforces the central role of the material electronic structure and structural defects, a reminder that only full understanding of light absorption and electron excitation can allow us to optimize photocatalytic systems.

3.1.2. Charge Carrier Dynamics

The carrier dynamics determine the efficiency of photocatalytic materials. Charge carrier formation, transfer, and recombination are determining factors in the overall efficiency of the reaction in photocatalysis. The principal process is light absorption by the photocatalyst, exciting electrons from the valence to the conduction band and generating electron-hole pairs. These charge carriers must be efficiently separated and transported to the surface in order to engage in redox reactions. Poor charge carrier dynamics can lead to high recombination rates, significantly lowering photocatalytic activity. Maximizing charge carrier lifetime is therefore important, and materials that have more efficient charge separation mechanisms by nature tend to offer improved photocatalytic activity. Research in structural alteration and doping has shown that manipulated materials are capable of exhibiting favorable charge dynamics, thus supporting photocatalytic uses in environmental remediation and energy conversion technology. The need for stable materials with charge carrier dynamics is emphasized by recent initiatives for creating enhanced photocatalysts. For instance, tin halide perovskites are a new solution because they are water stable and can accommodate enormous charge carrier dynamics change when they are subjected to some solvents. Recent research on crystals of DMASnI3 indicates that the environmental conditions lead to the formation of protective surface layers, thus enhancing stability and regulating photochemical properties. Interestingly, this *in situ* surface structure directly impacts the charge carrier pathways and allows for sustainable applications, such as hydrogen production. External factors offering means to control these dynamics highlight the key space for research in photocatalytic science, illustrating how knowledge of charge carrier behaviors at the atomic level can result in improved materials, thereby addressing the challenge for sustainable energy applications (Tanaka et al., 2025). Besides, appropriate charge separation is of prime significance not only for the facilitation of photocatalytic activity but also for minimizing the adverse effects of photocorrosion in metal oxide systems. Photocatalytic reactions are generally confronted by stability issues that can be due to inefficient management of charge carriers, leading to side product formations and reduced catalyst life. The application of heterojunctions is an attractive method to improve the charge carrier dynamics with enhanced separation efficiency. The heterostructures enable effective interfacial transfer of charges, thereby reducing charge collection at active sites, which is more likely to result in photocorrosion. By engineering such heterojunctions, researchers have achieved significant advancements in overcoming key bottlenecks in photocatalytic removal of NO_x, particularly reaction routes and selectivity (Hailili et al., 2025). This new method creates a pathway towards the creation of more efficient photocatalytic materials that are resistant to stresses of operation and can still remain effective for environmental use.

3.1.3. Reaction Pathways and Product Formation

The understanding of reaction mechanisms and product formation in photocatalysis is central to the optimization of degradation of organic pollutants. Photocatalysis relies on the generation of reactive species such as hydroxyl radicals and holes to catalyze contaminant destruction. Such photocatalytic processes are effective depending on the properties of the materials used and light absorption ability. For instance, such materials as BiOBr nanosheets have been used to improve charge carrier splitting under piezoelectric fields, furthering the generation of reactive radicals and consequently reaction channels for pollutant degradation. Interestingly, under combined solar and ultrasound conditions, BiOBr exhibited strikingly greater mineralization rates towards recalcitrant contaminants, suggesting greater energy transfer to the reactants, which leads to enhanced product formation (Falletta et al., 2023). Another important aspect in defining reaction pathways is the identification of the intermediate products formed during the degradation process. In the majority of photocatalytic studies, focus is placed more on the parent compounds rather than by-products, which can sometimes be toxic or less innocuous compared to the original pollutants. Identification of these intermediates is important in determining the overall efficiency and safety of photocatalytic systems. For instance, in studying the degradation of ibuprofen, advanced analytical techniques like ultraperformance liquid chromatography–mass spectrometry (UPLC/MS) have been useful for understanding transformation products. These analyses underscore the complex nature of the pathways taken when organic pollutants are mineralized into less harmful materials, providing critical information that can be applied to optimizing catalyst design and catalytic processes

(Afanasiev et al., 2023). Significance of reaction pathways extends far beyond degradation rates in and of themselves; they are integral to charting the larger-scale environmental impact of photocatalytic systems. Since the photocatalytic world is increasingly realizing the importance of by-products, future research focuses on minimizing the formation of toxic intermediates and maximizing the stability and recyclability of the catalyst. Challenges brought forth by the constraints in reaction rates and catalyst recovery are being addressed by new approaches and materials, which offer enhanced mineralization and reduced ecotoxicity. By continued enhancement in understanding reaction pathways and by-product formation, photocatalysis can emerge as a leading method in pollution abatement and encouraging environmental green practices (Afanasiev et al., 2023) (Figure 3.4).

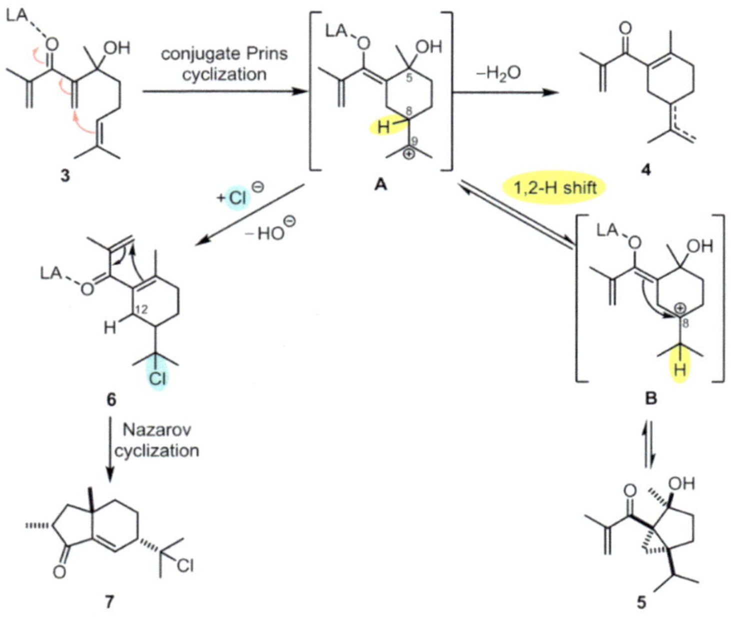

Figure 3.4. *Reaction pathways and product formation.*

Source: https://www.researchgate.net/figure/Proposed-reaction-pathways-for-the-formation-of-products-4-to-7-LALewis-acid_fig4_351111025.

3.2. TYPES OF PHOTOCATALYSTS

The photocatalyst environment is diverse, drawn from various structures, compositions, and mechanisms, which collectively provide improved photocatalytic performance. Semiconductors, and more specifically titanium dioxide (TiO_2), are the most investigated photocatalytic compounds since they are stable, nontoxic, and possess broad spectral response. Their activity is primarily based on the crystallographic structure and morphology, and they can be tailored for specific applications such as environmental remediation or energy conversion.

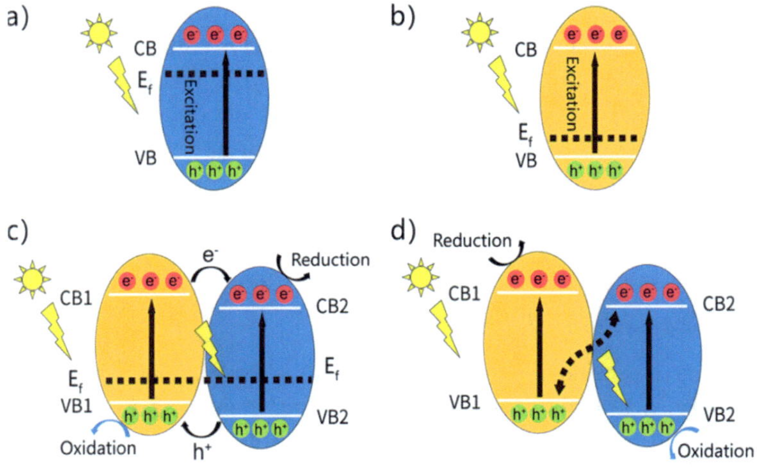

Figure 3.5. *Types of photocatalysts.*

Source: http://researchgate.net/publication/339834188/figure/fig1/AS:114312812 19309383@1706017509323/Schematic-diagram-of-different-types-of-photocatalysts-a-and-b-single-material.jpg.

The incorporation of metal nanoparticles, such as gold or silver, enhances the photocatalytic activity through localized surface plasmon resonance (LSPR). The mechanism facilitates enhanced light harvesting and better charge separation, hence enhancing photocatalytic activity (Deepak et al., 2024). The other important class of photocatalysts includes metal oxides, such as zinc oxide (ZnO), which possess unique properties that enhance their photocatalytic activity. ZnO nanostructures possess increased degradation activities against contaminants due to their superior light absorption and efficient electron-hole separation processes. For example, ZnO nanorods (ZnO-R) perform better than commercial ZnO by a considerable difference, removing 98.2% of carbaryl pesticide from wastewater compared to the removal of 87.3% of ZnO-C (Thi et al., 2024). The rod-like morphology for greater interaction with light is the reason for improvement. As photocatalysis continues, the research into many metal oxide structures is promising for future successes in the degradation of environmental contaminants and other technologies. Plasmonic photocatalysts are a novel collection of materials that incorporate metal and semiconductor systems and leverage both plasmonic and semiconducting processes. This hybrid approach is aimed at avoiding inherent deficiencies present with typical photocatalysts. For instance, the hybridization of TiO_2 with gold nanoparticles has shown high improvements in photocatalytic action due to the creation of a Schottky barrier at the heterojunction, which enhances charge separation (Deepak et al., 2024). Hybrid systems can effectively manage charge dynamics and improve other mechanisms for photocatalytic reactions. The distinct mechanisms utilized by plasmonic photocatalysts uncover the possibilities of new applications, such as the regeneration of important cofactors like NADH, thus providing promising paths for developing photocatalytic processes and materials in a wide range of areas, from energy to environmental science (Figure 3.5).

3.2.1. Semiconductor Photocatalysts

When it comes to semiconductor photocatalysts, the efficiency of the utilization of solar energy as a driving force for chemical reactions is primarily dependent on materials and structural types. Traditional photocatalysts such as titanium dioxide have been the norm for many years due to their stability and activity under UV light. However, material science breakthroughs have led to the research of complex structures, including heterojunctions and two-dimensional materials, which hold the promise for greatly enhancing light absorption and charge separation. These advanced semiconductor materials are crucial to engineer, as they aim to surpass the limitations posed by conventional photocatalysts. With continued evolution of the field of study, increased understanding of the mechanisms will be required to optimize photocatalytic processes and expand their range of applications in energy conversion and environmental remediation. The advent of 3D printing technology has made photocatalyst semiconductor synthesis a new, transformative field and allowed unprecedented complexity and composition control over photocatalytic materials. This new method streamlines the manufacturing process so that heterogeneous materials with varying photocatalytic properties can be integrated in a single substrate. Additionally, 3D printing provides an extremely low amount of material wastage during manufacture, potentially saving costs and increasing ease of access. Using 3D printing, researchers are able to tailor the properties of semiconductor photocatalysts to make them optimal for specific applications, such as cleaning water or converting carbon dioxide. Despite these advantages, in-depth reports detailing the use of 3D printing in photocatalytic research are still emerging, calling for additional targeted research into this promising nexus of technology and materials science (Khoo et al., 2024). Among recent materials in photocatalysis, carbon-nitrogen molecules, i.e., poly (heptazine imide) (PHI) and poly (triazine imide) (PTI), are of considerable attention due to their unique electronic structures and potential enhanced photocatalytic activity. The underlying formation mechanisms inherent to these materials induce energy band structures within them that can facilitate various photocatalytic reactions, hence enhancing their efficiency for solar energy applications. The synergy between PHI and PTI, along with emerging characterization techniques, facilitates the full picture of their photocatalytic performance. As outlined in current literature, the optimization of such photocatalysts for specific purposes can significantly improve their performance in a variety of reactions, highlighting their prospects for a promising future in furthering sustainable solutions to energy and environmental issues (Jing et al., 2024). With ongoing research, the full potential of PHI and PTI as effective photocatalysts might be unlocked in the near future (Figure 3.6).

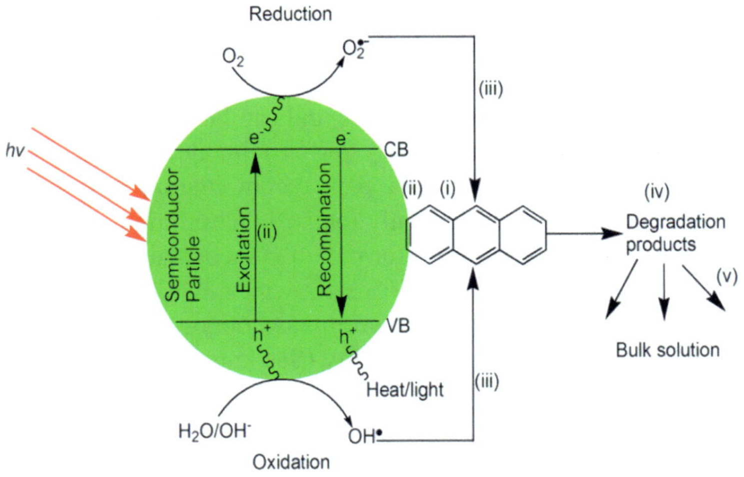

Figure 3.6. *Semiconductor photocatalysts.*

Source: https://duckduckgo.com/?q=Semiconductor+Photocatalysts&iar=images&t=h_&iai=https%3A%2F%2Fwww.researchgate.net%2Fpublication%2F303627507%2Ffigure%2Ffig2%2FAS%3A369883800981-505%401465198578545%2FKey-steps-in-semiconductor-photocatalysis.png.

3.2.2. Metal-Based Photocatalysts

Research on metal-based photocatalysts has become increasingly important in surmounting the energy dilemma and environmental concerns of the modern age.

Solar light is absorbed by metal-based photocatalysts to drive photochemical reactions, offering a clean method of energy conversion and pollutant degradation. Metal clusters are among the photocatalysts that have remained at the focal point of research over the past few years due to their unique physical and chemical behavior. Such clusters are also capable of enhancing the absorption of light and charge separation, which are both vital to overall improvement of photocatalytic systems. Also, novel application of artificial intelligence in both synthesis and characterization of the clusters is a significant milestone, which allows modification to be tailored in an effort to maximize their performance. Hence, the preparation of metal-based photocatalysts not only accelerates the advancement of materials science but also addresses the global energy and environmental problems urgently, in effect revolutionizing photocatalysis (Li et al., 2025). Among various metal-based photocatalysts, metal vanadates have garnered considerable attention owing to their excellent properties and versatility. These high semiconducting metal oxides also possess high redox potential and effective separation of charge carriers generated through light, thereby providing a best-suited candidate for a variety of applications like energy conversion and environmental remediation. The structure and physicochemical characteristics of metal vanadates are adjustable in nature and can be manipulated to enhance their photocatalytic activity, particularly for sustainable hydrogen generation. Factors like the quick electron-hole pair recombination still affect them, however. Extensive research is currently focused on optimizing these materials through methods like heterojunction formation and cocatalyst loading with the intent to enhance their photocatalytic performance. Ultimately, the ongoing betterment of metal vanadates holds promising opportunities for clean fuel production and environmental sustainability (Sasikumar et al., 2025). Merging metal-based photocatalysts with traditional and modern technologies opens up new horizons for new uses. For instance, combination of these catalysts with nanotechnology can lead to the development of multifunctional photocatalytic systems with increased efficiency and expanded applicability. In addition to hydrogen production and degradation of pollutants, metal-based photocatalysts may be of the utmost significance for carbon dioxide reduction and production of useful chemical feedstocks. With the growing demand for sustainable technologies, researchers are now placing greater focus on improving the fabrication techniques and scalability of these materials. Advanced synthesis methods coupled with insight from computational modeling can provide critical information in order to further enhance the design and activity of metal-based photocatalysts. By bridging the theory-practice divides, it is possible to unlock the full potential of such materials to provide solutions to the dual challenges of energy production and environmental conservation (Li et al., 2025) (Figure 3.7).

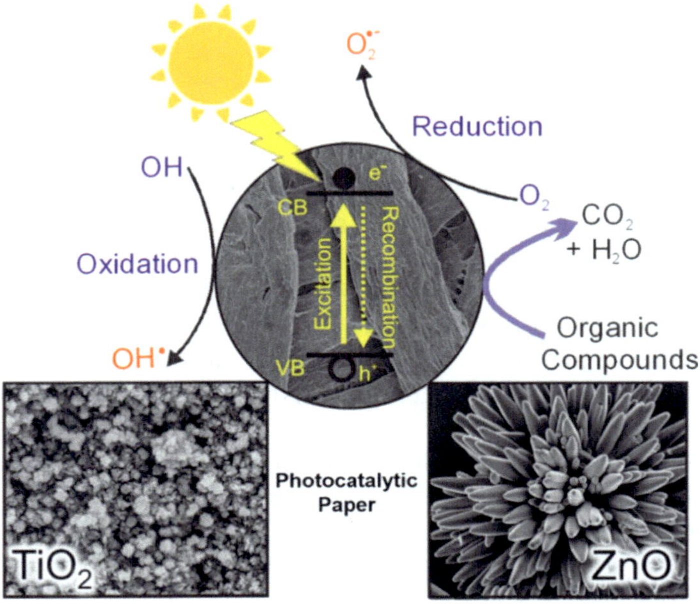

Figure 3.7. *Metal-based photocatalysts.*

Source: https://pub.mdpi-res.com/catalysts/catalysts-11-00504/article_deploy/html/images/catalysts-11-00504-ag.png?1618906404.

3.2.3. Hybrid Photocatalysts

The integration of hybrid photocatalysts in photocatalytic applications is a breakthrough in the pursuit of effective and efficient materials that are suitable for use in environmental remediation and energy conversion. Hybrid photocatalysts, which are typically made of two or more semiconductors and metals or other semiconductor compounds, have exhibited improved photocatalytic activity compared to their respective components. The interaction of different materials created by their union has a tendency to enhance charge separation and enhance the light absorption efficiencies. In particular, the hybrid systems exploit the individual optical and electronic properties of the components with resultant multifunctional applications such as CO_2 reduction and water treatment. As pointed out, Cu_2O-based composites are paradigm examples of this potential because of the display of desirable traits such as low cost and greenness, although challenges such as the high rate of charge recombination need to be addressed in real applications (Su et al., 2023). Furthermore, the application of hybrid semiconductors has also been driven by the necessity for reducing environmental problems caused by the consumption of fossil fuels and pollution. The unique characteristics of hybrid photocatalysts such as their increased charge carrier mobility and increased surface area enable them to be extremely efficient in most photocatalytic applications. Hybrid semiconductors, when structured with non-noble and noble metals or carbon-based materials, show improved activity in multidisciplinary fields like the generation of energy and the detoxification of the environment. These substances promise to resolve energy problems and mitigate environmental degradation with effective degradation of contaminants and recycling of CO_2 (Mishra et al., 2023). Cooperation among different materials provides a broader range of photocatalytic reactions compared to that in single-component systems. It is essential to overcome the difficulties of hybrid photocatalysts to develop them further and utilize them in practice. While they are perceived to bring significant advantages, stability and durability over long times in working conditions remain pre-eminent factors. Research is increasingly focused on surface-interface engineering for achieving optimum interaction among the components of hybrid systems to realize enhanced photocatalytic performance and stability. Stable hybrid photocatalysts with the new materials and novel fabrication techniques are important to overcome these challenges. Thus, with the growth of this area, the path to the development of more efficient and stable hybrid photocatalysts is significant not only for scholarly advancement but also for application in energy and environmental protection (Mishra et al., 2023; Su et al., 2023).

3.3. FACTORS INFLUENCING PHOTOCATALYTIC ACTIVITY

In photocatalysis, a number of factors significantly influence the efficiency and effectiveness of photocatalytic performance. Among them, the structure of the catalyst is a critical factor. The intrinsic structural characteristics such as the surface area and crystal facets significantly influence the behavior of photogenerated carriers. For instance, a work indicated that the photocatalytic activity of CeO_2, governed by different crystal facets, showed considerable differences. Specifically, the rod-like morphology of NR-CeO_2 exhibited a photocatalytic activity of approximately 19.7 times higher than CdS alone, demonstrating the role played by morphology and facet exposure in maximizing photocatalytic performance (Liu et al., 2023). The observation reaffirms the importance of tailored design and synthesis of the photocatalyst, wherein optimal morphology may generate enhanced efficiency in applications such as hydrogen evolution and degradation of pollutants. Another critical photocatalytic activity parameter is the existence and type of dopant in semiconductor materials. Doping is a thoroughly proven means to regulate the electronic properties of photocatalysts, thereby enhancing their photochemical efficiency. Dopants, which are capable of developing new levels within the band gap, facilitate easy separation of charge carriers and decrease the rate of recombination. As noted in literature, the addition of suitable dopants such as transition metals or nonmetals can significantly improve the photocatalytic activity of compounds like TiO_2 and ZnO. Controlling the doping type and concentration can customize the electronic structure so that light absorption and subsequent generation of resulting charge carriers can be enhanced (Gusain et al., 2020). This approach not only enhances photocatalytic activity but also expands the operating window of photocatalysts so that photocatalysts can be activated through visible light, which is generally a limitation in conventional systems. Finally, external parameters such as light intensity, temperature, and pH significantly regulate photocatalytic activity. The efficiency of a photocatalytic reaction is directly related to the intensity of incident light since it determines the amount of available energy for electron excitation in the photocatalyst. Temperature also affects the kinetics of the photocatalytic reaction where an increase will accelerate rates of reaction up to a point. Similarly, the pH of solution regulates ionization of photocatalysts as well as contaminants and hence impacts adsorption kinetics as well as the consequent efficiency of degradation. These are parameters that can be adjusted to maximize the photocatalytic reactions so that systems are used at maximum capability in real-world applications for environmental remediation and treatment of wastewater (Gusain et al., 2020). Therefore, considering these externalities is important when designing efficient photocatalytic systems to address contemporary environmental issues.

3.3.1. Light Wavelength and Intensity

The intricate relationship between the intensity and wavelength of light forms the pillar of photocatalysis that significantly influences photochemical reaction efficiency. The energy of the photons is determined by the wavelength of light, therefore electron excitation in photocatalyst materials. If photons of appropriate wavelengths strike a photocatalyst, they will be able to initiate charge separation, thereby accelerating the reaction kinetics. For instance, bandgap energy-modulated semiconductors for the absorption of a particular wavelength show maximum photocatalytic activity. At low intensities of light, charge carrier excitation may be brief, limiting reaction rates. Hence, wavelength versus intensity correlation is crucial in optimizing photocatalytic systems and facilitating efficient conversion of solar energy into chemical energy. Moreover, the light intensity not only dictates the excitation rate but also regulates the subsequent vibrational and electronic transitions of the photocatalytic material.

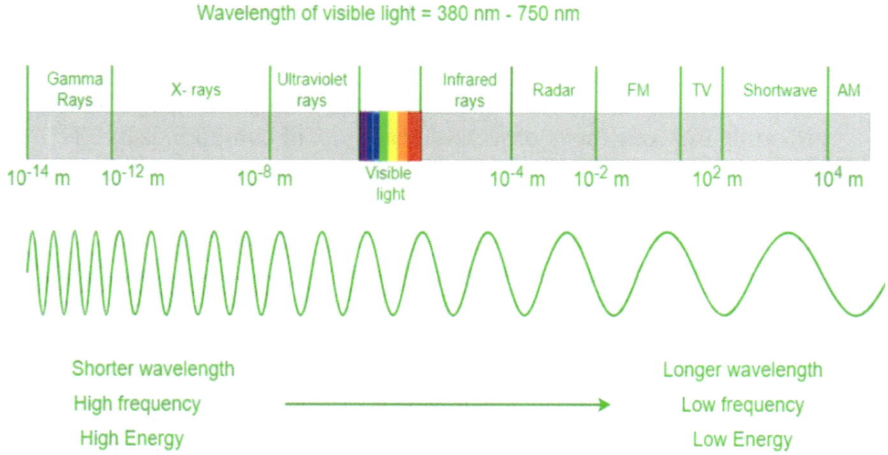

Figure 3.8. *Light wavelength and intensity.*

Source: https://media.geeksforgeeks.org/wp-content/uploads/20220511115054/spe1.jpg.

The association between optical property and solvent polarity reflects the extent to which the two facets are embedded in one another while they dictate photocatalytic performance. Recent studies have proven that altering the solvent conditions can alter both the size and intensity of emission peaks, which are significant for catalysis. As illustrated in research on the solvent polarity influence on circularly polarized luminescence, there exists a linear correlation between optical properties of this kind and reaction efficiencies, and consequently avenues for the optimization of photocatalytic performances via tailored solvent selection are presented (Liu et al., 2024). Such findings present new paradigms for the design of next generation photocatalysts that utilize light wavelength and intensity for amplified reactivity. In plasmonic photocatalysis, the sensitivity to light intensity is even more important. Localized surface plasmon resonance (LSPR) in metal nanoparticles enhances collective oscillations of charges that significantly enhance light-matter interactions and, therefore, strengthen the electric field intensity at the catalyst surface. The increase in intensity encourages electron transfer processes that are involved in photocatalytic reactions. However, there remain controversies in balancing the idea of the effect of temperature on reaction rates and light intensity, in relation to understanding contributions from thermal and non-thermal effects. Kinetic studies of photocatalytic reaction usually employ Arrhenius-type expressions but detect strong wavelength and reaction condition dependencies, highlighting the complexity in optimizing light intensity for optimal catalytic efficiency

(Verma et al., 2024). This general strategy opens avenues for future efforts toward the ultimate potential of photocatalytic systems (Figure 3.8).

3.3.2. Surface Area and Morphology of Photocatalysts

The surface area and morphology of photocatalysts largely contribute to their overall catalytic efficiency and performance in photochemical reactions. A greater surface area provides more sites for the adsorption of light and contact with the reactants, enhancing the photocatalytic process. For instance, ultra-porous materials such as metal-organic frameworks (MOFs) and covalent organic frameworks (COFs) facilitate efficient diffusion of the reactants and products while providing maximum light exposure at the catalyst's surface. The tunability and freedom of design of their characteristic structural features enable them to optimize photocatalytic reactions in a wide variety of applications, including CO_2 reduction and hydrogen evolution, as highlighted by recent studies exploring the fine interactions between structure and property (Tang et al., 2024). The morphology of the photocatalysts, from particle size and shape to structural texture, also plays an important role in governing their reactivity. For example, nanostructured material exhibits improved electron transport characteristics, which are beneficial in inhibiting electron-hole recombination pair and thus improving the photocatalytic activity. Advances in synthesis processes make possible selective tuning of these morphological characteristics and therefore the modulation of the photocatalytic activity of novel materials. Furthermore, some composite structures that merge different dimensionalities can potentially take advantage of synergism among different morphologies to improve photocatalytic activity. This has been key to the design of highly advanced photocatalysts able to make efficient use of solar energy in CO_2 and other impurities' conversion (Mishra et al., 2024). Ultimately, sustained research activities will attempt to unravel and take advantage of the relationships between morphology, surface area, and photocatalytic activity, with a view to designing improved materials. Complete understanding of the relationships is crucial for the advancement of photocatalyst technology. As the researchers enhance the surface properties and establish novel morphologies, new directions for improved charge carrier performance and light absorption are achieved. Along these lines, integration of functional components into the photocatalyst structure, e.g., cocatalysts or hybrids, provides promising directions for performance augmentation. The future of photocatalysis lies in this continued advancement of materials science, with controlling morphology and surface area strategically behind innovation for renewable energy use (Mishra et al., 2024; Tang et al., 2024).

3.3.3. Environmental Conditions (pH, Temperature)

Environmental conditions, particularly pH and temperature, play a big role in the photocatalysis processes. Alterations in pH could be strikingly effective in altering the surface charge and activity of photocatalytic materials. For instance, most photocatalysts will exhibit optimal activity at specific pH levels, which are typically related to ionization states of their active sites. Moreover, as in the case of research on *Staphylococcus aureus*, pH changes are responsible for extreme modifications in gene expression and molecular behavior and hence have a general impact on the efficacy of photocatalytic processes. Absorption and interfacial interaction of reactive species differ at various pH levels, ultimately affecting the rate of photocatalytic reactions and efficiency. Hence, optimization of pH is crucial to enhance the photocatalytic activity of materials in applications such as water purification and CO_2 reduction. Temperature is another critical environmental parameter that regulates the efficacy of photocatalytic reactions. The rate of most photocatalytic processes will increase with higher temperature due to enhanced molecular mobility and kinetic energy, which results in more favorable interactions between reactants and photocatalysts. But after the temperature exceeds an optimal value, it can lead to unwanted side reactions or deterioration of photocatalyst materials. The same phenomenon has been observed in glass fiber reinforced polymer composites in which different environmental conditions, temperature and acidic environments included, play an important role in influencing the material properties. Increased temperatures also produce effects on photocatalytic systems' structural reliability and durability. Therefore, a fine balance of temperature is of paramount importance in the design and deployment of photocatalysts to exploit their fullest potential while providing stability and efficiency of the material. In systems of photocatalysis, interactions between pH and temperature could generate complex dynamics on overall performance. Various materials involved in photocatalysts show varying degrees of sensitivity to these factors, and familiarity with their interaction is necessary for particular applications. For example, research indicates that acidic environments can reduce the wear resistance of composite materials while enhancing degradation reactions. In contrast, applying moderate temperatures in combination with optimal pH conditions can enhance photocatalytic efficiency through increased reactant interactions and less potential for degradation. These results highlight the need for intensive studies regarding the existence of these environmental factors and their effects on photocatalytic efficiency. Greater understanding of these interdependencies is likely to lead to new innovations in the design of photocatalysts that are more efficient and longer lasting for remediation of environmental challenges in energy conversion and pollutant degradation.

3.4. APPLICATIONS IN ENVIRONMENTAL REMEDIATION

The ongoing increase in environmental pollution, particularly in water bodies, necessitates new remediation processes. One such process that is gaining favor is photocatalysis, where light is employed to activate catalysts that accelerate the degradation of organic pollutants. The process harnesses solar energy for the production of reactive oxygen species (ROS), which effectively break down toxic compounds to less toxic products. Of the numerous materials used in photocatalytic processes, natural mineral materials (NMMs) have attracted interest based on their affordability, availability, and environmental friendliness. NMMs have been shown to greatly improve the effectiveness of advanced oxidation processes (AOPs), providing green solutions to long-lasting organic pollutants. With the inclusion of these materials in photocatalytic systems, researchers can actually increase pollutant degradation without overmuch hazardous byproducts, thereby answering the critical need for effective environmental remediation (Xu et al., 2025). With NMMs combined with them, catalytic carbon nanocomposites (CCNs) have also shown amazing potential in photocatalytic environmental processes. These high-surface-area and tunable reactivity materials significantly enhance the photocatalytic degradation of a wide range of contaminants, including organic dyes and heavy metals. The unique characteristics of CCNs facilitate sophisticated catalysis such as photocatalysis, electrocatalysis, and biocatalysis. Research presents the effectiveness of CCNs in various real-world applications and reveals their potential for large-scale application in continuous-flow and solar-powered systems. However, more research in the future will need to address problems such as leaching of nanoparticles and performance degradation in complex water matrices in order to use them effectively to maximize their application in environmental remediation (Mahesh et al., 2025). Integration of such advanced materials into the photocatalytic system is a step towards superior research, supporting the ongoing quest for green solutions to pollution. Furthermore, photocatalytic technology is not merely for water purification but also for air improvement and soil restoration. Photocatalysis for the breakdown of volatile organic compounds (VOCs) and other toxic air pollutants is a promising step toward maintaining cleaner urban air. Photocatalytic processes can be incorporated into building construction materials, allowing surfaces to actively degrade contaminants upon exposure to sunlight. Similarly, soil decontamination through photocatalysis aims to clean contaminated lands, employing advanced materials to break down contaminants embedded in the soil matrix. As studies on the mechanism and functioning of photocatalytic materials are continually evolving, there is vast potential in supporting these technologies for mitigating environmental challenges. Therefore, additional research into the optimization of these applications is needed to best utilize the potential benefits of photocatalysis in total environmental remediation strategies (Mahesh et al., 2025; Xu et al., 2025) (Figure 3.9).

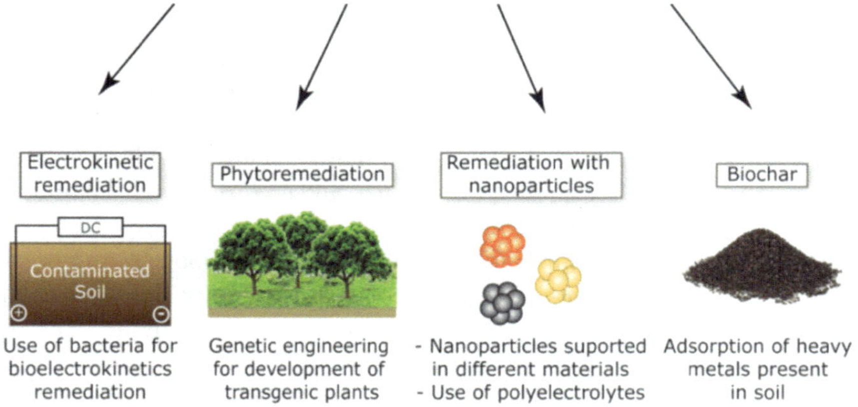

Figure 3.9. *Applications in environmental remediation.*

Source: https://www.researchgate.net/publication/339312252/figure/fig8/AS:944809816633352@1602271622137/The-new-approaches-of-the-techniques-for-soil-remediation.png.

3.4.1. Degradation of Organic Pollutants

Organic pollutant destruction has been a focus of environmental research, particularly with the development of photocatalysis as a premier remediation technology. Traditional wastewater treatment practices are unsuccessful in the management of complex organic pollutants, including drugs, personal care products, and industrial chemicals. Such pollutants can possess varying chemical stability and would remain resistant to conventional degradation operations. Photocatalysis employs semiconductor materials, mostly titanium dioxide (TiO_2), to decompose these organic pollutants effectively upon light irradiation. Photocatalytic activity initiates the formation of reactive oxygen species, which subsequently react with the pollutants, leading them to disintegrate into innocuous by-products. Researchers emphasize that the conditions under which photocatalysis takes place should be optimized such that best performance in pollutant degradation with little harmful by-products is obtained, addressing problems of secondary pollution in the process of treatment (Gatou et al., 2024). Optics in photocatalytic materials significantly enhances the ability of organic pollutants to degrade. For instance, structural modifications such as metal or nonmetal doping of TiO_2 can efficiently modulate its bandgap, leading to improved visible light absorption and improved photocatalytic performance. Additionally, nanocomposite photocatalysts are formed by combining the desirable properties of several materials, which lead to synergistic benefits accelerating pollutant degradation rates. Techniques such as ultrasound-assisted synthesis are now efficient methods for synthesizing such nanostructured photocatalysts, which can guarantee enhanced mass transfer as well as catalyst-pollutant interactions (Mahmoud et al., 2023). Tailoring the structural characteristics of TiO_2 and exploring alternative synthesis routes can assist the researchers to obtain optimized photocatalytic systems capable of degrading a broader range of organic pollutants to guarantee effective environmental remediation strategies. The integration of photocatalysis with newer approaches, i.e., sonophotocatalysis, provides new possibilities for the enhancement in the degradation of organic pollutants. The synergistic combination of ultrasound application along with photocatalytic processes makes the degradation more efficient by faster reaction rates and higher generation of reactive species. Sonophotocatalysis processes involve the generation of microjets, cavitation bubbles, and local heat, in addition to further breaking down the pollutants. In addition, all significant parameters such as the catalyst loading, the concentration of the pollutants, and the light intensity decide the overall efficiency of this process. Identifying these dynamics is necessary to design effective photocatalytic systems for given organic pollutants. By way of analysis and enhancement of these advanced techniques, the potential for wastewater treatment applicability grows, offering eco-friendly alternatives to the acute issue of organic pollutant breakdown (Gatou et al., 2024; Mahmoud et al., 2023).

3.4.2. Water Purification Techniques

The quest for effective water purification systems has driven significant progress in the field of photocatalysis that offers novel solutions for the degradation of pollutants as well as water safety. Traditionally, systems such as filtration, chlorine disinfection, as well as reverse osmosis have been employed in the purification of water, but these systems are prone to inefficiencies and potential ecological harms. Photocatalysis stands out in that it has the capability to utilize light energy to induce chemical reactions to degrade organic contaminants as well as destroy microbial pathogens. It is particularly helpful due to its lower cost of operation and capacity to eliminate various types of contaminants. As research continues to explore the mechanisms of photocatalysis, the development of innovative materials such as titanium dioxide (TiO_2) nanoparticles improves the method and widens its application in various water purification applications (Kapoor et al., 2024). The inclusion of nanotechnology in photocatalytic devices has also revolutionized the process of water purification. Nanoparticles, with large surface area and clear attributes, significantly enhance photocatalytic reactions.

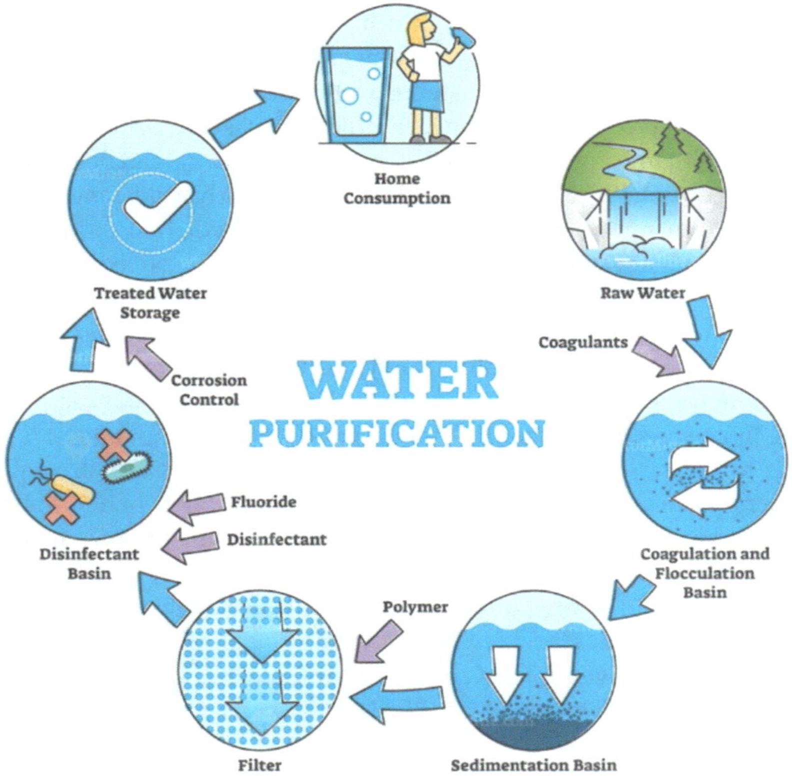

Figure 3.10. *Water purification techniques.*

Source: https://vectormine.b-cdn.net/wp-content/uploads/Water_Purification_V2_outline.jpg.

For instance, metal doping of TiO_2 creates increased photocatalytic activity visible to the naked eye, pushing the daylight available to be utilized in water purification beyond UV light. Innovative advances come in the form of novel photoreactor structures, which maximize the photon-catalyst-contaminant interaction for optimal degradation effectiveness. Among some of the improvements in recent years are portable and multi-purpose photoreactors that are able to effectively degrade some of the contaminants, such as drugs and bacteria, in different light conditions (Castro et al., 2024). Aside from making water purification technology more accessible, portability also allows it to be used in many settings where water quality is a problem. As water shortage and pollution persist to become vastly acute environmental issues, the imperative for scalable and effective purification methods becomes increasingly urgent. Photocatalysis, especially when married with nanotechnology, becomes a compelling means of tackling these problems by ensuring sustainable practices. The capacity to effectively remove toxic pollutants and pathogens places photocatalysis at the center of global initiatives to enhance water quality. Besides, the development of new portable photoreactors offers a new degree of flexibility in managing water treatment for laboratory and external use. When these advanced technologies are integrated in traditional water management systems, the capability to enhance public health as well as induce sustainability is significantly amplified, rendering photocatalysis the pillar in the future of water treatment (Kapoor et al., 2024) (Figure 3.10).

3.4.3. Air Quality Improvement

The deterioration of urban air quality has become a biting problem that has made it imperative to seek new remedies such as photocatalysis. The technology employs the energy of certain catalysts, most often from nanomaterials, to initiate chemical reactions by irradiation with light, disintegrating toxic air pollutants into harmless products. Photocatalysis involves mechanisms of electron-hole pair generation when exposed to light, promoting degradation reactions that can eliminate volatile organic compounds (VOCs) and particulates. Controlling air quality with photocatalytic processes not only increases environmental sustainability but also reduces health hazards from air pollution. The incorporation of advanced materials like TiO_2 and g-C_3N_4 reflects contemporary improvements in photocatalytic efficiency, which is a promising direction for city air cleanup (Prakash, 2025). Practically speaking, photocatalysis exceeds theoretical benefit, being used in novel systems with applications towards air cleaning. One promising development along this line is the inclusion of low-cost, modular modules that are efficient and sustainable. Additional advancements have emphasized solar-powered systems that are self-sufficient, hence reducing reliance on conventional energy sources that lead to pollution. One such example is the use of internet of things-enabled air purifiers, which shows the way photocatalytic technology can offer live air quality reports and dynamically adjust operating parameters in a bid to achieve optimal performance. These techniques ensure extensive accessibility and adaptability to fit numerous indoor environments, constituting an efficient solution to combat air quality problems in urban areas (Vakharia et al., 2025). Moreover, the integration of photocatalysis with other filtration technologies can result in synergistic effects, maximizing the overall management of air quality. As illustrated in the development of air-cleansing machines using combinatory filters like HEPA and polypropylene, photocatalytic materials significantly improve the filtration of particulate matter such as $PM_{2.5}$ and PM_{10}. Not only are the systems efficient in cleaning air, but their affordability is also achieved as they are designed to be cheap to use on a mass scale. The application of photocatalysis as a technology for improving air quality aligns with global intentions aimed at achieving sustainable urbanization and public health. Therefore, continued research on photocatalytic processes and materials will continue to play an important role in guiding effective strategies to clean air in increasingly polluted urban areas (Prakash, 2025).

3.5. APPLICATIONS IN ENERGY CONVERSION

The new science of photocatalysis is firmly placed at the crossroads of energy conversion and environmental sustainability. Its implications are most far-reaching in addressing the challenge of undoing the effects of climate change by enhancing the efficiency of clean energy technologies. Photocatalysis can help convert sunlight to chemical energy, thereby fulfilling the mandate of renewable sources of energy. This is especially important in the case of photocatalytic reduction of CO_2 since, aside from being a means to renewable fuels, it also helps to reduce levels of greenhouse gases in the environment. By the use of photocatalytic systems, one can then proceed to actually transform waste products into valuable resources according to principles of circular economy while, at the same time, offering cleaner energy solutions. Metal–organic frameworks (MOFs) are multidisciplinary materials that have reached photocatalysis and have shown high potential for application in energy conversion. The structural properties of MOFs, such as their porosity and electronic tunability, allow them to exhibit better photocatalytic performance without the need for additional catalysts, which introduce complexity and cost. In addition, the capability of MOFs to enable self-tuning by structural engineering lends it a prime place as effective photocatalysts in numerous applications, such as water splitting and processes of CO_2 conversion. Specifically, ongoing studies have shown how structural component modifications and defects could play a fundamental role in maximizing the photocatalytic activity of materials towards energy conversion processes such as organic transformation and environmental remediation processes that are vital in addressing energy challenges globally (Qian et al., 2023). Applications of layered double hydroxide (LDH) for photocatalytic reduction of CO_2 are another potential area for energy conversion. As 2D materials, LDHs are beneficial in that they couple metal hydroxide sheets with interlayer anions, facilitating facile light absorption and charge separation in photocatalytic processes. Mechanistic insights into the photocatalytic CO_2

reduction by LDHs reveal that they play multifaceted roles—as direct catalysts, precursors to other catalysts, and as composite components that enhance overall performance. The scope of research in the area underlines the need for new strategies, such as heterojunction formation and material modification, to further enhance the photocatalytic activity of LDHs. The advances collectively underpin the potential of LDHs to significantly impact energy conversion, particularly towards the generation of renewable energy and removal of carbon dioxide (Zhe et al., 2022) (Figure 3.11).

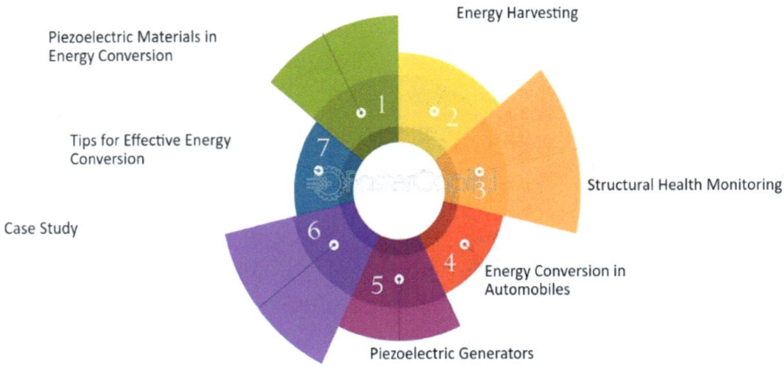

Figure 3.11. *Applications in energy conversion.*

Source: https://fastercapital.co/i/Piezoelectric-Materials-and-Hysteresis--Exploring-Energy-Conversion--Applications-of-Piezoelectric-Materials-in-Energy-Conversion.webp.

3.5.1. Hydrogen Production Through Water Splitting

The requirement for alternative energy sources has created special interest in photocatalytic water splitting to produce hydrogen. This process utilizes sunlight for the energy to power the chemical reaction of water splitting into oxygen and hydrogen. Photocatalysis is doubly advantageous: it can transform abundant sunlight into a clean fuel source while addressing environmental pollution by means of purification processes. Scientists have noted the potential of emulating natural photosynthesis using materials, as noticed in the study that illustrates the critical role played by photocatalytic systems in hydrogen generation. The ongoing investigation into the processes underscores the call for new materials and techniques to enhance efficiency, particularly in the translation of photocatalysis to practical applications in renewable energy fields (Chen, 2025). Since the material is central to photocatalytic performance, recent innovations have highlighted the promising potential of elemental phosphorus in maximizing the hydrogen production system. Specifically, the design of a cobalt single-atom-assisted photocatalytic structure shows the improvements that can be made through innovative material combinations. Through the transformation of amorphous red phosphorus into crystalline black phosphorus, researchers have unveiled an improved photocatalytic reaction space, ultimately making electrochemical water splitting possible. This novel architecture presents the capability of single-atom catalysts in ensuring high efficiency and durability. These record-breaking hydrogen production rates achieved, coupled with solar-to-hydrogen efficiency, illustrate the possibility of this approach in breaking current bounds and driving commercialization of photocatalytic devices (Zhai et al., 2025). The possibilities of photocatalytic water splitting outside fuel production are of significant concern in solving world energy needs. As hydrogen is brought into the spotlight as a clean carrier of energy, its economic and sustainable generation becomes increasingly important. Photocatalytic systems that are capable of using advanced materials most effectively have a prospective path towards meeting rising energy demands without exacerbating carbon emissions. The fusion of the technologies in existing infrastructures holds out the promise of reversing the paradigms of energy generation and storage. With the conduit of scholarly research and actual applications, the pursuit for effective hydrogen production via photocatalysis can revolutionize the Renewable Energy agenda and supply meaningful contributions to initiatives at curbing global warming (Chen, 2025).

3.5.2. CO_2 Reduction to Fuels

The photocatalytic reduction of CO_2 to renewable fuels is a matter of prime research in photocatalysis, particularly towards addressing the problem of climate change as well as energy needs. Solar radiation acts as a catalyst for chemical reactions for reducing CO_2 to valuable hydrocarbons and thus deliver a double benefit in terms of conserving greenhouse gases as well as generating renewable energy sources. The latest advances in catalytic materials, including the synthesis of single-atom catalysts (SACs), have been promising in maximizing CO_2 reduction efficiency. SACs possess higher atom utilization and better catalytic activity compared to their traditional heterogeneous counterparts with the potential to utilize lower reaction pressures and temperatures and greater selectivity for desired products. This new approach not only highlights the potential of tapping copious solar energy but also emphasizes the central role played by sophisticated materials in achieving efficient CO_2 utilization and conversion (Yang et al., 2024). The photocatalyst process is at the center of facilitating the CO_2 reduction reaction. Photogenerated charge carriers, electrons and holes, play a crucial role in catalyzing reduction reactions, but the processes' efficiencies largely rely on the properties of the photocatalytic material.

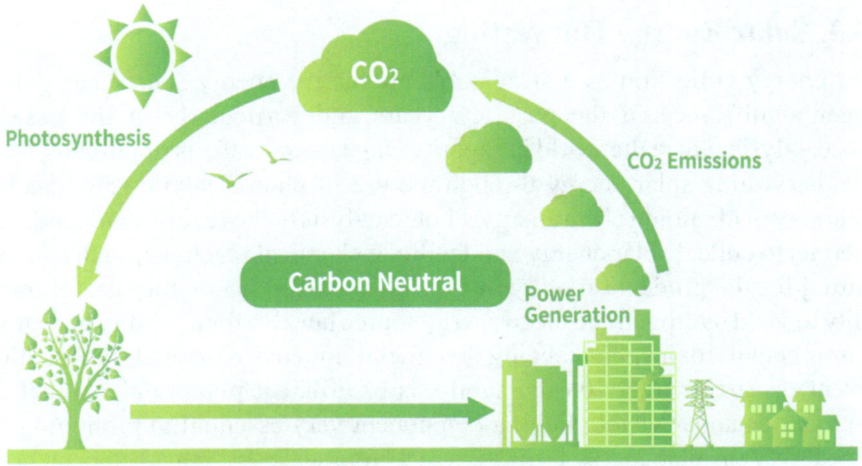

Figure 3.12. *CO_2 reduction to fuels.*

Source: https://www.erex.co.jp/en/common/images/business/img_pks_01.png.

Another key issue in current photocatalytic systems is poor CO_2 adsorption and activation on catalyst surfaces, which typically results in low conversion efficiencies. Advances in materials science, particularly with the implementation of MXenes such as $Ti_3C_2T_x$, have shown great promise in overcoming such constraints. The materials provide not only high surface concentrations of active sites for augmented catalytic activity but also better electronic properties to facilitate enhanced separation and transport of charge carriers, leading to further photocatalytic performance (Wang et al., 2024). With the continued research in the field of photocatalytic CO_2 reduction, the relation between the materials and the mechanism remains in spotlight. Though there is considerable development in high-performance single-atom and MXene-based catalysts, structuring their optimization and operational conditions to render them scalable as well as practically usable remains a challenge. The transition from small-scale laboratory experimentation to industrial application demands complete understanding of the reaction mechanisms involved and ongoing innovation in catalyst design. The ultimate goal is the effective conversion of CO_2 into fuels that easily integrate into existing energy infrastructures and can facilitate a sustainable future for energy. Through further overcoming the technical and practical barriers, research in this area could not only significantly reduce carbon emissions but also create a sustainable source of renewable energy generation (Yang et al., 2024) (Figure 3.12).

3.5.3. Solar Energy Harvesting

Solar energy collection as a sustainable alternative energy source has gained greater significance in the past few years and particularly in the case of photocatalysis. Since the world is experiencing adverse effects of utilizing fossil fuels, harvesting solar energy through the use of photocatalytic reactions is a viable way of obtaining clean energy. Photocatalysis makes use of semiconductor materials to collect solar energy and facilitate chemical reactions, such as water splitting for the production of hydrogen. This method is not only useful for its ability to yield hydrogen as a clean energy source but also for its ability to degrade environmental toxins, thus being beneficial in environmental reclamation. Current research centers on the synthesis of efficient photocatalysts that are able to work under visible light, a component very essential in promoting the applicability of solar energy utilization in a large scale. Recent advancements in materials science have brought single-atom catalysts (SACs) into photocatalyst systems, optimizing solar energy harvesting opportunities. Single-atom catalysts have fascinating coordination characteristics and electronic properties that enhance their reactivity and efficiency in photocatalytic reactions. By integrating SACs into semiconductor photocatalysts, researchers aim to address the key limitations of traditional photocatalytic systems, such as rapid charge recombination and poor exploitation of the solar spectrum. Understanding these mechanisms is significant, as it guides the design of improved photocatalytic materials and processes. Particularly, mechanistic understanding gained from SACs makes the gateway to new strategy in photocatalysis available, and the advancement of the science of solar energy harvesting to meet global energy demand is emphasized (Chen et al., 2025). Notably, recent work has shown promise in the synthesis of composite materials to optimize the efficiency of photocatalysis-induced solar energy harvesting. For instance, novel formulations involving nickel-doped cadmium sulfide have exhibited increased photocatalytic activity, particularly in the reaction of hydrogen evolution. The variations achieved through doping not only optimize the band gap energy but also enhance stability and catalytic reactivity under sunlight illumination. Research indicates that such doped materials achieve high rates of hydrogen production together with high degradation rates of organic pollutants, thereby playing twin functions of sustainable energy generation and pollution mitigation (Kumar et al., 2024). With advancing science, increasingly it is clear that the application of advanced materials in solar energy harvesting is a deciding element for the fate of photocatalytic technologies and by doing so, an indication of the potential direction for a greener energy future.

3.6. CHALLENGES AND LIMITATIONS

Photocatalysis is an area of science that has garnered a lot of attention with regard to its potential applications in environmental remediation and clean energy solutions. However, there are various limitations and challenges to the widespread application of photocatalytic material. One of them is the inherent inefficiency of light absorption, particularly under visible light, which constitutes a high percentage of solar radiation. The majority of traditional photocatalysts, such as TiO_2, absorb primarily UV light, severely limiting their practical application under actual natural sunlight conditions. Moreover, even when composites are engineered to induce visible light activity, they usually experience quick charge recombination, which negatively affects their photocatalytic activity. Advancements in the synthesis and preparation of photocatalysts by enhancing structural properties and constitutions are required to overcome these inherent limitations and achieve higher overall efficiency in real applications, as evident in the nanomaterials analysis illustrated in the literature (Prakash, 2025). Stability of the photocatalytic material is another crucial factor that requires overcoming in order to facilitate practical use. Some materials, such as $CsPbBr_3$ perovskite, are promising due to their advantageous band positions and visible-light absorptivity but are notoriously unstable in aqueous environments. This limitation poses a key challenge when the aforementioned materials are employed in photochemical reaction, particularly those for water purification or hydrogen evolution, which often involve aqueous solutions. Degradation-induced loss of photocatalyst activity is not only efficiency-biased but also generates environmental safety and material durability problems. Techniques such as heterostructure engineering have shown potential in resolving such stability problems, optimizing performance and extending photocatalyst lifespan (Lv et al., 2025). It remains important to address these stability challenges if one hopes to move towards more stable and dependable photocatalytic systems. Lastly, the scalability of photocatalysis is an important issue for practitioners to deal with in terms of real-world applications. Success in the lab is not always scalable nicely to factory-scale production and implementation, largely due to economic and engineering considerations. The economics of synthesizing high-performance photocatalysts are sometimes uneconomic, particularly for advanced nanomaterials requiring highly specific fabrication processes. Also, large-scale production of these processes can introduce variability in material properties and lead to uneven photocatalytic activity. For making photocatalysis a reality as a solution to global energy and environmental challenges, meticulous studies should focus on optimizing synthesis protocols to bring reproducibility, efficiency, and cost-effectiveness. By overcoming these drawbacks, the potential of photocatalysis can be accessed in practical applications (Prakash, 2025).

3.6.1. Stability and Reusability of Photocatalysts

The stability and reusability of photocatalysts are crucial parameters that can influence their performance across an array of applications, from environmental remediation to energy conversion. To exploit the intrinsic characteristics of photocatalysts, researchers hope to develop materials that will have long-lasting performance. For instance, photocatalysts with good structural stability combined with resistance against deactivation processes such as photocorrosion and leaching are essential. Recently prepared donor-acceptor (D-A) structured materials, including polymeric carbon nitride (PCN), are stable under working conditions. According to Li et al. (2025), the materials possess not only excellent physicochemical stability but also a favorable electronic band structure. The research on D-A structures aims at the development of photocatalysts that can withstand numerous environmental stress factors while maintaining their high reactivity, which maximizes their functional lifetime. In addition, the recyclability of photocatalysts adds great value to their industrial and environmental applications. A recyclable photocatalyst that can be reused without significant loss of catalytic activity is preferred. This aspect not only saves economic expenditures on offering fresh catalysts but also avoids environmental disruptions resulting from the disposal of catalysts. The changes in metal halide perovskites (MHPs) have exhibited encouraging impacts on enhancing their stability and photocatalytic properties, a crucial element in real-world usage. As noted by Wang et al. (2025), doping and the construction of heterostructures are some of the methods employed to stabilize the materials under working conditions, such as issues regarding lattice and band structure engineering. These advances lead the way to sustainable photocatalytic systems that facilitate long-term reusability and operational sustainability. The interaction between stability and reusability brings situations of dual optimization in photocatalytic systems. Surface modification, composite development, and hybrid system approaches are becoming increasingly used to enhance both characteristics. The design of tandem D-A type structures illustrates a new direction in this context, where the photocatalyst is designed to not only be stable but also have synergistic enhancement in its recyclability. As discussed by Li et al. (2025), such novel designs have the potential to address universal issues of photocatalysts in real application, thus unveiling new horizons for efficiency in energy transformation and environmental purification. Together, these innovations point the way to a more sustainable future as an example of how improvements in photocatalyst stability and reusability can result in significant strides toward addressing urgent global problems like energy shortages and pollution (Figure 3.13).

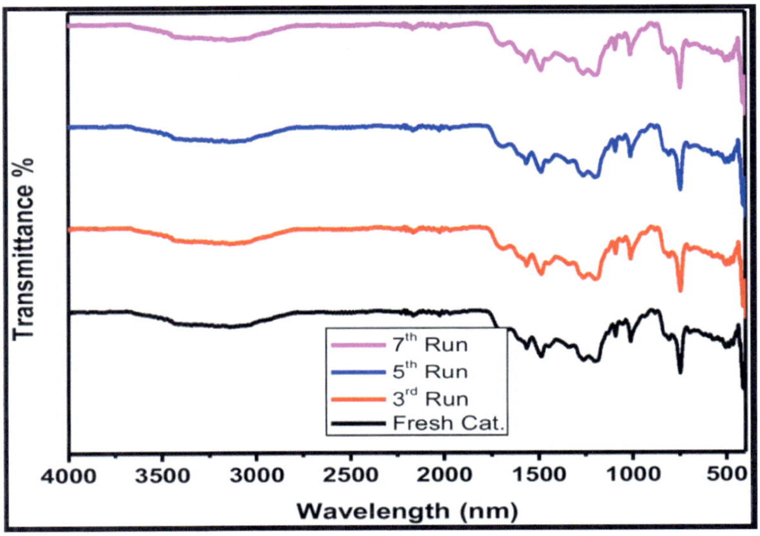

Figure 3.13. *Stability and reusability of photocatalysts.*

Source: https://www.researchgate.net/publication/353025588/figure/fig8/AS:1143 1281169366052@1687314343551/Stability-and-reusability-of-the-photocatalyst-CFPBIL.png.

3.6.2. Efficiency and Cost-Effectiveness

Efficiency and cost-effectiveness are critical considerations in the evaluation of photocatalytic processes in environmental remediation, with particular focus on wastewater treatment. Photocatalysis, dominated primarily by titanium dioxide (TiO_2), is a dominant technology due to its ability to degrade a wide range of organic pollutants under visible light. The operational efficiencies of TiO_2 are maximized through the exploitation of its high oxidizing ability, which causes the rapid degradation of pollutants. Of recent, new TiO_2 composites synthesis and surface engineering towards enhanced visible light-induced photocatalytic activity have been at the center of addressing the limitations presented by traditional UV-pumped systems. The shift towards visible light not only advances the usability of photocatalysis but also complements sustainability goals via reduced energy uptake and operational costs associated with traditional photoreactive materials (Chauke et al., 2025). Furthermore, the economic feasibility of photocatalytic technologies is intrinsically connected to the life cycle cost engineering of the material used and its resulting activity in real-world applications. Traditional wastewater treatment processes are economically burdensome with high-scale operation maintenance and energy expenditures. In contrast, photocatalysis provides a potentially attractive option, such as the application of low-cost and easily available TiO_2 minimizing material expenses, with its non-toxic nature reducing associated environmental liabilities. Additionally, approaches such as photocatalyst immobilization on various supports enable reusability and reduce the need for constant replenishment of materials. By integrating photocatalysis in existing wastewater treatment systems, operators can save money significantly while addressing major issues of pollution elimination and environmental compliances (Chukwunonso et al., 2025). Use of photocatalysis as a component of more extensive environmental treatment processes must also take into account scalability, as this parameter significantly influences efficiency and economy. While laboratory efficacy of TiO_2 photocatalysis has been extensively documented, actual application to beneficial, scale-up operations is difficult. Parameters of reactor design, penetration of light through the treatment medium, and durability of materials must be taken care of in order to realize the total potential of the technology. In parallel with this, advancements in the synthesis of the photocatalysts, such as the development of hybrid materials and optimization of reaction conditions, have been promising the overall efficiency. The outcomes of recent research indicate that the multi-disciplinary approach towards photocatalysis involving the merging of novel material science and engineering ideas can yield cost-effective as well as efficient systems for large-scale water treatment operations (Chauke et al., 2025).

3.6.3. Scale-Up and Industrial Application Issues

The upscaling from laboratory test-scale experiments to industrial-scale production in photocatalysis is beset with issues that can retard the larger application of this technology. Scaling up photocatalytic processes is one of the most significant challenges, and cost-effectiveness is a key issue. Laboratory results have a tendency to emphasize the efficiency and effectiveness of photocatalysts, but actual implementation can be wildly less efficient due to heterogeneity of environmental conditions and material quality. Issues concerning high production cost, material low durability, and low thermal and mass transfer efficiency must be addressed. Such issues are particularly pertinent in the case of integrating photocatalysis with solar-driven interfacial evaporation (SDIE) systems where material synergy like graphene must overcome pragmatic constraints concerning large-scale application such as system design optimization and material compatibility (Zhang et al., 2025). The other significant field of photocatalytic applications concerns the demand for effective waste disposal and byproduct removal in the process of industrial-scale operations.

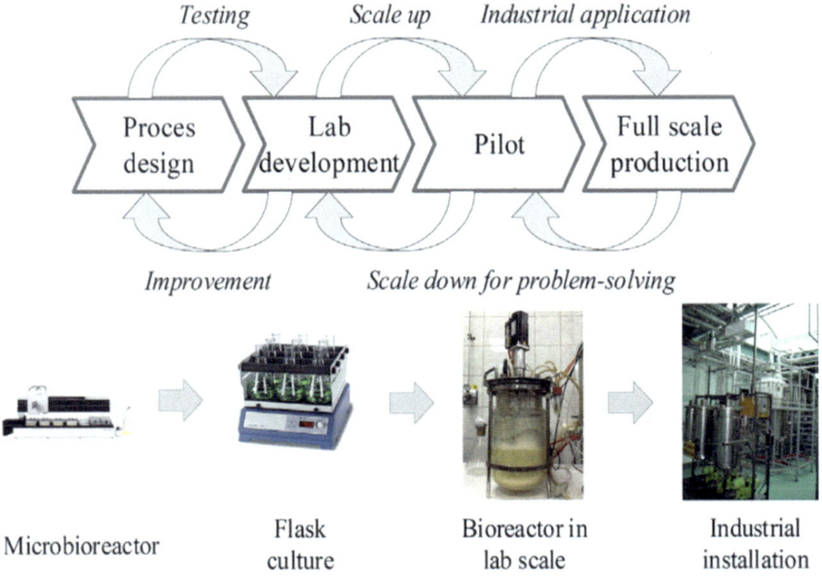

Figure 3.14. *Scale-up and industrial application issues.*

Source: https://www.researchgate.net/profile/Aleksandra-Mironczuk/publication/339194013/figure/fig1/AS:860917478748161@1582270130531/ndustrial-process-scale-up.jpg.

With the areas of photocatalysis further developing into high-impact technologies such as water purification and degradation of organic pollutants, the treatment of residual material becomes crucial. Nanoscale material synthesis from biomass, as attested to by biosorbent use, offers a promising trend to halt waste creation while retaining material performance (Arantes et al., 2024). Process designs in such instances not only must be effective in pollution removal but also eco-friendly to meet legal standards as a measure against environmental degradation. This combination of efficiency and environmental concerns typically renders scale-up strategies difficult to execute, as it demands the investment of research to maximize processes and develop useful products that may be appropriately scaled. Furthermore, realistic application of photocatalytic systems is also within the bounds of technical limitations that are enhanced under scale-up. For instance, even illumination of larger reactor surfaces is necessary for the best photocatalysis; nevertheless, this is more difficult with scale. In addition, the risk of catalyst deactivation with extended operating time is an area of concern for long-term material stability and economic viability. Resolution of these issues involves a synergistic approach consisting of sophisticated reactor design, effective catalyst regeneration methods, and integration of sophisticated materials with improved stability and activity under real conditions. These are crucial not only for the efficiency of photocatalytic systems but also for the competitiveness of such systems in international markets that are increasingly focused on sustainability and environmental integrity (Zhang et al., 2025) (Figure 3.14).

SUMMARY

In summarizing the study of photocatalysis, it is evident that such a process has vast potential in many applications, especially environmental sustainability. The photocatalytic processes are quite coupled with the respective properties of photocatalytic materials that facilitate degradation of organic pollutants and harvesting of solar energy. As has been discussed herein, advancements in materials such as titanium dioxide and novel nanocomposites have enhanced the efficiency of photocatalytic reactions significantly. These developments not only aid in environmental remediation but also hold the potential for effective energy solutions, reinforcing the imperative role of photocatalysis to solve contemporary ecological issues. Moreover, as technologies advance to incorporate photocatalysis into feasible applications, sustainable materiality is increasingly vital. The exploitation of biowaste, like sericin-based silk, as a precursor to the synthesis of nanoparticles is a manifestation of innovative ways towards green alternatives. Silver nanoparticle synthesis (Scn-AgNPs) using sericin, as highlighted, demonstrates the versatility and efficiency of natural polymers as catalysts. Their potential anti-inflammatory and antibacterial properties confirm their potential use in biomedical applications and ecological conservation, promoting the status of sustainable materials in photocatalysts. This double benefit not only minimizes waste but also promotes the circular economy. Lastly, the research of nanomaterials, particularly graphene, also highlights the multi-dimensionality of the potential of photocatalysis for contemporary applications. The unique structural characteristics of graphene enhance the efficiency of photocatalytic processes to address vital challenges such as water pollution with novel treatment technologies. Through utilizing the supportive nature of graphene in combination with photocatalytic processes, researchers are able to create cost-effective water treatment schemes that are both efficient and environmentally friendly. The future of photocatalysis therefore appears promising since it combines cutting-edge material science with environmental stewardship, providing a model for innovative solutions to the societal environmental dilemma, as noted in ongoing research on nanomaterials and graphene's involvement in this topic (Das et al., 2025; Song, 2024).

The research into photocatalysis has become increasingly imperative in the context of the twin challenges of energy demand and environmental pollution. Central to the research is an understanding of photocatalytic mechanisms, with water splitting being most prominent. Water splitting offers a process for the conversion of solar light to chemical energy. It is important for the manufacturing of hydrogen, the primary component of renewable energy. The importance of using appropriate materials in the enhancement of photocatalytic performance cannot be overstated, considering that some materials exhibit enhanced light absorption as well as catalytic performance. Consequently, studies have focused on the development of new photocatalytic materials that not only enhance performance but also durability under operating conditions, as articulated by Chen (2025). This intricate relationship of material properties and photocatalytic reactions emphasizes the need for a multidisciplinary approach to material development towards sustainable energy technologies. Photocatalyst design innovation has also been key to developing significant degradation of contaminants, paving the way for pragmatic uses in environmental remediation. Heterogeneous photocatalysis has emerged as a promising means of efficient degradation of novel pollutants in water by leveraging the ability of specially engineered photocatalysts to get excited under visible light. Various strategies like band gap engineering and heterojunction formation have been employed to facilitate better electronic properties of the photocatalysts, thereby resulting in enhanced degradation efficacy. Further, the introduction of plasmonic materials has also improved light absorption and catalytic efficiency. These technologies are crucial to addressing water quality issues and are explored comprehensively through multipurpose reactor designs evidencing the adaptability of photocatalytic systems for real-world applications (Paiu et al., 2025). Further advancement of this technology signifies future breakthroughs in environmental as well as energy domains. The future of photocatalysis will benefit from multipurpose methods that combine materials engineering, system design, and data-driven methodologies. Trends today observe a growing coordination between machine learning and artificial intelligence to automate catalyst discovery and rationalize photocatalytic processes. This development is significant in the designing of systems not only that function well but also remain sustainable, particularly in decentralized or constrained resource environments. The potential of hybrid systems which integrate photocatalysis with other treatment steps such as membrane filtration and biological treatments enhances the operational efficiency and overall effect of water treatment processes. Pathbreaking research in this line is sensitive to the dire need for new solutions for environmental issues all over the world, reasserting the pivotal role of photocatalysis for the stimulation of water purification activities and solar energy production from solar resources (Paiu et al., 2025).

REFERENCES

1. Afanasiev, P., Anastasescu, C., State, R., Vasile, A., Papa, F., & Balint, I. (2023). Photocatalytic degradation of organic and inorganic pollutants to harmless end products: Assessment of practical application potential for water and air cleaning. *Catalysts, 13*, 380–380. https://doi.org/10.3390/catal13020380 (accessed on 28th September 2025).
2. Ahmed, M. A., & Mohamed, A. A. (2023). Advances in ultrasound-assisted synthesis of photocatalysts and Sono photocatalytic processes: A review. *I Science, 27*, 108583–108583. https://doi.org/10.1016/j.isci.2023.108583 (accessed on 28th September 2025).
3. Arantes, M. S. T., Miranda, T. M. P., Magalhães, W. L. E., Helm, C. V., & da Silva, V. R. (2024). Development of cellulose-based biosorbents from the residue of the peach palm agribusiness and simulation of industrial scale-up for copper removal from Brazilian spirit. *Journal of the Science of Food and Agriculture.* https://www.semanticscholar.org/paper/2d5d144d96e698d88ae25bdbe1736a053a37c9c7 (accessed on 28th September 2025).
4. Bi, Z. X., Guo, R. T., Hu, X., Wang, J., Chen, X., & Pan, W. (2022). Research progress on photocatalytic reduction of CO_2 based on LDH materials. *Nanoscale, 14*, 3367–3386. https://doi.org/10.1039/d1nr08235c (accessed on 28th September 2025).
5. Bystrov, V. S., Avakyan, L. A., Paramonova, E., & Coutinho, J. (2019). Sub-band gap absorption mechanisms involving oxygen vacancies in hydroxyapatite. *The Journal of Physical Chemistry C, 123*, 4856–4865. https://doi.org/10.1021/acs.jpcc.8b11350 (accessed on 28th September 2025).
6. Castro, N., Queirós, J. M., Alves, D. C., Fernandes, M. M., Lanceros-Méndez, S., & Martins, P. (2024). A 3D-printed portable UV and visible photoreactor for water purification and disinfection experiments. *Nanomaterials, 14.* https://www.semanticscholar.org/paper/37d10e55de92113f12997e5732aca7c68f86a0a7 (accessed on 28th September 2025).
7. Chauke, N., Ngqalakwezi, A., & Raphulu, M. (2025). Transformative advancements in visible-light-activated titanium dioxide for industrial wastewater remediation. *International Journal of Environmental Science and Technology.* https://www.semanticscholar.org/paper/9b57407582f530365b690275c7ae6f6a6f6660e5 (accessed on 28th September 2025).
8. Chen, S. (2025). Learn from photosynthesis: A summary of photocatalysis mechanisms and applications. *Applied and Computational Engineering.*

https://www.semanticscholar.org/paper/0e9053d944dc6fc14fa9994812dd896449ad9b5f (accessed on 28th September 2025).

9. Chen, Z. Q., Ngan, A., Chan, C., Lee, J., Sanjaya, D. D., & Gu, F. X. (2025). Lighting the path to practical applications of single-atom catalysts in photocatalysis: The role of platinum group single atoms in enhancing catalytic activity. *Solar RRL, 9*. https://www.semanticscholar.org/paper/2e3a3dfb9c27f607178d18089f7fc29610ebfd77 (accessed on 28th September 2025).

10. Das, G., Shin, H. S., Yang, I. J., Nguyen, L. T. H., & Patra, J. (2025). Silk biowaste protein mediated silver nanoparticles synthesis and analysis of anti-inflammatory, wound healing, antidiabetic, antioxidant, tyrosinase inhibition, and antibacterial mechanism of action. *International Journal of Nanomedicine, 20*, 6551–6571. https://www.semanticscholar.org/paper/25d284aeb9900c34f289e59ee762e17eb8a1b1d6 (accessed on 28th September 2025).

11. Deepak, N., Jain, V., & Pillai, P. P. (2024). Metal-semiconductor heterojunction accelerates the plasmonically powered photo regeneration of biological cofactors. *Photochemistry and Photobiology, 100*, 1000–1009. https://www.semanticscholar.org/paper/3bb52c0cc0f13d715dce6ef00b3887dd40780c5b (accessed on 28th September 2025).

12. Du, A., Sun, X., Dong, M., Liu, Y., Chen, M., Wang, Y., Zhang, Y., et al. (2025). Modelling the effects of temperature, pH and osmotic shifts on the autofluorescence of *Staphylococcus aureus in vitro*. *Future Microbiology*. https://www.semanticscholar.org/paper/9a78bebda58e4a0fe579718e4b82a7085a11290e (accessed on 28th September 2025).

13. Falletta, E., Galloni, M. G., Mila, N., Roslan, M. N. B., Ghani, N. A., Cerrato, G., Giordana, A., et al. (2023). Fast and efficient piezo-photocatalytic mineralization of ibuprofen by BiOBr nanosheets under solar light irradiation. *ACS Photonics, 10*, 3929–3943. https://doi.org/10.1021/acsphotonics.3c00724 (accessed on 28th September 2025).

14. Gatou, M. A., Syrrakou, A., Lagopati, N., & Pavlatou, E. A. (2024). Photocatalytic TiO_2-based nanostructures as a promising material for diverse environmental applications: A review. *Reactions, 5*, 135–194. https://doi.org/10.3390/reactions5010007 (accessed on 28th September 2025).

15. Gusain, R., Kumar, N., & Ray, S. S. (2020). Factors influencing the photocatalytic activity of photocatalysts in wastewater treatment. https://doi.org/10.1002/9781119631422.ch8 (accessed on 28th September 2025).

16. Hailili, R., & Gan, Y. (2025). Tailoring multiscale interfaces in heterojunction photocatalysis for NOx removal. *ACS Applied Materials & Interfaces.* https://www.semanticscholar.org/paper/fc616cd6476f9035e7b928affef7bf6deef3abf2 (accessed on 28th September 2025).
17. Huang, Q., Canady, T. D., Gupta, R., Li, N., Singamaneni, S., & Cunningham, B. T. (2020). Enhanced plasmonic photocatalysis through synergistic plasmonic–photonic hybridization. *ACS Photonics, 7*, 1994–2001. https://doi.org/10.1021/acsphotonics.0c00945 (accessed on 28th September 2025).
18. Jiang, Y., Fan, Y., Li, S., & Tang, Z. (2022). Photocatalytic methane conversion: Insight into the mechanism of C(sp^3)–H bond activation. *CCS Chemistry, 5*, 30–54. https://doi.org/10.31635/ccschem.022.202201991 (accessed on 28th September 2025).
19. Jing, L., Li, Z., Chen, Z., Li, R., & Hu, J. (2024). Engineering polyheptazine and polytriazine imides for photocatalysis. *Angewandte Chemie (International Ed. in English), 63.* https://www.semanticscholar.org/paper/652b7e86807cd5146a24f22934e9b169b971a31f (accessed on 28th September 2025).
20. Kapoor, N., Bhardwaj, R., Kotnala, S., Kandwal, P., Jain, D., Gambhir, L., & Chandra, S. (2024). Nano-resolutions for environmental salvation: Leaping to sustainability. *Current Green Chemistry.* https://www.semanticscholar.org/paper/5edd3f8693ed216018db2e77c8e4d494364a2895 (accessed on 28th September 2025).
21. Khoo, V., Ng, S. F., Haw, C., & Ong, W. J. (2024). Additive manufacturing: A paradigm shifts in revolutionizing catalysis with 3D printed photocatalysts and electrocatalysts toward environmental sustainability. *Small.* https://www.semanticscholar.org/paper/43860f5a36daba8eb6ba09656517ef37c13b3fa8 (accessed on 28th September 2025).
22. Korku, M., Ilhan, R., & Feyzullahoğlu, E. (2024). Investigation of effects of environmental conditions on wear behaviors of glass fiber reinforced polyester composite materials. *Polymer Composites.* https://www.semanticscholar.org/paper/c1c7f1d8d55a80a39e1c7d65d8df3cba247334a8 (accessed on 28th September 2025).
23. Kumar, V., Raj, B., Kommu, P., Paswan, S. K., & Singh, G. P. (2024). Growth of Ni loaded CdS in nanorods structure for photocatalytic and dye degradation applications under solar irradiation. *Nano Express, 5.* https://www.semanticscholar.org/paper/658620110f7b276c26583d5d2344699fe7d0ed64 (accessed on 28th September 2025).

24. Li, L., Zhou, Z., Shi, Y., Tang, R., Li, W., Deng, Y., & Huang, Y. (2025). Donor-acceptor type carbon nitride photocatalysts in photocatalysis: Current understanding, applications and challenges. *Small.* https://www.semanticscholar.org/paper/cebfbad3a5312fa8e21fa52a2cb1e91cc5a2dc62 (accessed on 28th September 2025).

25. Li, T., Zhang, R., Fang, N., Shi, Y., Li, J., He, C., & Chu, Y. (2025). Metal cluster-mediated photocatalysis: Synthesis, characterization and application. *Nanoscale.* https://www.semanticscholar.org/paper/f57bf4993af57773bbbd0237da19605313dd0f93 (accessed on 28th September 2025).

26. Liu, Y., Xu, Q., Wang, L., Gao, A., Li, Q., Chen, S., Zhao, Y., et al. (2024). Rational control of maximum EMI/CPL intensity and wavelength of Bora[6]helicene via polarity and vibronic effects. *The Journal of Physical Chemistry Letters.* https://www.semanticscholar.org/paper/5a3515437b410f8770b4757018d95e4a1cd0403f (accessed on 28th September 2025).

27. Liu, Z., Zhuang, Y., Dong, L., Mu, H., Li, D., Zhang, F., Xu, H., et al. (2023). Enhancement mechanism of photocatalytic hydrogen production activity of CeO_2/CdS by morphology regulation. *ACS Applied Energy Materials, 6,* 7722–7736. https://doi.org/10.1021/acsaem.3c01255 (accessed on 28th September 2025).

28. Lv, K., Li, Z., Huang, X., Cheng, Z., Wang, Z., & Zhao, H. (2025). $CsPbBr_3$ perovskite-based heterostructures in photocatalysis: Mechanisms, stability, and multifunctional performance. *Advanced Science.* https://www.semanticscholar.org/paper/2f0ca417b3b12ecd16b7dc3fbfa936bec7d64a87 (accessed on 28th September 2025).

29. Mahesh, N., Balakumar, S., Kamaraj, M., Palanisamy, S. B., & Aravind, J. (2025). Sustainable innovative approaches on catalytic-carbon nanocomposites for environmental remediation. *Environmental Quality Management.* https://www.semanticscholar.org/paper/bc31c073ce44f876227bae9c59fc28a34f685719 (accessed on 28th September 2025).

30. Mishra, B., Alam, A., Chakraborty, A., Kumbhakar, B., Ghosh, S., Pachfule, P., & Thomas, A. (2024). Covalent organic frameworks for photocatalysis. *Advanced Materials.* https://www.semanticscholar.org/paper/90e6e8f6bef69075c163a8b02b7b84a3fe2c12c0 (accessed on 28th September 2025).

31. Mishra, K. B., Devi, N., Siwal, S. S., Gupta, V. K., & Thakur, V. K. (2023). Hybrid semiconductor photocatalyst nanomaterials for energy

and environmental applications: Fundamentals, designing, and prospects. *Advanced Sustainable Systems, 7*. https://doi.org/10.1002/adsu.202300095 (accessed on 28th September 2025).

32. Mulliqi, N., Blilie, A., Ji, X., Szolnoky, K., Olsson, H., Boman, S. E., Titus, M., et al. (2025). Foundation models–A panacea for artificial intelligence in pathology? https://www.semanticscholar.org/paper/516d76eb4547667ff9191853afc407fedb0564e4 (accessed on 28th September 2025).
33. Nematov, D. (2024). Titanium dioxide and photocatalysis: A detailed overview of the synthesis, applications, challenges, advances and prospects for sustainable development. *Journal of Modern Green Energy*. https://www.semanticscholar.org/paper/b3d250b2eb309ac34c0467cf571685be6ce930e7 (accessed on 28th September 2025).
34. Onyenanu, C. N., & Onyenanu, I. U. (2025). Advanced treatment technologies for heavy metal removal from water and wastewater: A comprehensive review. *International Journal of Applied and Natural Sciences*. https://www.semanticscholar.org/paper/443a2e36c92da166609a3b952e46919d84053421 (accessed on 28th September 2025).
35. Paiu, M., Lutic, D., Favier, L., & Gavrilescu, M. (2025). Heterogeneous photocatalysis for advanced water treatment: Materials, mechanisms, reactor configurations, and emerging applications. *Applied Sciences*. https://www.semanticscholar.org/paper/d7c94cc298d3ccbfc13da431a96f3c18f8d71f7f (accessed on 28th September 2025).
36. Pham Thi, H., Chu, T. T. H., & Nguyen, M. (2024). Improved photocatalytic decomposition of carbaryl pesticide in wastewater using ZnO nanorods. *Journal of Environmental Science and Health, Part B, 59*, 758–766. https://www.semanticscholar.org/paper/cb4cb0c64abb6d01bdbf83c0d95d98bad0916468 (accessed on 28th September 2025).
37. Prakash, T. B. (2025). Review on nanomaterial-based photocatalysis: Mechanisms, materials, and environmental application. *International Journal of Applied Engineering and Management Letters*. https://www.semanticscholar.org/paper/cb36182953e9923239fb5a35979cc41a66cc6d99 (accessed on 28th September 2025).
38. Qian, Z., Zhang, R., Xiao, Y., Huang, H., Sun, Y., Yang, C., Ma, T., et al. (2023). Trace to the source: Self-tuning of MOF photocatalysts. *Advanced Energy Materials, 13*. https://doi.org/10.1002/aenm.202300086 (accessed on 28th September 2025).

39. Sasikumar, K., & Ju, H. (2025). Recent advances in vanadate-based materials for photocatalytic hydrogen production. *Molecules, 30.* https://www.semanticscholar.org/paper/1aa9a3ec6e0bf4641913d82d8d1a815c3a033366 (accessed on 28th September 2025).

40. Sharma, R., & Bisht, K. (2024). Development of metal oxide-based catalysts for CO_2 capture and conversion. *International Journal of Behavioral Sciences.* https://www.semanticscholar.org/paper/76f95e1b535dccff2a146f934be997b42827cfdd (accessed on 28th September 2025).

41. Song, C. (2024). Application prospects of graphene in environmental science. *Applied and Computational Engineering.* https://www.semanticscholar.org/paper/58f7ef543e0bb23b7e9e9ace8216842e64ef4811 (accessed on 28th September 2025).

42. Su, Q., Zuo, C., Liu, M., & Tai, X.–S. (2023). A review on Cu_2O-based composites in photocatalysis: Synthesis, modification, and applications. *Molecules, 28*, 5576–5576. https://doi.org/10.3390/molecules28145576 (accessed on 28th September 2025).

43. Sun, D., Zhang, Y., Zhou, Y., Nie, Y., Ban, L., Wu, D., Yang, S., et al. (2025). Photocatalytic and electrochemical synthesis of biofuel via efficient valorization of biomass. *Advanced Energy Materials, 15.* https://www.semanticscholar.org/paper/a7db630f2c1c10faa25d3b1c05af89e30821ea07 (accessed on 28th September 2025).

44. Tanaka, N., Kumabe, Y., & Tachikawa, T. (2025). Impact of solvent-induced surface restructuring on charge carrier dynamics in *$DMASnI_3$. ACS Applied Materials & Interfaces.* https://www.semanticscholar.org/paper/24f880d9501b410cf3e1f3845416a30cdba8204f (accessed on 28th September 2025).

45. Tang, C., Li, X., Hu, Y., Du, X., Wang, S., Chen, B., & Wang, S. (2024). Porphyrin-based metal-organic framework materials: Design, construction, and application in the field of photocatalysis. *Molecules, 29.* https://www.semanticscholar.org/paper/9177e306f87a5d1ceab79ad94a507a9e5e699c7b (accessed on 28th September 2025).

46. Vakharia, A., & Chavan, A. (2025). Development of a compact IoT-enabled air purification system for indoor air quality improvement. *2025 5th International Conference on Trends in Material Science and Inventive Materials (ICTMIM).* https://www.semanticscholar.org/paper/d53f089ba27e31446e1a2c4b048f5eadab00997f (accessed on 28th September 2025).

47. Verma, R., Sharma, G., & Polshettiwar, V. (2024). The paradox of thermal vs. non-thermal effects in plasmonic photocatalysis. *Nature Communications, 15.* https://www.semanticscholar.org/paper/2e58a2f91f3e51d738e7db60abe0c36c6d0537db (accessed on 28th September 2025).
48. Wang, C., Guo, Q., & Liu, J. (2025). Perspectives of soft lattice halide perovskite photocatalysts, mechanisms and applications. *Chemistry.* https://www.semanticscholar.org/paper/57858fe0e6422df1f7aa77c98ef8a37b2c9bc267 (accessed on 28th September 2025).
49. Wang, Z., Al Jitan, S., AlNashef, I. M., Tardy, B. L., & Palmisano, G. (2024). Recent progress of MXene as a cocatalyst in photocatalytic carbon dioxide reduction. *Chemical Engineering Journal Advances, 18*, 100593–100593. https://doi.org/10.1016/j.ceja.2024.100593 (accessed on 28th September 2025).
50. Xu, D., Yang, Y., & Gan, L. (2025). Utilization of natural mineral materials in environmental remediation: Processes and applications. *Minerals.* https://www.semanticscholar.org/paper/ba6f4592a068f2f5d56584b42861220a572dd3ff (accessed on 28th September 2025).
51. Yang, G., Wang, Q., Kuwahara, Y., Mori, K., & Yamashita, H. (2024). Recent progress of studies on photoconversion and photothermal conversion of CO_2 with single-atom catalysts. *Chem & Bio Engineering, 1*, 289–311. https://doi.org/10.1021/cbe.3c00110 (accessed on 28th September 2025).
52. Yang, J. L., Wang, H. J., Qi, X., Zheng, Q. N., Tian, J. H., Zhang, H., & Li, J. (2024). Understanding the behaviors of plasmon-induced hot carriers and their applications in photocatalysis. *ACS Applied Materials & Interfaces.* https://www.semanticscholar.org/paper/cbbc10fd252db9bb98b17830b89e04e1237bb4a1 (accessed on 28th September 2025).
53. Yang, Z., Vrielinck, H., Jacobsohn, L. G., Smet, P. F., & Poelman, D. (2024). Passive dosimeters for radiation dosimetry: Materials, mechanisms, and applications. *Advanced Functional Materials, 34.* https://www.semanticscholar.org/paper/7822d65f1bc5672f56b592419427b25aeabb7065 (accessed on 28th September 2025).
54. Yaqoob, A. A., Parveen, T., Umar, K., & Ibrahim, M. N. M. (2020). Role of nanomaterials in the treatment of wastewater: A review. *Water, 12*, 495–495. https://doi.org/10.3390/w12020495 (accessed on 28th September 2025).

55. Zhai, X., Wei, Z., Lu, Z., Zhang, X., Chen, X., Liu, Y., Deng, J., et al. (2025). Cobalt single atom-enhanced photocatalysis: Hetero-phase elemental phosphorus for visible light hydrogen production from pure water splitting. *Advanced Functional Materials*. https://www.semanticscholar.org/paper/5ddc4c016a40216ca15fdd4d27b0667919fa25ed (accessed on 28th September 2025).
56. Zhang, Y., Wang, H., & Zhang, J. (2025). Application of graphene-based solar driven interfacial evaporation-coupled photocatalysis in water treatment. *Catalysts*. https://www.semanticscholar.org/paper/d14e1a557877e41132dd2fdd7256a2545ce92711 (accessed on 28th September 2025).

CHAPTER 4

ADVANCED OXIDATION PROCESSES (AOPs) IN ENVIRONMENTAL REMEDIATION

Contents

Introduction 133
4.1. Mechanisms of AOPs 138
4.2. Types of Advanced Oxidation Processes (AOPs) 143
4.3. Applications of AOPs in Water Treatment 148
4.4. AOPs in Soil Remediation 154
4.5. Advantages and Limitations of AOPs 159
4.6. Future Trends and Innovations in AOPs 163
Summary 169
References 171

INTRODUCTION

The emergence of advanced oxidation processes (AOPs) is a significant advancement in environmental cleanup. The processes employ high oxidizing power reagents, the foremost among them being hydroxyl radicals, to degrade stable organic pollutants and contaminants in various environmental media, predominantly water. AOPs are distinct from conventional treatments in their ability to undertake complete mineralization of harmful substances and thus convert them into harmless compounds such as carbon dioxide and water. With the increasing occurrence of harmful chemicals within ecosystems and their probable impact on human health, AOPs present a promising solution for addressing water pollution and calling for innovative remediation technologies during this age of environmental stewardship. An efficient implementation of AOPs includes the combination of ultrasonication, a process that triggers chemical reactions by creating hydroxyl radicals. The application of ultrasonic waves not only increases the destruction of toxic organic molecules but also maximizes the efficiency of oxidants used in the degradation process. For instance, the application of ultrasound in combination with zero valent iron (ZVI) and hydrogen peroxide to significantly enhance the mineralization rates of sodium dodecylbenzene sulfonate (DBS) and dodecyl pyridinium chloride (DPC) under optimal temperatures. Experiments reveal that under controlled temperature and pH conditions, the enhancement offered by ultrasound may lead to excellent degradation results, with a maximum of 93% mineralization of DBS and 87% of DPC, thereby revealing the actual environmental potential of AOPs to handle intricate environmental problems (Amat et al., 2011). Besides degradation, the AOP variability is also such that they can be made to treat specific pollutants by adjusting operational parameters. For instance, variations in initial concentration levels, ultrasonic frequencies, and addition of oxidizing agents or inorganic particles may influence degradation significantly. Recent research has demonstrated efficacy of such modulations in opening the door for more selective environmental remediation strategies. Its continuous development alongside sonochemistry advances and research for novel approaches emphasizes their vital role in pollution control. With growing interest in treating recalcitrant pollutants, integration of AOPs into existing water treatment technologies will play a pivotal role in successful sustainable environmental management (Bremner et al., 2011) (Figure 4.1).

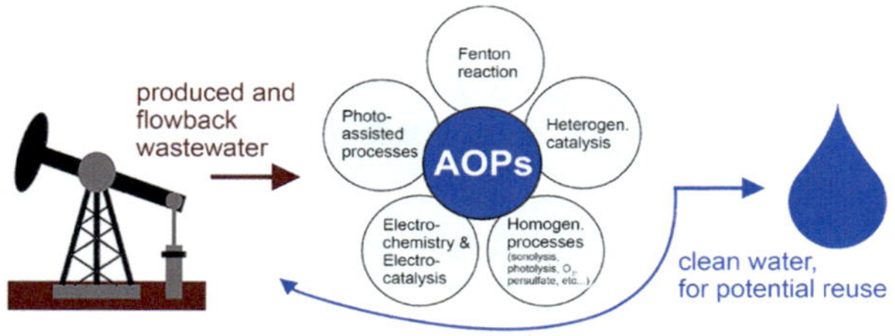

Figure 4.1. *Advanced oxidation processes (AOPs).*

Source: https://aimequip.com.au/wp-content/uploads/2025/02/Advanced-Oxidation-Processes-AOPs1.jpg.

Definition and Overview of AOPs

Advanced Oxidation Processes (AOPs) are a group of next-generation treatment methods that are designed to destroy recalcitrant organic contaminants in various environmental matrices, with particular interest in water and soil. AOPs are characterized by the generation of hydroxyl radicals, highly reactive species that can degrade nearly all pollutants, even highly resistant ones to conventional treatment. The significance of AOPs lies in their ability to remediate contaminants that are resistant to degradation by traditional means, such as chlorination. Hence, scientists and professionals are familiar with AOPs as potential candidates to remediate complicated contaminants, especially where process kinetics are slow or the contaminants are extremely recalcitrant (Gencheva et al., 2025). With the combined use of various oxidation methodologies, AOPs are being employed more frequently in laboratory and field-scale operations for the efficient remediation of contaminated environments. The regime of operation of AOPs is diverse, encompassing a range of methodologies such as Fenton's reagent, heterogeneous photocatalysis, ozonation, and electrochemical oxidation. Based on the method implemented, all these processes have varying mechanisms for radical generation and can be optimized to treat specific contaminant matrices. For instance, Fenton's method utilizes the iron salt-hydrogen peroxide reaction to generate hydroxyl radicals that oxidize a wide variety of organic pollutants (Gencheva et al., 2025). Photocatalysis employs light, on the other hand, to trigger catalysts to enhance contaminant degradation by producing reactive species under regulated conditions. The inbuilt flexibility in AOPs not only allows tailoring based on contaminants but also facilitates reactor design and operation conditions, thereby securing maximum removal efficiencies for recalcitrant wastewater streams. Moreover, AOPs are increasingly viewed as being a part of general environmental management practice, particularly wastewater treatment and reuse. As urbanization intensifies and freshwater resources become scarce, reclamation of wastewater becomes a most significant challenge towards sustainability. The conventional treatment processes are generally inadequate to deal with the complexity of effluent organic matter (EfOM) in wastewater. AOPs become the driving force in this direction, effectively treating endocrine-disrupting chemicals and pharmaceuticals that are toxic to human health and also the natural ecosystem. Employing advanced oxidation processes, the performance of removal of such recalcitrant pollutants can be significantly enhanced, contributing to the general goal of achieving safe and reusable water resources. Thus, application of AOPs in environmental remediation contexts highlights their significance as protectors for both public health and local environment.

Historical Development and Evolution

The history of Advanced Oxidation Processes (AOPs) in the past is strongly based on the necessity to effectively clean environmental pollutants. Early environmental cleaning focused more on conventional types such as physical or biological treatment, which were not very effective against recalcitrant organic pollutants. The understanding of oxidation phenomena shifted significantly towards the close of the 20th century when researchers identified the advantages of using hydroxyl radicals as aggressive oxidants that may demolish a wide variety of pollutants. This fact laid the groundwork for the creation of AOPs, which utilize highly reactive species generated by a variety of processes, including photolysis and chemical injection. Developments in this field have continuously moved AOPs from a niche subject of study to a widely embraced method of treating effluents, with various applications across industrial processes that are interested in environmental sustainability and protection. As AOPs gained acceptance across the field of environmental engineering, the variety of methodologies further expanded with the addition of new methods with the focus on greater effectiveness and efficiency.

Most strikingly, the finding of employing the use of ultrasound as a means to improve the generation of hydroxyl radicals reflects the emerging trends in the field. Not only did the utilization of ultrasonic waves speed up chemical reactions but also break down complex organic molecules that conventional techniques struggled to address. These advancements brought forth a wave of research examining the synergistic effect of ultrasound when combined with various additives and conditions, thus enhancing the capabilities of AOPs further. For instance, experiments have demonstrated how varying pollutant concentration or using varying frequencies of ultrasonic waves can have a considerable influence on degradation rates, making it obvious in process optimization why special strategies are needed (Bremner et al., 2011). Subsequent developments in AOPs have revealed the potential of plasma-liquid systems, which have found extensive interest in the past due to their multiple applications from environmental treatment to biomedicine and food conservation. The emerging technology exploits the interaction between the plasma and liquid phases to generate reactive species that can efficiently break down pollutants. The multidisciplinary nature of plasma-liquid interactions has driven extensive research into their mechanism and application, revealing a complicated landscape of possibilities beyond the traditional means of oxidation. As researchers continue to explore the kinetics of plasma as well as various environmental media, AOPs will develop even further to satisfy contemporary demands for advanced, efficient, and environmentally friendly remediation strategies (Geyter et al., 2019). The integration of state-of-the-art methodologies reflects the historical trend of AOPs, which clearly expresses a strong focus on addressing hard environmental issues in an increasingly emergent world.

Importance in Environmental Remediation

The significance of environmental cleaning has remained increasingly imperative in the face of intensifying concerns regarding water contamination and its detrimental effects on the environment and human health. Advanced Oxidation Processes (AOPs) are key in eliminating a wide range of pollutants, particularly micropollutants that traditional treatment methods often prove ineffective in eliminating efficiently. AOPs employ highly reactive radicals to facilitate the breakdown of organic pollutants into less harmful substances, therefore improving water safety and quality. By means such as Fenton oxidation, which has improved degrading ability for recalcitrant compounds, AOPs facilitate more effective cleanup operations in contaminated environments. The comparative analysis of different AOP methods shows a promising future for their application in environmental management practices and emphasizes their importance in mitigating the hazards presented by chemical pollutants and providing sustainable water supplies. The application of AOPs in environmental remediation becomes more evident while analyzing particular contaminants such as methyl tert-butyl ether (MTBE). This compound poses unique problems due to its high solubility and water-loving nature, which makes it difficult to be eliminated from wastewater using conventional processes. Studies have indicated that emerging treatment processes, including the combination of granular activated carbon (GAC) with hydrogen peroxide, significantly enhance MTBE removal efficiency from model wastewater. The findings highlight the imperative need to optimize operating conditions in AOP processes. Specifically, determining the ideal pulsation time of the cycle and air flow rates can enhance remediation efficiency for multipollutant contaminants, illustrating how AOPs can be tailored to respond to a specific environmental problem while maintaining effective pollutant removal and resource efficiency. The efficacy of AOPs is easily scalable from the lab phase to actual applications in diverse environments. Literature on the efficiency of $FeSO_4^-$ and zero-valent iron (Fe^0)-catalytic Fenton oxidation shows the suitability of such processes for degrading extensive types of emerging micropollutants, from pharmaceutical byproducts to agricultural products. Fe^0-catalytic Fenton processes, specifically, have been distinguished by greater efficiency in pollutant degradation and lower sludge production, which is crucial in minimizing the environmental footprint of treatment processes. With as much as 70% elimination of DOC, such processes work toward the restoration of freshwater ecosystems adversely affected by pollution. Continuing advancements in AOP technologies signify a growing appreciation for their contribution to environmental remediation, defining the path toward creative solutions to intricate environmental problems and enhancing our capacity to preserve water resources for future generations.

4.1. MECHANISMS OF AOPS

Advanced Oxidation Processes (AOPs) are characterized by the ability to generate very reactive species, primarily hydroxyl radicals (•OH), which are the most important species in the degradation of organic contaminants in environmental cleanup. AOP mechanisms vary and encompass processes such as photolysis, electrochemical oxidation, and heterogeneous photocatalysis. Such answers enable effective degradation of persistent pollutants, like perfluoroalkyl and polyfluoroalkyl substances (PFASs), which have significant environmental and health effects through their stability and bioaccumulative tendency. By leveraging their oxidative potential, AOPs enable the oxidation of complex organic molecules into less complex, less harmful substances, significantly enhancing water treatment effectiveness (Deeb et al., 2016). Further, the versatility of AOPs allows for their integration with other treatment technologies in order to create comprehensive treatment strategies for different contaminants. The operational parameters which influence the effectiveness of AOPs are such parameters as pH, temperature, concentration of the reactants, and the catalysts used in the treatment process. Each AOP variant has specific operating conditions that can influence hydroxyl radical formation, thereby impacting overall efficiency and completeness of pollutant removal. For instance, pH modification may optimize the reactivity and stability of catalysts, while temperature fluctuations may regulate reaction kinetics. Additionally, catalyst choice, such as titanium dioxide used in photocatalysis, can potentially contribute significantly to speeding up the oxidation process through an enhancement in the production of reactive oxygen species. As research continues to advance, the identification and control of these parameters are important in minimizing cost of operation and maximizing treatment efficiency, particularly of complex mixtures found in contaminated water sources (Agüera et al., 2019). In addition, the multi-functionality of AOPs makes them suitable for co-treatment with other technologies to create synergistic results for enhanced remediation efficiency. For example, treating wastewater using AOPs in combination with biological treatment can release even more energy-efficient and environmentally friendly wastewater management. The integration of advanced oxidation technologies with microbial degradation not only enhances contaminant destruction but also enables treatment of recalcitrant substances that are resistant to conventional processes. The integrated approach is required to address the complexity demands of contaminant mixtures and co-contaminants associated with environmental matrices. While the research community continues to explore potential combinations of AOPs with alternative strategies, there is increasingly clear potential for more efficient and lower-cost remediation strategies for pollutants, including persistent organic pollutants, to grow (Agüera et al., 2019; Deeb et al., 2016) (Figure 4.2).

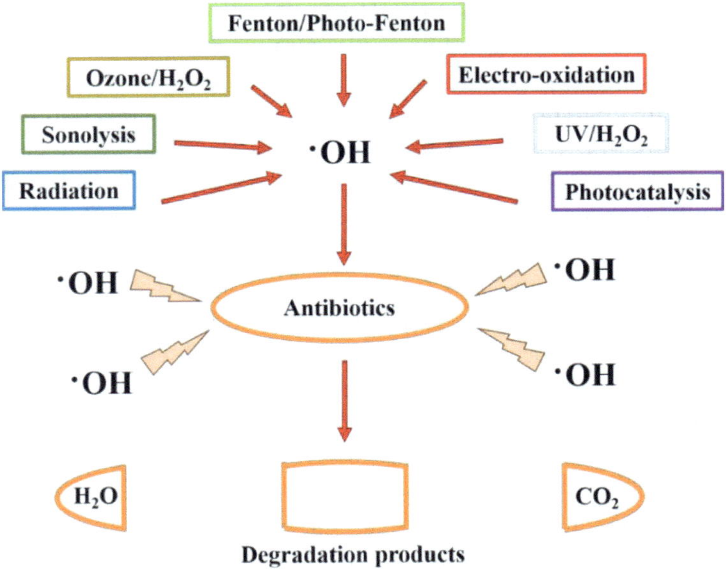

Figure 4.2. *Mechanisms of advanced oxidation processes (AOPs).*

Source: https://www.researchgate.net/profile/Tharindu-Trishan-Darshana-Senarathna-Dapana-Durage/publication/356429474/figure/fig1/AS:1094567705219073@1637976685899/General-mechanism-of-AOPs-to-degrade-antibiotic.jpg.

4.1.1. Generation of Hydroxyl Radicals

The generation of hydroxyl radicals (•OH) is a key to advanced oxidation processes (AOPs), which are being increasingly adopted for their efficacy in environmental remediation. Hydroxyl radicals are extremely reactive species capable of oxidizing many organic contaminants, breaking them down into less harmful compounds. One of the primary methods through which these radicals are formed is by the use of hydrogen peroxide (H_2O_2) and transition metals such as iron under conditions replicated after Fenton's reaction. Such reactions can be catalyzed under acidic conditions, often at a pH of around 2.5, that increases the rate of formation of hydroxyl radicals, and therefore the degradation of most impurities, for example, surfactants such as sodium dodecylbenzene sulfonate (DBS) and dodecyl pyridinium chloride (DPC) as in some studies (Amat et al., 2011). Ultrasound treatment is another sophisticated technique for the generation of hydroxyl radicals, which effectively converts liquid media into reactive conditions conducive to pollutant degradation. Ultrasound application causes cavitation processes that create localized high pressure and temperature regions, followed by generation of •OH from water molecules. This treatment significantly enhances the speed of oxidation reactions, providing a powerful avenue towards the degradation of recalcitrant organic pollutants, including polycyclic aromatic hydrocarbons and chlorinated compounds (Bremner et al., 2011). Therefore, the synergy of ultrasound with chemical oxidants like H_2O_2 enhances the efficiency of degradation, as reflected in high mineralization levels achieved in laboratory experiments, such as up to 93% mineralization of DBS under optimized conditions. Moreover, the role played by hydroxyl radicals in AOPs is similar inasmuch as their ability to interact with a wide variety of organic pollutants, making it possible to extend the range of contaminants amenable to degradation. The versatility may be influenced by a variety of factors, including the concentration of the initial substrate, the amplitude of the ultrasonic employed, as well as the type of inorganic particles or salts that may prolong the degradation process. Recent research has isolated how the variation of these parameters can significantly increase the efficiency of AOPs, ultimately making them a viable solution for the treatment of recalcitrant organic compounds in water. The combination of a number of oxidation reactions, as demonstrated in comparative studies, suggests the potential for future strategies for environmental remediation based on hydroxyl radicals as a cornerstone in the oxidation of difficult organic pollutants (Amat et al., 2011; Bremner et al., 2011).

4.1.2. Role of Oxidants in AOPs

The role of oxidants in Advanced Oxidation Processes (AOPs) is the secret to their efficacy as a method of environmental cleanup. AOPs utilize highly reactive hydroxyl radicals that have the capacity to degrade a wide range of organic contaminants and are particularly useful for wastewater treatment. These oxidants facilitate the chemical reactions that destroy persistent pollutants, such as chlorinated solvents and polycyclic aromatic hydrocarbons, into less toxic by-products. For instance, use of ultrasound with chemical oxidants can enhance the generation of these radicals, thus maximizing the oxidative potential of the system. Studies have demonstrated that alterations in operating conditions, such as ultrasonic frequency and concentration of the oxidizing species, are capable of significantly influencing the rates of degradation of organic pollutants, underscoring the necessity of optimizing these conditions to enhance the effectiveness of AOPs in environmental applications (Bremner et al., 2011). Addition of different oxidants into AOPs can also increase the effectiveness of degradation processes. For example, with the integration of zero-valent iron (ZVI) and hydrogen peroxide in a sono-amplified Fenton process, the oxidative degradation of contaminants is significantly improved. This technology utilizes both the synergistic potential of ZVI to generate reactive iron species and the exploitation of hydrogen peroxide as a supplementary source of hydroxyl radicals.

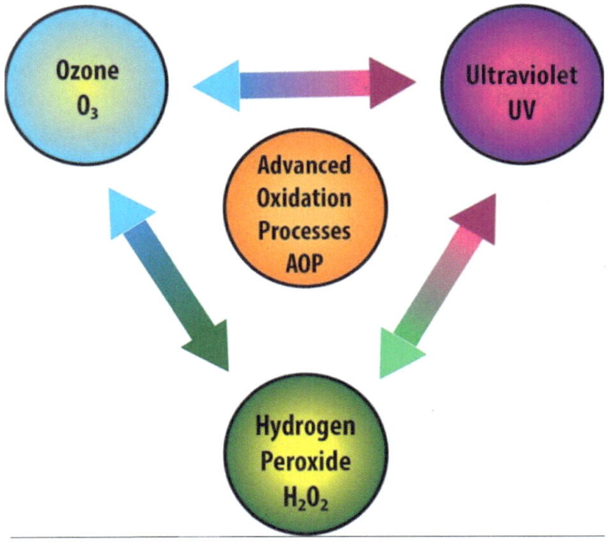

Figure 4.3. *Role of oxidants in AOPs.*

Source: https://poolchemistrytraininginstitute.com/wp-content/uploads/2020/10/AOP-Triangle-graphic-color-061617-768x693.jpg.

The addition of ultrasonic irradiation in this system not only stimulates the activation of ZVI but also maintains high oxidation capacities even after the exhaustion of the initial oxidant. Interestingly, studies have concluded that this enhanced methodology is able to recover nearly 90% total organic carbon (TOC) removal in a 24-hour period, and this shows the determinantal impact oxidants have on keeping AOPs efficient and lengthy in silent degradation phases of treatment procedures (Bremner et al., 2012). Overall, the regulation of oxidant levels and types plays an integral role in maintaining the effectiveness of AOPs in environmental remediation. Through understanding the specific roles of various oxidants, researchers can design improved treatment systems for a variety of pollutants. Through the integration of innovative methods, such as the use of ultrasound, overall reactivity and ability to degrade are enhanced. With their continuous advancement, AOP technologies' ability to address complex environmental challenges is dependent on relentless discovery of oxidant methods in line with sustainable development and economic feasibility. Facilitating such advancements, AOPs will be a critical solution for effective degradation of toxic organic pollutants in contaminated environments (Figure 4.3).

4.1.3. Reaction Pathways and Degradation Processes

Within the framework of advanced oxidation processes (AOPs), understanding mechanisms of reaction and degradation is crucial to environmental remediation success. AOPs are characterized by their ability to generate very reactive species, i.e., hydroxyl radicals, which facilitate enhanced degradation of the complex organic pollutants into less harmful products. The reactions typically entail multiple mechanisms, including hydroxylation, dehalogenation, and mineralization. Through such processes, toxic substances, like polycyclic aromatic hydrocarbons and phenols, are transformed into lower molecular weight substances that can ultimately lead to complete mineralization. The performance of such processes can be regulated by a range of operating conditions, including pH, temperature, and the presence of catalysts or additives, which all have the potential to significantly influence the kinetics of the reaction and overall potential for degradation. The efficacy of certain AOPs can also be evaluated through certain case studies, which throw light on their use in the degradation of contaminants. For instance, the application of ultrasound as an accelerator in AOPs, whose potential in enhancing chlorinated organics and refractory pollutant degradation has been demonstrated. The generation of hydroxyl radicals by ultrasonic cavitation not only accelerates the oxidation process but also enables an increased distribution of reactants, which promotes reaction rates. In addition to this, recent studies have highlighted the effects of varying ultrasonic frequencies and additives, revealing the aspect that these factors can increase the efficiency of degradation of toxic organic substances as well. This multimethod approach to the maximization of the degradation processes demonstrates the dynamic interaction between different AOP methods and the conditions for their operation, thereby establishing their flexibility in diverse environmental applications (Bremner et al., 2011). Reaction pathways can also be one of the most significant problems of degrading recalcitrant pollutants like 1,4-dioxane. Recent findings indicate that while ozone (O_3) oxidation has been widely viewed as insufficient as a solo process for treatment of such substances in the past, the process can achieve remarkable degradation rates under optimal operating conditions. For 1,4-dioxane wastewater treatment, studies have confirmed that maintaining the pH above 9 helps improve the degradation processes with the formation of ethylene glycol as a principal intermediate. Conversely, under acidic conditions, there is slow degradation with ethylene glycol diformate as the most prevalent degradation product. By investigating the different pathways of degradation with on-line Fourier transform infrared spectroscopy, researchers can gain more insight into how various AOPs are efficient in treating hazardous chemicals and assist in maximizing improved remediation methods in environmental engineering (Barndok et al., 2014).

4.2. TYPES OF ADVANCED OXIDATION PROCESSES (AOPs)

The case of advanced oxidation processes (AOPs) describes a series of methodologies with different applications and effectiveness in environmental cleanup. Evident among them is the application of hydrogen peroxide (H_2O_2) that is capable of producing hydroxyl radicals through homogeneous and heterogeneous catalytic processes. This non-selective radical possesses the ability to oxidize organic pollutants effectively, leading to their mineralization into harmless end products. Additionally, catalyst use significantly enhances the effectiveness of H_2O_2 in AOPs. Catalysts such as nanoscale zero-valent iron (nZVI) have shown great promise since they facilitate long-term activation of hydrogen peroxide that enhances the generation of reactive oxygen species (ROS) needed for effective remediation. By tapping such catalytic capabilities, researchers aim to accelerate treatment methods to manage the longtime problems of organic contamination in various ecosystems, ultimately to sustainable environmental use. Another well-known class of AOP employs peroxymonosulfate (PMS) and peroxydisulfate (PDS) as oxidants for oxidative treatment. Such oxidants generate sulfate radicals, which are powerful contributors to the degradation of recalcitrant organic contaminants. The PMS and PDS promiscuity and ability to be activated by various catalysts render them effective agents for pollutant remediation. Different mechanisms of activation, which include thermal, photochemical, and catalytic, can be employed to generate sulfate radicals from PMS and PDS to provide a robust platform for oxidative treatment. Research continues to examine the effectiveness of hydrochar-based catalysts in facilitating PMS and PDS activation processes, specifically through the type of modifications like heteroatom-doped or metal-doped hydrochar. Such advancements imply the potential of superior catalytic performance within AOPs, addressing limitations identified in traditional methods (Manickavasagam et al., 2025). The role of ozone (O_3) in advanced oxidation processes is an indicator of its harsh ability to non-selectively oxidize a wide range of organic contaminants. Ozone-producing systems have become a significant part of AOPs because of their ability to produce hydroxyl radicals when they decompose. Besides, combined with other oxidants or catalysts, the effectiveness of ozone can be significantly enhanced, leading to improved contaminant destruction. The synergy of ozone with catalysts such as nZVI, which allows for the generation of ROS, also puts into perspective the importance of optimized AOPs in specific environmental applications. Current research is focusing on describing the mechanisms of action when ozone acts with various types of pollutants and the subsequent stable degradation product formation. This general understanding of ozone's use in AOP systems is the cornerstone in advancing effective remediation steps that ensure environmental well-being and foster sustainable development (Liu et al., 2023) (Figure 4.4).

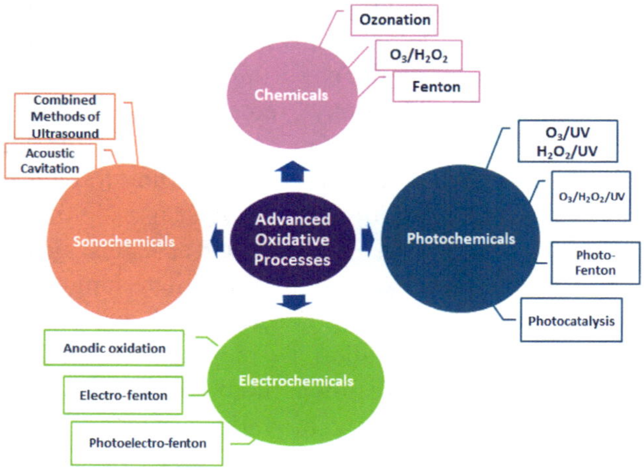

Figure 4.4. *Types of advanced oxidation processes.*

Source: https://www.researchgate.net/profile/Francielli-Monteiro/publication/366313741/figure/fig1/AS:11431281119061803@1675962116958/Classification-of-types-of-advanced-oxidative-processes.png.

4.2.1. Ozone-Based AOPs

Ozone-mediated Advanced Oxidation Processes (AOPs) are an important advancement in environmental pollution control technologies. These processes, which involve the use of ozone as a potent oxidizing agent, effectively oxidize a wide variety of organic pollutants in wastewater. Oxidation is achieved by the creation of hydroxyl radicals, which are highly reactive molecules that possess the capability to break down complex molecules into simpler, harmless compounds. Such an approach is most hopeful for the treatment of recalcitrant pollutants that are resistant to conventional treatment processes. Currently, the application of ozone has come into the limelight with the perspective of enhancing the effectiveness of such processes towards placing them as an acceptable solution for the elimination of persistent organic pollutants from contaminated water sources (Agüera et al., 2019). The combination of ozonation with other oxidation processes even further increases the efficiency of AOPs towards the destruction of various pollutants. For instance, a combination of zero-valent iron (ZVI) and hydrogen peroxide within a new Fenton-like system can appreciably increase the mineralization rates of surfactants as well as other organic pollutants. It has been established through research that the incorporation of ultrasound in combination with these combined processes significantly increases the oxidation rates and overall treatment efficiency. Specifically, mineralization of pollutants such as sodium dodecylbenzene sulfonate and dodecyl pyridinium chloride under optimized conditions exhibits marked improvement with mineralization efficiency of nearly 93% and 87%, respectively. These findings demonstrate the adaptability of AOPs employing ozone in addressing complex wastewater problems and affirming their viability in green remediation processes at the environmental level (Amat et al., 2011). Moreover, the AOPs scientific branch based on ozone is dynamic and characterized by continuous scientific cooperation among leading world institutions. Outstanding contributions from countries such as China, Spain, and the USA have pushed the area to create new methods of water treatment. Systematic review of the current pool of literature suggests that different modifications of AOPs, including those with ozone, are being investigated in various water matrices, most significantly industrial and urban. By recognizing patterns of research and collaboration threads, the findings illuminate the emergence of ozone-facilitated AOPs as a critical cog in the grand puzzle of sustainable wastewater treatment. The more the environment is recognized and prioritized by the populace, the more essential it becomes to include these new processes within regulatory guidelines for water treatment in guarding water resources for generations to come (Agüera et al., 2019).

4.2.2. Fenton and Fenton-Like Reactions

Fenton and Fenton-type reactions are excellent advancements in the field of advanced oxidation processes (AOPs) for environmental remediation. The reactions predominantly employ the hydroxyl radicals generated as a result of the oxidation of ferrous iron with hydrogen peroxide to decompose a very wide variety of organic pollutants. The efficiency and versatility of these processes are unveiled by various studies demonstrating their capacity to mineralize complex organic compounds that are generally refractory to normal treatment. Specifically, the application of zero-valent iron (ZVI) and hydrogen peroxide under optimized conditions is highly efficient in realizing significant reductions in total organic carbon (TOC) that reflect the degradation of target compounds. The process can also be optimized by adjusting parameters like temperature and pH, providing a specific remediation technique targeted at specific pollutants in the environment.

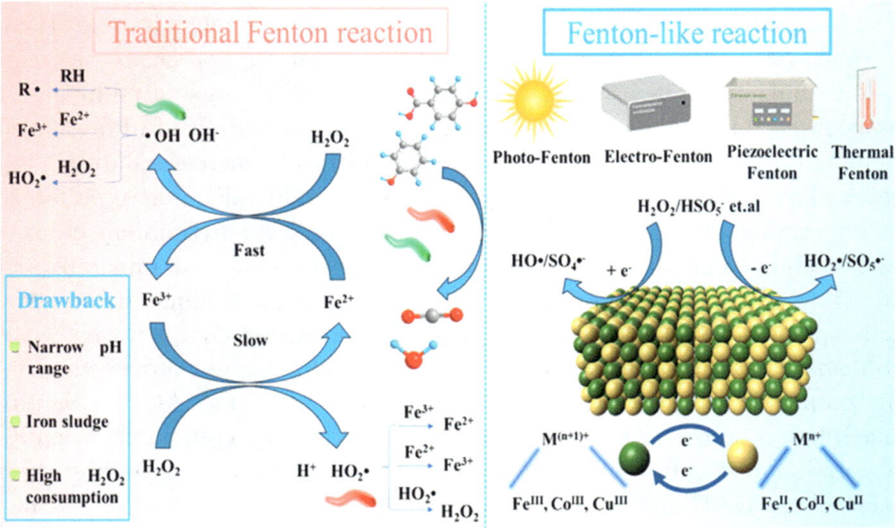

Figure 4.5. *Fenton and Fenton-like reactions.*

Source: https://pubs.acs.org/cms/10.1021/accountsmr.3c00289/asset/images/large/mr3c00289_0007.jpeg.

Adaptability is one of the unique features of Fenton-like reactions, involving the adjustment through amendments, for instance, the introduction of ultrasound or hydrodynamic cavitation. These changes enable the generation of other hydroxyl radicals, which are capable of breaking down organic matter at a faster and more efficient rate. For instance, the application of ultrasound in conjunction with the improved Fenton process (AFP) has been found to significantly improve mineralization rates, as noted for degradation of surfactants like sodium dodecylbenzene sulfonate (DBS) and dodecyl pyridinium chloride (DPC). Through reaction condition optimization, such as the regulation of some pH levels and addition of various amounts of ZVI and hydrogen peroxide, researchers have achieved impressive TOC reduction improvements—93% for DBS at optimal conditions (Amat et al., 2011). These advancements demonstrate how effective Fenton-like reactions are at defeating stable environmental pollutants. Furthermore, the ability of Fenton and Fenton-like processes to treat specific pollutants, i.e., 2,4-dichlorophenoxyacetic acid (2,4-D), signifies their potential effectiveness for specific applications in environmental management. It has been demonstrated in laboratory experiments that the effectiveness of mineralization may enhance significantly with the addition of hydrogen peroxide together with zero-valent iron, leading to enhanced hydroxyl radicals production, crucial for degradation. Cumulatively, these are effects leading to increased removal of TOC up to a level of 64% under certain conditions (Bremner et al., 2008). Hydrodynamic cavitation combined with Fenton chemistry is also an emerging technology, which makes it possible for wastewater contaminated with recalcitrant organic compounds to be continuously remediated. As there are more stringent environmental regulations, the optimization and optimization of Fenton and Fenton-like reactions become pivotal factors in the arsenal of techniques for effective water treatment and cleaning processes (Figure 4.5).

4.2.3. Photocatalytic AOPs

The application of photocatalytic Advanced Oxidation Processes (AOPs) has been a pioneering technique in environmental remediation, particularly in countering the issues caused by persistent organic pollutants. Photocatalysis takes advantage of light energy, typically sunlight, to activate a photocatalyst, typically titanium dioxide (TiO_2), to generate reactive species, such as hydroxyl radicals. These radicals, in turn, induce the degradation of harmful pollutants in wastewater, like phenolic compounds and other harmful chemicals (Romero et al., 2024). The use of photocatalytic AOPs not only assists in effective contaminant removal but also proves energy efficient, with solar power usage helping in applying green methodologies to water treatment technologies. Due to the urgent demand for having efficient environmental remediation technologies, photocatalyst system development remains a critical research and application area. Photocatalytic AOP efficiency can significantly be enhanced by the addition of various types of nanostructured materials. Current studies have projected the photocatalytic potential of nickel sulfide (NiS) nanostructures due to their superior optical and electronic characteristics, and therefore they are fine materials for photocatalysis (Sarkar et al., 2024). These nanostructured materials can harvest sunlight in an efficient manner as well as catalyze oxidation reactions at a comparative lower cost than standard noble-metal photocatalysts. Besides, controlled NiS modification, i.e., metal doping or heterojunction formation, has the ability to further enhance their photocatalytic activity and achieve higher organic pollutant degradation efficiency. Continued research into new semiconductor materials suggests improving photocatalytic methods further to make integrated and feasible environmental cleansing technologies possible. In practice, the synergistic action of photocatalytic AOPs with other technologies has also increased their efficacy in environmental remediation. For instance, application of hybrid systems that couple photocatalysis with ozonation or high-performance filtration techniques can achieve a better removal of pollutants. Through the use of techniques that allow for the generation of more reactive oxidative species, the combined system is capable of lowering the concentration of toxic pollutants by a substantial extent in water bodies. Effective removal of environmental pollutants not only alleviates ecologic distress but also aligns with emerging regulatory requirements for water quality and safety. Therefore, sustained R&D efforts for photocatalytic AOPs supported by cutting-edge material science and complemented by other treatments are essential in creating a cleaner and greener environment (Figure 4.6).

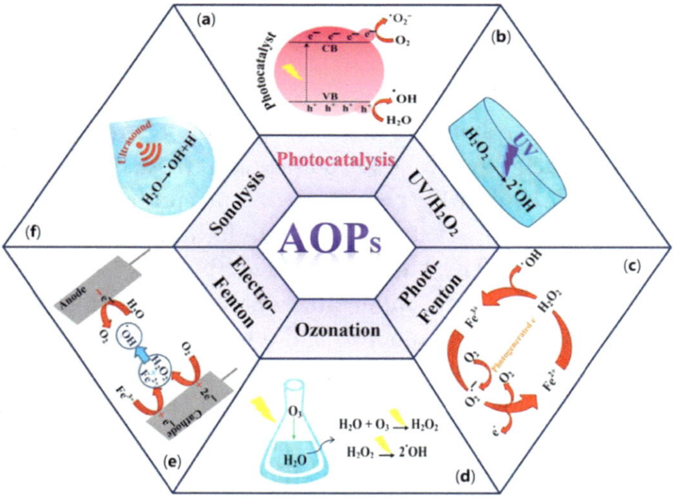

Figure 4.6. *Photocatalytic advanced oxidation processes (AOPs).*

Source: https://www.researchgate.net/publication/372385929/figure/fig4/AS:1143 1281174909802@1689430459068/Different-types-of-AOPs-processes-a-photocatalysis-b-UV-H2O2-processes-c.ppm.

4.3. APPLICATIONS OF AOPs IN WATER TREATMENT

Advanced Oxidation Processes (AOPs) are one of the most potential technologies for the treatment of water, particularly in the annihilation of recalcitrant organic pollutants and microbes. These technologies utilize highly reactive species, predominantly hydroxyl radicals, to oxidize refractory contaminants that cannot be eliminated by conventional water treatment technologies. Industrial chemicals, pharmaceuticals, and surfactants are some of the pollutants involved here that are of greater concern to the public health system as well as aquatic biota. Application of AOPs can enhance the efficacy of water treatment plants through the ability to break down complicated molecules into simpler units that are simpler to remove.

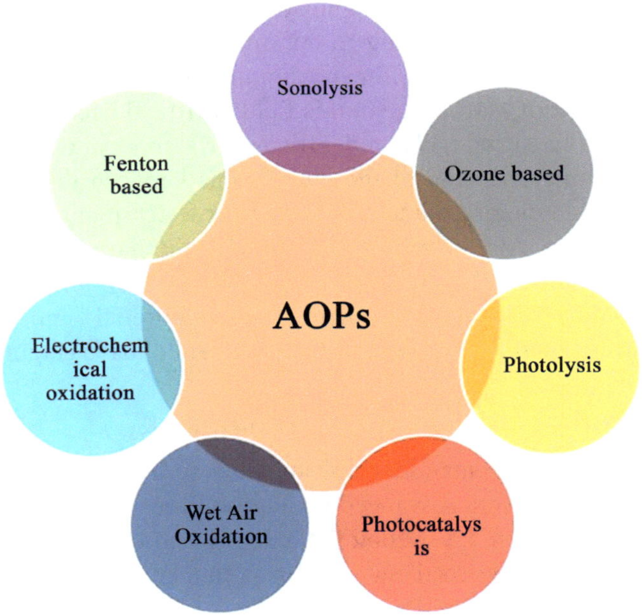

Figure 4.7. *Applications of AOPs in water treatment.*

Source: https://www.researchgate.net/profile/Pratibha-Gautam-3/publication/356423085/figure/fig9/AS:1122161901215751@1644555654804/Classification-of-a-major-class-of-AOPs-used-to-treat-complex-wastewater.png.

As highlighted in recent studies, countries like China, Spain, and the USA are leading research on AOP techniques, showing global interest in advancing this technology towards sustainable water management (Agüera et al., 2019). Precisely, the integration of AOPs with biological processes has led to significant improvements in the treatment of wastewater for various water sources such as industrial, urban, and drinking water. The application of such integrated strategies ensures higher removal rates for recalcitrant compounds with higher levels of mineralization. For instance, the Fenton-like system with zero valent iron and hydrogen peroxide has proved to be remarkably efficient in surfactant degradation like sodium dodecylbenzene sulfonate. Specifically, studies have reported a remarkable improvement in total organic carbon (TOC) removals when ultrasound is combined with these agents, allowing the breakdown of contaminants under the most favorable conditions (Amat et al., 2011). Such synergy not only beefs up the performance of AOPs but also poses potential cost-saving adaptations within treatment systems worldwide. More so, the scientific community has turned its eye to the scalability of AOPs, resolving the problems of their practical application within varying environments. Research advances further, seeking to optimize working parameters like pH, temperature, and oxidant dosing for maximum contaminant degradation. The steady advancement of knowledge in this field is critical because it illuminates the design of more sustainable and efficient water cleanup processes. By facilitating the comprehension of the interactions between disparate oxidants and pollutants, researchers work toward particular water qualities and treatment goals in optimizing AOP applications. Lastly, the large-scale application of AOPs will significantly help improve water quality and environmental sustainability and emphasize the potential for additional innovation in this valuable environmental treatment field (Agüera et al., 2019) (Figure 4.7).

4.3.1. Removal of Organic Contaminants

The removal of organic contaminants from water sources is a daunting task in environmental remediation. Advanced Oxidation Processes (AOPs) have been hailed as a ready solution, leveraging the oxidative power of hydroxyl radicals to degrade even the most resilient organic pollutants. Current research puts emphasis on the ability of AOPs to remove most forms of wastewater, from industrial to domestic wastewater, through the elimination of a wide variety of organic pollutants such as pharmaceuticals and personal care products. Such pollutants are particularly problematic due to their persistence in the environment as well as potential toxicity, necessitating the creation of effective treatment systems. Analytical examination shows that countries like China, Spain, and the USA have been at the forefront of research into these new methodologies, making it possible for future breakthroughs in the field of wastewater treatment that will encourage efficient management of water (Agüera et al., 2019). One of the significant classes of AOPs involves the introduction of chemical reagents such as hydrogen peroxide and zero-valent iron (ZVI), which has been successful in enhancing the degradation of organic pollutants. For instance, the enhanced Fenton process has shown capability in optimizing reaction conditions and attaining high mineralization levels for various pollutants. It has been demonstrated in research that use of ZVI under hydrogen peroxide at acidic pH will result in drastic decreases in TOC levels, or essentially surfactant destruction for compounds such as sodium dodecylbenzene sulfonate (DBS) and dodecyl pyridinium chloride (DPC). Inclusion of ultrasound in such processes will further enhance the mineralization efficiency, to a maximum of 93% for DBS and 87% for DPC under optimized conditions. This combination of technologies is an expression of the interdisciplinary approach necessary in the removal of persistent organic pollutants in modern wastewater treatment (Amat et al., 2011). Lastly, the continued evolution of AOPs holds great potential in overcoming the challenge of organic pollutants in the environment. Research efforts have managed to maximize the effectiveness of these processes in various applications, leading to improved methodologies in pollutant degradation. The international collaborative nature of research work in this area is crucial in shaping innovation and best practice sharing between countries. As this field continues to develop, the integration of emerging technology and optimization of procedures will be central to enhancing the cost-effectiveness and productivity of AOPs. Lastly, the successful application of these new technologies will result in effective water management and protect aquatic ecosystems from hazardous organic pollutants. Through sustained commitment to R&D, AOPs can become a focal point in environmental remediation in the years to come (Agüera et al., 2019) (Figure 4.8).

Figure 4.8. *Removal of organic contaminants.*

Source: https://www.mdpi.com/nanomaterials/nanomaterials-12-02148/article_deploy/html/images/nanomaterials-12-02148-g001.png.

4.3.2. Treatment of Industrial Wastewater

Industrial wastewater treatment is a multifaceted challenge with severe environmental health implications. Contemporary industrial manufacturing processes tend to yield effluents containing a diversity of pollutants, including heavy metals, organics, and pathogenic microbes. Traditional wastewater treatment technologies also tend to be ineffective, being unable to handle the toxicity and intricacies of these pollutants. In response to this, Advanced Oxidation Processes (AOPs) meet an urgent need for new technologies that enhance the efficiency of wastewater treatment by employing powerful oxidative agents to effectively degrade recalcitrant contaminants. By creating extremely reactive species, such as hydroxyl radicals, AOPs are able to oxidize and mineralize organic pollutants of wide range, significantly improving wastewater quality and preventing potential public health hazards from untreated effluent. AOPs are characterized by the ability to create reactive oxygen species through a variety of mechanisms, including chemical reactions involving catalytic use. Some of the novel materials that have emerged are metal–organic frameworks (MOFs). These multicomponent materials facilitate adsorption of contaminants and possess antimicrobial properties, enhancing their effectiveness in wastewater treatment. When integrated into AOPs, MOFs enhance the formation of oxidant radicals, hence accelerating the degradation of complex organic contaminants. Recent studies indicate that the inclusion of MOFs in electro-Fenton and photocatalytic processes is able to provide remarkable improvements in efficiency for treatment. The tunability of such frameworks represents a significant breakthrough in the ongoing quest for effective industrial wastewater treatment technologies. Additionally, their structure can be modified readily to fine-tune performance to specific pollutants, with material design being at the forefront of AOP applications (Terrón et al., 2025). Whereas the promising advancements ushered in by AOPs and the employment of MOFs give hope to anticipate, challenges remain that constrain their widespread application in industry. Catalyst recoverability limitations, secondary byproduct formation, and overall process costs are limiting scalability of such technologies. Homogeneous AOPs are highly efficient in degradation but release sludge making it difficult for disposal. Heterogeneous AOPs possess better reusability and recoverability but with comparable limitations of potential mass transfer issues. Researchers now focus more on developing hybrid systems by joining AOPs with add-on treatment processes such as membrane filtration or conventional biological treatments. This integration is done to establish a more efficient, sustainable wastewater treatment system, overcoming the limitations of traditional systems while ensuring effective removal of pollutants and safeguarding the environment (Satyam et al., 2025) (Figure 4.9).

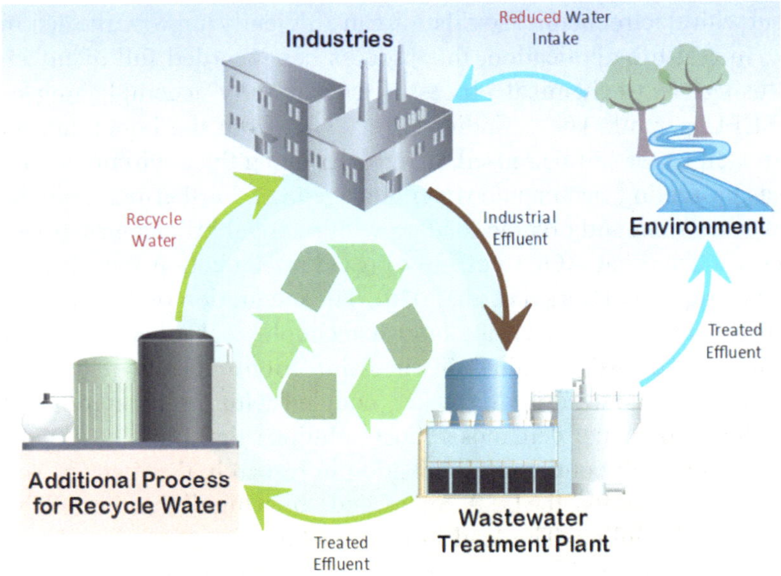

Figure 4.9. *Treatment of industrial wastewater.*

Source: https://www.albeadvance.com/wp-content/uploads/2022/10/Recycle-Water-Concept-FOR-WEBSITE-2.png.

4.3.3. Disinfection and Pathogen Control

The significance of pathogen disinfection and control during environmental cleanup cannot be overstated as water shortages emerge as a very serious problem on the international level. Advanced Oxidation Processes (AOPs) are proving to be great potential as a tool for water purification by effectively killing detrimental microorganisms, like bacteria and viruses. AOPs utilize powerful oxidants, such as hydroxyl radicals, to oxidize organic pollutants in water, thus offering a greener method of disinfection. The increasing prevalence of antibiotic-resistant microorganisms calls for new approaches, which AOPs solve in a unique way through the degradation of such toxic entities using oxidative mechanisms. The variability of the abilities of AOPs enables them to be used for the treatment of heterogeneous wastewaters, either industrial, urban, or agricultural in nature, and thus presents a reliable solution to complex disinfection issues in water treatment facilities (Agüera et al., 2019). Current research has shown the effectiveness of certain AOP technologies in treating pathogen control and disinfection. An example is the Closed Loop Advanced Sanitation System (CLASS) that highlights how the combination of biological treatment with electrochemical oxidation can efficiently improve the disinfection process. In real-life application, this process has recorded full disinfection of *E. coli* using electrochemical processes, successfully accomplishing a count of zero CFU per ml. These findings are vital within the home environment where treated water can be reused, thereby reducing the environmental impact as well as promoting sustainable water management. Furthermore, biochemical removal of nutrient and organic loading with removal efficiencies greater than 85%–95% means that AOPs not only provide disinfection but also improve overall water quality (Barani et al., 2020). The integration of AOPs in drinking water treatment programs is not only a technological advancement; it is a paradigm move towards sustainability in disinfection and pathogen control. The global demand for pure drinking water, coupled with the inherent challenges of emerging pathogens, demands robust solutions that must be efficient yet nature friendly. With continuing innovation in research, the focus of scientific activities in countries like the USA, Spain, and China highlights the collaborative nature of this discipline. The lead in the development of advanced oxidation methodologies is taken by these countries, with special emphasis on different wastewater types and the treatment of organic pollutants. The balance between practice and research is vital for advancing disinfection technologies, putting AOPs at the forefront of future environmental remediation (Agüera et al., 2019).

4.4. AOPs IN SOIL REMEDIATION

The use of Advanced Oxidation Processes (AOPs) in soil cleaning has been a groundbreaking approach to treating extensive contamination in the soil, particularly by harmful contaminants from industrial waste. AOPs, which generate highly reactive species like hydroxyl radicals, are distinguished by their efficiency and rapidity in breaking down organic contaminants. This process is particularly useful in the remediation of persistent pollutants that are difficult to destroy with conventional methods due to their complex chemical structures, such as dyes and heavy metals present in textile wastewater industry (TIWW) and oil refinery sludge. The use of AOPs not only enhances the rates of degradation of such pollutants but also eliminates the risk of secondary pollution, hence making it an environmentally friendly method for soil remediation processes. Understanding these operations and their efficiency is crucial in the development of holistic environmental programs aimed at the restoration of soil health. In recent times, the textile industry has come under criticism for its significant ecological impact, primarily due to the dumping of toxic dyes and other chemicals-impregnated wastewater. The total organic carbon (TOC) and chemical oxygen demand (COD) contents of TIWW are among the attributes that necessitate advanced treatment methods to be utilized for effective remediation. AOPs are an appropriate choice, considering that they have the ability to oxidize complex dye molecules and other toxic components in TIWW to facilitate their subsequent degradation into non-toxic byproducts. The exploration of novel AOP configurations, i.e., the combination of Fenton's reagents with ultrasound use, has also been shown to be viable in significantly enhancing the efficiency of degradation of some dye compounds present in TIWW, with levels of color strength and toxicity achieved. AOPs are hence a major component in addressing the pressing issues from textile wastewater while promoting environmental sustainability. Besides the textile industry, AOPs are also applied to counteract oil refinery contaminations, to be more precise oily sludge contaminations, which might impact soil and aquifers. Using AOPs in treating oily sludge, particularly when integrated with other technologies such as bioremediation, enhances TPH reduction, showing the diversified use of these technologies. According to recent studies, combinations of ultrasound with Fenton's reagent have achieved outstanding TPH removal rates of up to 88.1% when optimized. The combined process not only destroys harmful materials but also recovers valuable resources from waste products, illustrating the dual benefit of AOPs for both remediation and resource recovery. Hence, this new technology is a cornerstone in modern soil rehabilitation processes, treating both short-term and long-term environmental problems in numerous industrial applications (Figure 4.10).

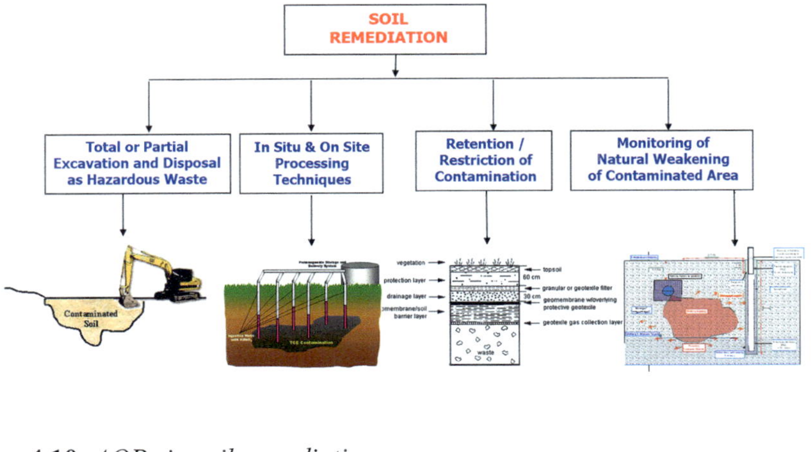

Figure 4.10. *AOPs in soil remediation.*

Source: https://intergeo.gr/wp-content/uploads/2019/03/katigories_metron_01-copy.jpg.

4.4.1. In Situ versus Ex Situ Remediation Techniques

For environmental remediation, a terminological difference is considerable between ex situ and *in situ* approaches. *In situ* remediation refers to activities applied directly on the contaminated site, modulating environmental conditions to degrade pollutants without digging. It is typically less costly and less disruptive to the ecosystem since it minimizes site disturbance and transport of hazardous materials. Advanced Oxidation Processes (AOPs) greatly enhance *in situ* processes through the generation of effective oxidants such as hydroxyl radicals, thus facilitating the degradation of recalcitrant organic pollutants. This has demonstrated tremendous potential towards the remediation of biorefractory materials that traditional remediation practices often struggle to degrade, thus meeting the urgent need for novel approaches in environmental remediation processes (Carlos et al., 2023).

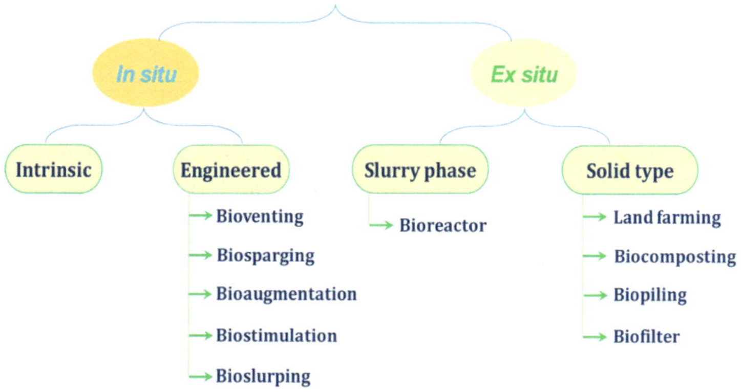

Figure 4.11. *In situ versus ex situ remediation techniques.*

Source: https://www.researchgate.net/publication/374783987/figure/fig4/AS:1143 1281199612058@1697648454039/Bioremediation-processes-are-mainly-classified-into-two-categories-in-situ-and-ex-situ.tif.

On the other hand, ex situ remediation techniques involve the removal of the contaminated material from the site and treatment off-site. Physical removal, treatment within facilities with specific equipment, and disposal or recycling of the treated material make up the techniques. Ex situ techniques can offer more control and efficiency in degrading contaminants but typically cost more and are logistically more difficult, involving having to manage waste transportation as well as potential additional contamination during transportation. Yet, AOPs can be integrated into ex situ processes, offering a robust platform for effective treatment of toxic wastes. Electrochemical technologies and material science advancements, particularly with the application of carbon nitride as an effective photocatalyst, indicate how the adoption of AOPs can improve ex situ remediation techniques by improving electron transfer and contaminant degradation (Deng et al., 2023). In conclusion, the choice between *in situ* and ex situ remediation techniques is a function of site-specific characteristics, pollutant properties, and economic factors. There must be an evaluation of the effectiveness and practicality of each technique, especially taking into account emerging technologies in AOPs with promises of increased pollutant degradation. By comparing these two techniques, practitioners can optimize remediation strategies based on efficiency, cost, and environmental factors. The continuous innovation of AOPs informs future approaches so that *in situ* and ex situ approaches also keep evolving, best resulting in synergistic strategies that harness the strengths of both to successfully address complex environmental issues. It is thus integration of state-of-the-art advanced oxidation processes into clean-up strategies that represents a paradigm shift in environmental remediation practice (Figure 4.11).

4.4.2. Effectiveness in Treating Contaminated Sites

The effectiveness of remediation of polluted sites is greatly enhanced with the use of Advanced Oxidation Processes (AOPs), which is one of the major environmental remediation methodologies. AOPs employ powerful oxidants in order to effectively break down harmful pollutants, rendering them less toxic and easier to treat. Traditional remediation techniques are typically incapable of satisfactorily handling complex mixtures of pollutants, leading to incomplete degradation and secondary pollution. AOPs, however, have been shown to be more efficient in remediating organic contaminants like petroleum hydrocarbons and other contaminant of emerging concerns. This is particularly the case when it comes to soil contaminated with petroleum because conventional technologies like excavation and incineration come with environmental disruptions, whereas AOPs offer a less invasive and more environmentally acceptable strategy (Rizal et al., 2025). Another significant advantage of AOPs is that they can be tailored to treat against specific contaminants and environmental factors. This versatility is crucial when the sites to be treated are contaminated with long-term organic pollutants and emerging contaminants, which may require specialty remediation procedures. AOPs can effectively tackle threats from resistant pollutants that cannot be degraded by standard wastewater treatment practices. Processes like electrochemical advanced oxidation processes (EAOPs) demonstrate this capability, generating reactive oxygen species which enable the breakdown of complex organic structure (Miguel et al., 2024). Furthermore, the scalability of AOPs means that they can be applied in a beneficial way to a wide variety of settings, from industrial effluent streams to agricultural runoff, thereby expanding their application in real-world remediation processes. Despite the promising results associated with AOPs, there are several drawbacks and difficulties that need further study and improvement. The cost of energy associated with sustaining the advanced oxidation approaches can be substantial, and ongoing research is necessary to enhance efficiency and reduce operating costs. In addition, the long-term environmental impacts of AOPs are underresearched, particularly regarding potential degradation byproducts generated during treatment. Future research endeavors should focus on enhancing AOP technologies, studying their integration with biological remediation technologies, and making sure they attain their maximum potential in different environments (Miguel et al., 2024; Rizal et al., 2025). The integrated approach not only offers the possibility of more effective remediation of polluted areas but also lays the framework for sustainable and eco-friendly practice in environmental remediation.

4.4.3. Case Studies of Successful Soil Remediation

Contamination of soil remains a high-profile environmental problem, particularly where industrial activity is high. In treating contaminated soil, advanced oxidation processes (AOPs) have emerged as very effective treatment for contaminated soils. A case study that testifies to the efficacy of employing AOPs for degradation of stable organic pollutants (POPs) in soil within industrial areas demonstrates just that. Through mechanisms such as Fenton processes and ozonation, researchers were able to decrease the concentration of pollutants by considerable amounts within a relatively short span of time. The study underscored the efficiency of AOPs in not only decontaminating water solutions but also in cleansing intricate soil matrices, highlighting their potential as powerful tools for environmental remediation strategies (Adebayo et al., 2022). A second good example of AOP applications includes oil refinery sludge remediation. Petroleum hydrocarbons toxicity necessitates the development of strong treatment strategies against potential dangerous environmental impacts. In an extensive study, a synergistic mixture of bioremediation techniques and advanced oxidation processes efficiently reduced the total petroleum hydrocarbons (TPH) in contaminated soil. The use of ultrasound and Fenton's reagents gave very good results with high TPH reduction rate of 88.1%. This case justifies the versatility of AOPs application with other techniques, finally resulting in an integrated solution to oily sludge waste management (N/A, 2013). Besides, the long-term success of AOP applications to clean up soil suggests major issues of concern for future environmental practice and regulation. Given the established ability of AOPs to oxidize harmful compounds with decreased secondary waste generation, their further use could totally revolutionize remediation processes at contaminated sites. However, the need to keep working on improving process conditions and establishing the environmental impact of by-products is simply imperative. As demonstrated in various successful case studies, the integration of AOPs into overall remediation strategies not only increases the level of pollutant degradation but also aids in reinstating balance in disturbed ecosystems (Adebayo et al., 2022).

4.5. ADVANTAGES AND LIMITATIONS OF AOPs

The advantages of Advanced Oxidation Processes (AOPs) are that they are very efficient in degrading stable organic contaminants that conventional processes cannot effectively target. AOPs' oxidizing abilities allow the formation of very reactive species, including hydroxyl radicals, which can mineralize a large range of contaminants, including pharmaceuticals, pesticides, and industrial chemicals. AOPs may be tailor-made for a given pollutant or a specific environmental situation so that they are more effective and versatile. For instance, the mineralization of phenol was achieved efficiently by a sono-enhanced Fenton process involving zero-valent iron and hydrogen peroxide to achieve excellent total organic carbon (TOC) removals efficiently, as per one study. This flexibility is an opportunity for developing cost-effective and large-scale applications of AOPs for wastewater treatment, potentially overhauling wastewater treatment processes for numerous industrial processes (Bremner et al., 2012). AOPs also possess some limitations that must be addressed for further use in environmental clean-up. A significant problem is the potential formation of deleterious by-products during the oxidation reaction, which could be of more dangerous implications than the benefits of pollutant degradation. By-products formed during the progression of AOPs in some instances can exhibit higher toxicity than the parent pollutant, and thus serious monitoring and regulation of the treatment process are warranted. Furthermore, the working conditions required for effective AOPs, such as regulation of reagent concentration and reaction time, can complicate and increase operation costs. More significantly, AOP utilization may not be sufficient on a large scale due to these economic and technical concerns. Hence, it is essential to understand such limitations so that AOP technologies not just work but are also safe and feasible in real-world scenarios (Gencheva et al., 2025). Another specific characteristic of AOPs is that they can be employed along with other methods of remediation, which can enhance their performance. By integrating AOPs with physical or biological approaches, professionals can deal with tricky pollution cases involving different kinds of pollutants. For instance, while AOPs ensure the rapid oxidative treatment, there may be additional biological processes that allow the destruction of residual by-products to ultimately result in complete mineralization of pollutants. Additionally, new AOP designs, e.g., hybrid systems involving coupling of photocatalysis with other oxidants, exhibit improved reaction selectivity and kinetics toward the target pollutants. This synergistic method is worth exploring for solving the issue presented by more resilient contaminants like munitions and other toxic compounds. Hence, creative application of AOPs along with supporting methods can lead to a more holistic remediation approach, thus forming a suitable scope of future research and implementation in the domain of environmental science (Bremner et al., 2012).

4.5.1. Benefits Over Traditional Treatment Methods

The efficiency of advanced oxidation processes (AOPs) in environmental cleaning offers a significant competitive advantage against conventional treatment processes, particularly in the management of complex pollutants in wastewater. Conventional practices such as biological treatment and primitive chemical processing fail to significantly degrade persistent pollutants, leading to poor water quality outcomes. AOPs, however, yield highly reactive hydroxyl radicals and other oxidants that can mineralize a wide variety of organic compounds. This radical degradation is particularly beneficial for the removal of recalcitrant substrates, thus yielding a more efficient removal of industrial wastewater that can be contaminated with emerging contaminants. Besides, the flexibility of AOPs allows for modulating based on pollutant patterns, thus enhancing their applications in different environmental conditions (Manickavasagam et al., 2025). Another added advantage of AOPs is its efficiency and flexibility during operation. Compared to traditional processes that may necessitate extensive pre-treatment or heavy operational cost, AOPs are likely to minimize the requirement for such auxiliary processes. For instance, heterogeneous hydrochar catalysts have been said to offer promise in the advancement of AOPs, improving catalytic efficiency and the rate of pollutant degradation. These catalysts, which are also diverse in nature like heteroatom-doped hydrochar and metal-heteroatom co-doped hydrochars, allow the production of reactive species under less aggressive conditions, thus rendering the remediation process cheaper and energy-efficient. In addition, the integration of AOPs with secondary treatments like membrane filtration can lead to synergisms, yielding an enhanced aggregate treatment performance (Manickavasagam et al., 2025). Besides this, AOPs also address most of the environmental problems associated with traditional wastewater treatment processes. Traditional processes produce sludge as a by-product, which may lead to disposal issues and secondary pollution. In contrast, mineralization characteristics of AOPs can significantly minimize the generation of solid waste and provide a more sustainable wastewater management strategy. Since there is an increasing demand for sources of more pure water, the ability of AOPs to kill pollutants without producing toxic byproducts positions them as a visionary method in environmental remediation. Through the use of new technologies and engineered catalysts, AOPs not only circumvent the current challenges in wastewater treatment but also provide the building blocks for a stronger and more sustainable future in environmental management (Satyam et al., 2025) (Figure 4.12).

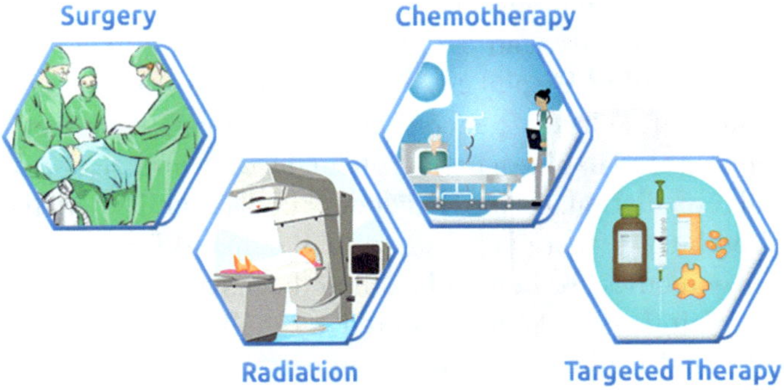

Figure 4.12. *Benefits over traditional treatment methods.*

Source: https://prolifecancercentre.co.in/wp-content/uploads/2019/12/liver-cancer-5.jpg.

4.5.2. Economic Considerations and Cost-Effectiveness

In environmental remediation, the financial factors concerning Advanced Oxidation Processes (AOPs) are a significant consideration in determining their mutual usability and viability. AOPs, which are celebrated for generating highly reactive radicals for contaminant destruction, are often preceded by high initial capital and operational costs. Such economic barriers call for an honest cost-effectiveness assessment, particularly when set against traditional remediation approaches. As industries contend with rigorous regulatory frameworks and increasing public attention, the need for effective and financially savvy solutions increases. Accordingly, an examination of AOPs' long-term worth—i.e., their likelihood of minimizing environmental impact and furthering public health—is paramount. Financing AOP technologies is an extremely lucrative venture by way of liability cost avoidance and improved environmental regulation compliance, an expense warranting the initial capital investment in their implementation. Moreover, inclusion of AOPs as part of a comprehensive environmental management system can enhance their economic feasibility. An interdisciplinary approach that synergizes AOPs with other technologies, such as bioremediation, can enhance full utilization of resources and reduce overall costs. For example, AOPs' capital costs can be recouped upon subsequent biological treatment, where AOP-treated residues serve as substrates for microbial decomposition. This reciprocally beneficial relationship not only heightens the efficiency of remediation treatments but also raises cost savings throughout the overall treatment process. Furthermore, collaborative effort among stakeholders, including between government agencies, private sector, and the science community, can lead to cost-effective AOP solutions that are tailored to specific environmental environments, which optimizes the aggregate economic viability of such next-generation technologies. Regulatory support and public opinion also tend to play a major role in the cost-effectiveness of AOPs in environmental remediation. As more is understood about the negative effects of conventional pollutants, there is greater public demand for new approaches to offer healthier outcomes for human health and ecosystems. Regulatory bodies are now recognizing the potential offered by AOPs for addressing hard-to-treat pollutants, as done in recent studies on petroleum products and persistent organic pollutants (Atlas et al., 2015; AEA Technology Environment et al., 2010), to promote the use of such technologies with a bias towards incentives. Through balancing economic interests with society's requirements, AOPs find themselves poised as not only viable alternatives, but more as components of future remediation planning. Ultimately, economic viability for AOPs in environmental remediation relies on multidimensional analyses encompassing technological advancements, popular opinion, and shifting regulatory landscapes, facilitating a sustainable and effective mechanism for combating environmental degradation.

4.5.3. Challenges and Limitations in Implementation

The application of Advanced Oxidation Processes (AOPs) in the clean-up of the environment is full of challenges and limitations that can prove counterproductive to their success and widespread application. The highest among these is the economy of these technologies, rendering it often costly to many remediation projects, particularly those with a limited budget. The operational costs that are associated with reagents applied to chemicals, power, and advanced devices contribute to these costs. In addition, the complexity of AOP technologies requires expert knowledge for their implementation and maintenance to be effective. Lack of experienced staff increases the likelihood of inefficiency, which may lead to partial degradation of contaminants, thereby rendering remediation useless. These financial and technical problems are key to address to make AOPs more feasible to apply in actual projects, as illustrated in a number of remediation initiatives aimed at addressing long-term liabilities such as those NASA is undertaking (Greene et al., 2025). Another significant shortcoming is the inconsistency of site-specific conditions that affect the performance of AOP systems. Each contaminated site would possess some unique characteristics like variations in contaminant composition, concentration, and environmental conditions like pH and temperature. Heterogeneities might complicate the application of AOPs because various contaminants would need varied oxidative methods to effective treatment. Also, as established in electrodialytic remediation research, operating parameters like liquid-to-solid ratio and current density have a great impact on treatment efficiency (YB A et al., 2019). Thus, making AOPs adaptable to these differences is unavoidable but potentially costly and technically challenging. The need to perform comprehensive preliminary tests towards identifying the most suitable working parameters similarly highlights the complexities in the effective use of AOP technology for environmental remediation. Last but not least, the risk of by-product formation is an obstacle to the effective use of AOPs. Even as these processes are particularly aimed at the destruction of harmful contaminants, the supporting reactions will, in doing so, lead to the formation of harmful secondary waste products that need to be treated separately. This is particularly a concern when oxidants and precursors react to form poisonous intermediate products that ultimately cloud the entire plan for waste treatment. Understanding of the degradation routes and that the end products are harmless to the environment is important but in most instances requires extensive research and monitoring. As such, their resolution is key to the safe and sustainable application of AOPs in real applications, balancing the questions of efficient remediation against environmental stewardship. Consequently, a thorough analysis of both the expected advantages and possible disadvantages is needed in order to properly assess the effectiveness of AOPs in environmental clean-up activities.

4.6. FUTURE TRENDS AND INNOVATIONS IN AOPs

The landscape of Advanced Oxidation Processes (AOPs) for the remediation of the environment is shifting with complex pollutants, particularly pharmaceuticals and synthetic dyes, taking the center stage. The evolution of AOP techniques is increasingly becoming crucial since industries are coming under the microscope in terms of waste disposal and sustainability in the environment. For instance, dye discharge in the textile industry poses the challenge of environmental pollution, and thus the necessity for efficient treatment processes that would be able to reduce contamination without an impact on operational performance.

Research now focuses on integrating AOP with other treatment processes to ensure maximization of dye molecule degradation since physicochemical and biological treatments are not sufficient to provide complete solutions. The trend towards hybrid systems combining AOPs with biological treatment strategies is a revolutionary approach to wastewater management, with enhanced oxidation strategies enabling more environmentally friendly industrial processes as well as meeting regulatory compliance standards (Alves et al., 2012).

Figure 4.13. *Future trends and innovations in AOPs.*

Source: https://greenmindai.blog/wp-content/uploads/2024/02/Image-01.jpg.

Additionally, the persistence of pharmaceutical contaminants in freshwater bodies underscores the imperative for the development of new AOP applications. The discovery of pharmaceuticals in different ecosystems has raised concerns regarding their environmental and human health effects, with much research being done to look for efficient degradation technologies. AOPs, which generate extremely active radicals, have tremendous potential to degrade recalcitrant chemicals resistant to conventional treatment strategies. Studies have proven the feasibility of advanced oxidation technologies in the removal of such compounds as carbamazepine and diclofenac that generally pollute wastewater effluent due to an absence of sanitation infrastructure (Olawade et al., 2025). Upcoming trends indicate the focus on the optimization of AOP conditions, such as exploring the use of new catalysts and incorporating machine learning algorithms in real-time monitoring. These advances have the potential to facilitate the application of treatment strategies tailored to the specific contaminant mixes prevalent in local conditions, potentially enhancing the performance of AOPs in environmental remediation. In addition to simplifying treatment technologies, growing use of decentralized AOP strategies reflects a trend towards localized environmental management. Regions with limited or no centralized wastewater treatment capacity are apt to experience much higher contaminant concentrations, necessitating effective and low-cost treatment options. New technologies such as solar-powered or low-cost material-based portable AOP systems present viable interventions for on-site remediation operations. This transition to decentralized approaches also acknowledges the importance of community involvement in environmental health as the systems can allow local communities to have control over their own water quality. Moreover, the integration of biochar-based systems in AOP structures is a groundbreaking advancement in the utilization of locally available resources for pollution management. Utilizing local materials, these systems not only enhance contaminant removal efficiency but also promote environmental sustainability (Olawade et al., 2025). The future advancements in AOPs therefore hold the promise of a more resilient and responsive system for countering the pressing challenges of environmental pollution (Figure 4.13).

4.6.1. Emerging Technologies and Research Directions

The research on new technologies employed in advanced oxidation processes (AOPs) is one of the most important areas for enhancing environmental remediation strategies. As new contaminants of concern (CECs) start invading aquatic ecosystems, new AOPs are emerging as a solid solution as these can eliminate these recalcitrant pollutants. Notably, the use of new catalysts and reactive oxidants such as peracetic acid (PAA) displays a promising path towards oxidation reaction enhancement. Research indicates that PAA exhibits excellent oxidative power after activation in conjunction with transition metals or carbon materials. Catalytic synergy not only enhances the formation of reactive species but also assists with improving the overall efficiency of pollutant degradation, paving the way for environmentally friendly water treatment technologies that preclude the release of harmful by-products into the environment (Jaime et al., 2025). Furthermore, the integration of photocatalytic systems within AOP schemes offers a new way to harness natural light for powering environmental remediation. Solar photocatalysis, with lower efficiencies of degradation under natural light than artificial sources, offers huge benefits through lower energy needs and smaller environmental footprint. New studies highlight the promise of photocatalysts like titanium dioxide, which can be activated by sunlight to induce degradation reactions in chemicals like metformin, a key pollutant in waters (Jaime et al., 2025). This focus on clean energy-driven technologies is a turning point towards more environmentally friendly remediation technologies, addressing ecological concerns directly while driving cost-effective and universally scalable solutions. Following these technological advancements, ongoing research must fill the essential gaps regarding awareness of the by-products generated through AOP applications. The risk of toxicity in intermediates formed during the degradation of CECs remains a pertinent challenge, similar to that of extensive hazard assessments in AOP applications. APO strategies to contain harmful by-products—mainly those associated with the fashionable techniques like ozonation—can involve controlling reaction conditions, such as pH and catalyst concentrations. Further, as researchers keep investigating the environmental consequences of AOP-generated intermediates, there is scope for enhancing existing strategies of resilience towards a myriad of pollutants. Keeping these recent lines of study in check will make AOPs not only more efficient but also lead to overall environmental security and sustainability goals (Bouzayani et al., 2025) (Figure 4.14).

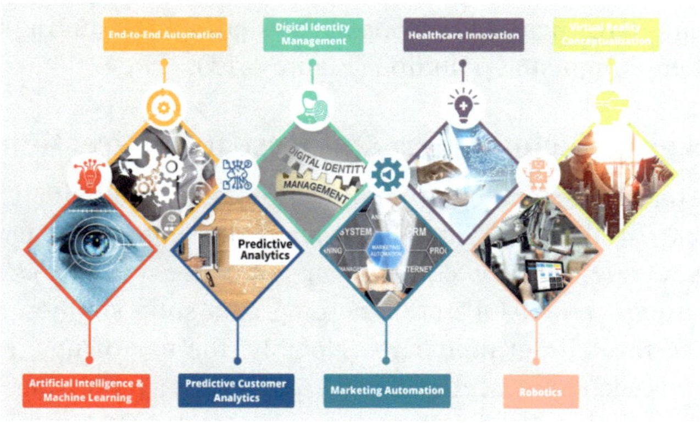

Figure 4.14. *Emerging technologies and research directions.*

Source: https://infigroup.com/images/services/emerging-technologies-chart.png.

4.6.2. Integration with Other Remediation Techniques

The combination of Advanced Oxidation Processes (AOPs) with other remediation techniques plays a key role in enhancing the efficiency of environmental remediation processes. AOPs are highly renowned for their ability to produce extremely reactive hydroxyl radicals, which are efficient in breaking down a wide range of organic pollutants. Isolation of AOPs alone, however, may not be sufficient to treat even more complex contaminant matrices such as textile industry wastewater (TIWW) or petroleum-contaminated environments. By combining AOPs with biological treatment processes, including activated sludge processes, the best features of both chemical and biological systems can be leveraged. This synergistic process not only boosts the overall treatment efficiency but also prevents secondary pollution by enabling efficient mineralization of recalcitrant compounds that are present in high quantities in TIWW, for example, permanent dyes and heavy metals, which are well-documented to cause widespread environmental harm (Bharagava et al., 2021). Combining AOPs with sorption processes also enhances the removal of various toxic compounds from contaminated waters and soils. For instance, inclusion of biochar materials in AOP systems can effectively adsorb organic pollutants and, at the same time, serve as a catalyst to promote advanced oxidation processes. Such a dual utilization is highly relevant with respect to oil-contaminated resources, where biochar has shown promise in the removal of petroleum hydrocarbons through both sorption and subsequent oxidative degradation. Such hybridization not only supports the decontamination process but also increases reusability of the used sorbents, which in turn leads to reduced overall operating costs associated with remediation (Liu et al., 2024). Hybrid systems achieved by combining AOPs with novel materials amplify the overall efficiency and sustainability of waste treatment. Lastly, combining AOPs with other emerging approaches—membrane filtration, electrochemical oxidation, or adsorption—is an active approach to overcoming multicomponent contaminant profiles. Every remediation technology can address some specific challenges, e.g., particle size, reactivity, or pollutant concentration, and hence offer a better remediation approach. Such integrated approaches tend to provide enhanced results in the decrease of chemical oxygen demand (COD) and total organic carbon (TOC) values, which are of crucial significance in the determination of water quality in highly polluted situations. The extremely intricate nature of pollutant properties demands a precise understanding of remedial operations, supporting the reality that no single approach can prove to be sufficient. With more research, the identification of further combinations has the potential to open up new paths for better environmental remediation outcomes, ultimately safeguarding ecological and human health.

4.6.3. Policy and Regulatory Considerations

The implementation of Advanced Oxidation Processes (AOPs) in environmental remediation is not only a technical matter but an actual policy and regulatory challenge. It is the responsibility of policymakers to make sure that there are accurate frameworks regulating the use of AOPs such that they are safe and environmentally friendly while they are efficient in eliminating such contaminants as per- and polyfluoroalkyl substances (PFAS). Current policies are always behind technology and therefore open up a gap that can hamper the implementation of new treatment procedures. A positive approach to policy development is therefore crucial in a way that it can promote research and implementation of AOPs, thereby promoting a positive climate for clean water initiatives and public protection of health. These guidelines can also enable stakeholder coordination across the likes of government agencies, private industries, and academe, leading to a more effective integration of AOPs into existing environmental management practices.

The evolving contaminant scenario, including the increasing role of PFAS, demands regulatory guidelines to incorporate the latest advances in AOPs and other treatment methods. The regulatory bodies must take note not only of establishing standards for the use of these technologies but also in addressing the particular issues involved with the new contaminants. For example, policy must dictate suitable levels of treatment efficacy as well as monitoring frequencies for AOPs employed for the removal of PFAS from water sources. In the predictive modeling arena, the incorporation of novel technologies and methods creates an opportunity for more effective regulatory decision-making, as highlighted in studies on the effectiveness of various AOP methods. Building on these conclusions, regulatory agencies can create evidence-based policies that retain the best of existing AOP solutions while also maintaining robust public health protections.

Finally, it is essential to promote a culture of continuous improvement regarding regulatory approaches to AOPs in order to ensure successful environmental remediation. Policymakers have to be in touch with scientific and engineering communities in order to stay updated on emerging technologies and their relevance for public health and safety. Regular communications among regulators and researchers are important in order to discern best practices, resolve potential risks, and maximize deployment of AOPs in environmental applications.

As highlighted in recent overviews of predictive PFAS remediation models, ongoing transformation of regulations is informed by advancements in treatment technology and understanding contaminant behavior. This coupling of policy with technological drivers will not only enhance regulatory frameworks but also spur up sustainable processes in environmental remediation, ensuring AOPs are optimally utilized in contamination challenges while protecting human health and the environment.

SUMMARY

In summary, advanced oxidation processes (AOPs) represent a paradigm shift in environmental remediation strategy, specifically the treatment of recalcitrant pollutants found in industrial wastewater. The performance of AOPs, as exemplified by processes such as the Photo-Fenton process, represents a revolutionary synthesis of chemical oxidation and photochemistry toward the extensive mineralization of organic pollutants. As illustrated in textile wastewater treatment with reactive dyes, AOPs have the potential to enhance biodegradability considerably, transforming the typical non-biodegradable chemicals into chemicals of adequate biodegradability for subsequent biological treatment. In addition to showing a high COD and TOC reduction, this activity highlighted the relationship between decolorization and general biodegradability, referencing the ability of photochemical treatments to maintain environmental regulation compliance (Ghaly et al., 2009). Furthermore, the comparison of different AOPs yields useful information about their treatment efficiencies and usage potentials under realistic conditions. For instance, a study comparing the $UVC/S_2O_8^{2-}$ process with the traditional UVC/H_2O_2 system demonstrates progress toward optimizing treatment processes. Findings indicated that the $UVC/S_2O_8^{2-}$ pathway indicated a cost-effective route, particularly against recalcitrant matrices frequently found in wastewaters from industry. This research not only explained the activation kinetics of oxidants but also identified the superior performance of UVC/H_2O_2 under various conditions and thus marking the direction of future studies. Such comparative assessments are critical in planning optimized treatment processes that not only maximize removal capacities but also minimize residual concentrations of oxidants to ensure residue-free discharge processes (Anipsitakis et al., 2015). The importance of these findings goes beyond technical advancement; they are very important to the environment and regulation. AOPs, for their ability to treat challenging-to-degrade pollutants at a high efficiency, can mark the breakthrough of the treatment of environmental pollution problems and regulatory acceptability issues. Future studies should cover integration of AOPs with conventional technologies and find synergies enabling simplification of the process. As the demand for efficient remediation strategies continues to grow globally, ongoing improvement and refinement of AOPs will be critical to addressing the challenges of today's environment. Overall, AOPs provide a promising direction for future research and development that will satisfy tight environmental standards while sustaining the management of contaminated water resources in a responsible manner.

AOPs are a ground-breaking technology in environmental remediation, especially in wastewater treatment. One strength of AOPs is their capability to effectively destroy organic pollutants, especially those that are toxic and bio-resistant in nature. This is achievable because of the generation of highly active radicals that have the ability to oxidize a wide range of pollutants, such as pesticides and drugs. The growing menace of water scarcity and the declining quality of water resources has been the driving force for scientists to explore AOPs as more sustainable options than traditional treatment processes. A literature review on AOPs shows dominance of some nations, notably China, Spain, and the USA, in leading the way in this area, which is a testament to an international drive for seeking cost-effective solutions for water purification (Agüera et al., 2019). Collaborative international research has been instrumental in advancing the discipline of AOPs. High research synergy has been observed among such countries as Spain, China, the USA, Italy, and France, pointing to an intersimilar consciousness of the water contamination threat. This collaborative platform has facilitated new ways of integrating AOPs with biological oxidation in order to enhance treatment efficiency. The keyword analyses conducted in a number of studies have identified 15 clusters of research targeting unique types of wastewater and pollutants, from urban to industrial settings. The outcomes demonstrate the multi-dimensionality of AOPs and the necessity for targeted strategies for unique contaminants (Agüera et al., 2019). The implications of these scientific advances extend beyond the mere elimination of pollutants; they align with global sustainable measures aimed at protecting water resources for future generations. AOPs not only prove to be effective but can also play a central role in the realization of many of the Sustainable Development Goals (SDGs), notably the sanitation and clean water ones. It is necessary to know how these processes function, what kinds of lead contaminants various AOP methods are addressing, in order to maximize treatment. As scientific investigation continues, the focus needs to be on maximizing efficiency, reducing cost, and providing access to these processes on a greater scale to address the emergent issues associated with environmental pollution.

REFERENCES

1. Adebayo, M. A., & Daramola, I. O. (2022). Recent developments in the application of advanced oxidative processes for remediation of persistent organic pollutants from water. https://core.ac.uk/download/492702391.pdf (accessed on 28th September 2025).
2. AEA Technology Environment, Ahel, M., Ahel, M., Ahel, M., APE Research Council, Ashfield, L. A., Ball, H. A., et al. (2010). A critical review of the formation of mono- and dicarboxylate metabolic intermediates of alkylphenol polyethoxylates during wastewater treatment and their environmental significance. https://core.ac.uk/download/20665860.pdf (accessed on 28th September 2025).
3. Agüera López, A. M., Esteban García, A. B., Garrido Cárdenas, J. A., & Manzano Agugliaro, F., et al. (2019). Wastewater treatment by advanced oxidation process and their worldwide research trends. https://core.ac.uk/download/286590742.pdf (accessed on 28th September 2025).
4. Alves, M. M., & Pereira, L. (2012). Dyes—Environmental impact and remediation. https://core.ac.uk/download/55625293.pdf (accessed on 28th September 2025).
5. Amat, A., Santos, A., Singh, A., Chakinala, A. G., Naldoni, A., Schiboula, A., Yim, B., et al. (2011). Mineralization of surfactants using ultrasound and the advanced Fenton process. https://core.ac.uk/download/228176854.pdf (accessed on 28th September 2025).
6. Anipsitakis, G., Antoniou, M., Antoniou, M., Antoniou, M., Antoniou, M., Antoniou, M., et al. (2015). Comparison of UVC/$S_2O_8^{2-}$ with UVC/H_2O_2 in terms of efficiency and cost for the removal of micropollutants from groundwater. https://core.ac.uk/download/20274599.pdf (accessed on 28th September 2025).
7. Atlas, R. M., Cunningham, C. J., Elkin, A. A., Ivshina, I., et al. (2015). Oil spill problems and sustainable response strategies through new technologies. https://core.ac.uk/download/42591507.pdf (accessed on 28th September 2025).
8. Barani, V., Basil, M., Cid, C. A., Grego, S., et al. (2020). Field testing of an onsite sanitation system on apartment building blackwater using biological treatment and electrochemical disinfection. https://core.ac.uk/download/288435725.pdf (accessed on 28th September 2025).
9. Barndok, H., Cortijo, L., Hermosilla, D., Negro, C., et al. (2014). Removal of 1,4-dioxane from industrial wastewaters: Routes of decomposition under different operational conditions to determine the ozone oxidation

capacity. https://core.ac.uk/download/33102935.pdf (accessed on 28th September 2025).

10. Bharagava, R., Bilal, M., Bharagava, R., et al. (2021). Ecotoxicological and health concerns of persistent coloring pollutants of textile industry wastewater and treatment approaches for environmental safety. https://core.ac.uk/download/573846659.pdf (accessed on 28th September 2025).

11. Bouzayani, B., Elaoud, S. C., & Sanromán, M. (2025). Current progress in advanced oxidation processes for the removal of contaminants of emerging concern using peracetic acid as an effective oxidant. *Catalysts*. https://www.semanticscholar.org/paper/e4e807e5e6e1a421783c30e2467a93dd5c8dc2d8 (accessed on 28th September 2025).

12. Bremner, D. H., Burgess, A. E., & Chand, R. (2011). The chemistry of ultrasonic degradation of organic compounds. https://core.ac.uk/download/228176790.pdf (accessed on 28th September 2025).

13. Bremner, D. H., Chakinala, A. G., Cravotto, G., Di Carlo, A., et al. (2008). Mineralisation of 2,4-dichlorophenoxyacetic acid by acoustic or hydrodynamic cavitation in conjunction with the advanced Fenton process. https://core.ac.uk/download/228176603.pdf (accessed on 28th September 2025).

14. Bremner, D. H., Chand, R., Martínez, F., Melero, J., et al. (2012). Enhancement of the advanced Fenton process (Fe^0/H_2O_2) by ultrasound for the mineralization of phenol. https://core.ac.uk/download/228176871.pdf (accessed on 28th September 2025).

15. Castañeda-Sánchez, J. M., Silerio-Vázquez, F., Villanueva-Fierro, I., García-Prieto, J. C., González-Burciaga, L. A., & Proal-Nájera, J. (2025). Metformin degradation by advanced oxidation processes: Performance, limitations, and environmental concerns. *International Journal of Molecular Sciences, 26*. https://www.semanticscholar.org/paper/ebdc805f2f670eb4b48f06e107f45219b1ce885a (accessed on 28th September 2025).

16. De Geyter, N., Morent, R., Nikiforov, A., Rezaei, F., et al. (2019). Applications of plasma-liquid systems: A review. https://core.ac.uk/download/288817520.pdf (accessed on 28th September 2025).

17. Deeb, R. A., Hawley, E. L., Hoffmann, M. R., Mahendra, S., et al. (2016). Degradation and removal methods for perfluoroalkyl and polyfluoroalkyl substances in water. https://core.ac.uk/download/77929091.pdf (accessed on 28th September 2025).

18. Deng, A., Sun, Y., Gao, Z., Yang, S., Liu, Y., He, H., Zhang, J., et al. (2023). Internal electric field in carbon nitride-based heterojunctions for photocatalysis. *Nano Energy, 108*, 108228–108228. https://doi.org/10.1016/j.nanoen.2023.108228 (accessed on 28th September 2025).
19. Fapohunda, O., Ige, A. O., Fapohunda, O., et al. (2025). Predictive modeling of PFAS behavior and degradation in novel treatment scenarios: A review. https://core.ac.uk/download/647986103.pdf (accessed on 28th September 2025).
20. Gencheva, P., & Jeleva, N. (2025). Innovative *in situ* technologies for remediation of energetics. https://core.ac.uk/download/658857332.pdf (accessed on 28th September 2025).
21. Ghaly, M., Haseneder, R., & Härtel, G. (2009). Integrated coupled systems as biodegradability enhancement of textile wastewater by photo-Fenton process. https://core.ac.uk/download/39152477.pdf (accessed on 28th September 2025).
22. Goswami, A., & Jiang, J. Q. (2019). Comparative performance of catalytic Fenton oxidation with zero-valent iron (Fe(0)) in comparison with ferrous sulphate for the removal of micropollutants. https://core.ac.uk/download/293866552.pdf (accessed on 28th September 2025).
23. Greene, B., Olsen, W., Romeo, J., Sorkin, E., et al. (2025). Remediation technology collaboration development–A compendium. https://core.ac.uk/download/pdf/162000436.pdf (accessed on 28th September 2025).
24. Huang, H., Huang, H. B., Leung, Y. C., Lu, X., et al. (2016). Recent development of VUV-based processes for air pollutant degradation. https://core.ac.uk/download/84930094.pdf (accessed on 28th September 2025).
25. Liu, M., Ye, Y., Xu, L., Gao, T., Zhong, A., & Song, Z. (2023). Recent advances in nanoscale zero-valent iron (nZVI)-based advanced oxidation processes (AOPs): Applications, mechanisms, and future prospects. *Nanomaterials, 13*. https://www.semanticscholar.org/paper/4a49472016e542ecca569ed251035531a29085f6 (accessed on 28th September 2025).
26. Liu, Y., Niu, S., Park, J. H., Wang, Z., et al. (2024). Biochar-based materials as remediation strategy in petroleum hydrocarbon-contaminated soil and water: Performances, mechanisms, and environmental impact. https://core.ac.uk/download/623809867.pdf (accessed on 28th September 2025).
27. Manickavasagam, G., Saaid, M., & Oh, W. D. (2025). Hydrochar as sustainable redox catalyst for advanced oxidation processes-based wastewater treatment. *Pure and Applied Chemistry*. https://www.semanticscholar.org/paper/3a4349939d869e91046dc8efd093543e821c1ce6 (accessed on 28th September 2025).

28. Martínez-Huitle, C. A., Rodrigo, M. A., Sirés, I., & Scialdone, O. (2023). A critical review on latest innovations and future challenges of electrochemical technology for the abatement of organics in water. *Applied Catalysis B: Environment and Energy, 328*, 122430–122430. https://doi.org/10.1016/j.apcatb.2023.122430 (accessed on 28th September 2025).

29. Mihaltan, A. I., Naghiu, A., Petcu, C., Varaticeanu, A., et al. (2010). Study of the combined effects of oxidation and adsorption process in removing MTBE from wastewater. https://core.ac.uk/download/pdf/159179239.pdf (accessed on 28th September 2025).

30. N/A. (2013). Treatment of refinery oily sludge using ultrasound, bio-surfactant, and advanced oxidation processes. https://core.ac.uk/download/84875039.pdf (accessed on 28th September 2025).

31. Olawade, D. B., Wada, O. Z., Olawade, D. B., et al. (2025). Recent occurrence of pharmaceuticals in freshwater, emerging treatment technologies, and future considerations: A review. https://core.ac.uk/download/649343312.pdf (accessed on 28th September 2025).

32. Rizal, A., Nordin, R., Navarro, A. R., Reyes, J. C., Maragathavalli, S., Kristanti, R. A., Wulandari, R., et al. (2025). Microbial bioremediation of petroleum-contaminated soil: A sustainable approach. *Tropical Aquatic and Soil Pollution*. https://www.semanticscholar.org/paper/e5154c4d2930c8d9936f7a440912aef350422c7e (accessed on 28th September 2025).

33. Romero, M., Hernández-Anzaldo, S., Vázquez-Lima, H., & Reyes-Ortega, Y. (2024). Probing the photocatalytic pathways of iron (III) protoporphyrin IX-functionalized TiO_2: A study on oxo-iron(IV) porphyrin π-cation radical formation. *ECS Meeting Abstracts*. https://www.semanticscholar.org/paper/b1a00adb456f8aff763882f7e6cacc17cbbcfc33 (accessed on 28th September 2025).

34. Sandoval, M. A., Salazar, R., Calzadilla, W., Vidal, J., & Brillas, E. (2024). Advancements in electrochemical advanced oxidation processes: Addressing emerging contaminants in Latin America. *ECS Meeting Abstracts*. https://www.semanticscholar.org/paper/13595dbff0b86d2e4f37fd46bf105034ad2c258a (accessed on 28th September 2025).

35. Sarkar, N., Mishra, S., Gadore, V., Panigrahi, B., & Ahmaruzzaman, M. (2024). Nanocosmos of catalysis: A voyage through synthesis, properties, and enhanced photocatalytic degradation in nickel sulfide nanocomposites. *Nanoscale Advances, 6*, 2741–2765. https://www.semanticscholar.org/paper/1eb6718f92da0f4192d1ff4b6de202f02d24bfa4 (accessed on 28th September 2025).

36. Satyam, S., & Patra, S. (2025). The evolving landscape of advanced oxidation processes in wastewater treatment: Challenges and recent innovations. *Processes*. https://www.semanticscholar.org/paper/1f1be4f0cf3779138726057b898c43b2a8cfab19 (accessed on 28th September 2025).
37. Shon, H. K., Snyder, S. A., & Vigneswaran, S. (2006). Effluent organic matter (EfOM) in wastewater: Constituents, effects, and treatment. https://opus.lib.uts.edu.au/bitstream/10453/3626/3/2006003988.pdf (accessed on 28th September 2025).
38. Terrón, D., Sanromán, Á., & Pazos, M. (2025). Metal–organic frameworks: Next-generation materials for environmental remediation. *Catalysts*. https://www.semanticscholar.org/paper/a59e5413a68a0d538b66f0c2ab637c2dc690a4d1 (accessed on 28th September 2025).

CHAPTER 5

PHOTOCHEMICAL APPLICATIONS IN WATER TREATMENT

Contents

Introduction 177

5.1. Principles of Photochemistry 182

5.2. Photocatalysis in Water Treatment 187

5.3. Advanced Oxidation Processes (AOPs) 193

5.4. Solar Water Treatment Technologies 198

5.5. Comparison with Traditional Water Treatment Methods 203

5.6. Challenges and Limitations 208

Summary 213

References 216

INTRODUCTION

Water pollution poses a significant threat to public health and environmental integrity, and as a consequence, needs novel treatment processes. Conventional treatment methods such as coagulation-flocculation and sedimentation often do not efficiently remove such long-lasting organic pollutants such as pesticides and other synthetic chemicals common in agricultural runoff. Conventional treatment methods are usually very costly to operate and can generate secondary pollution such as sludge formation, which further complicates waste management. With growing water pollution demand, especially in urban regions under the weight of industry effluent and population booms, sophisticated treatment technologies prove to be an area worth exploring. Photochemical processes, though, are a very viable alternative that harvests light energy for encouraging chemical alterations, with the aim of degrading undesirable pollutants effectively and sustainably. The latest photochemical technology developments, including Advanced Oxidation Processes (AOPs), have been reported to be effective in cleaning hard-to-treat contaminants. AOPs employ reactive species generated under methods such as UV irradiation in combination with hydrogen peroxide to oxidize recalcitrant organic pollutants, such as pesticides that taint water sources.

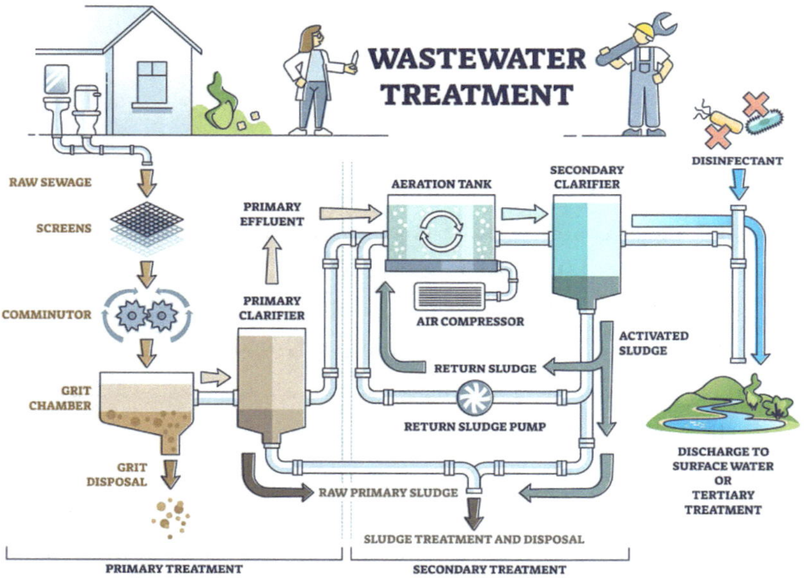

Figure 5.1. *Photochemical processes in water treatment.*

Source: https://www.elementar.com/fileadmin/_processed_/6/f/csm_wastewater-treatment-process_0d5e62d3d0.jpg.

The technologies have been proven to significantly enhance degradation of the pollutants without generating problematic secondary waste. Applicability of pesticide contamination treatment should not be exaggerated; these kinds of chemicals have a tendency to accumulate in biological organisms and, upon exposure, can lead to severe health risks. Photochemical application in water treatment is turning out to be a relevant pathway to pursue in the mitigation of adverse impacts that accompany water pollution, hence improving water quality. In exploring the future potential for photochemical applications, it is necessary to recognize the increasing sense of urgency to counter the multi-dimensional problems of water purification. With shrinking water supplies globally, the need for efficient and sustainable purification grows by the day. Research has established that most of the water body pollutants, including those emanating from illegal landfills and pesticide runoff, have severe health effects, leading to both carcinogenic and non-carcinogenic effects on human populations (Siddiqua et al., 2022; Syafrudin et al., 2021). Therefore, capital investment in photochemical technologies is not only supporting sustainable development but also addressing the painful consequences of environmental pollution. As we delve into the various photochemical processes and their uses in water treatment, the importance of implementing these processes in creating a healthier, sustainable future becomes apparent (Figure 5.1).

Definition of Photochemical Processes

Photochemical reactions refer to those chemical reactions that are induced or catalyzed by the energy that is absorbed by light. Photochemical reactions play a vital role in water treatment to destroy poisonous pollutants by converting chemicals into non-toxic or inactive compounds through photolysis or photosensitization under the use of sun's energy. During photochemical processes, molecules absorb photons and get excited with the subsequent formation of reactive intermediates such as free radicals or reactive oxygen species (ROS). The intermediates are involved in breaking down complex chemicals such as persistent pollutants such as industrial chemicals and pharmaceuticals, hence making photochemical processes an essential component in today's water purification technology. The specific mechanisms involved in photochemical processes can vary significantly depending on the nature of contaminants as well as environmental parameters. For instance, some pollutants, when exposed to ultraviolet (UV) or visible light during contact with engineered nanomaterials (ENMs), can be broken down at a higher rate. Recent studies highlight the degree to which degradation rates of perfluoroalkyl acid (PFAAs) precursors were tremendously enhanced under light-stimulated ENMs, highlighting the prospect for site-specific degradation of the hazardous compounds in aquatic systems (Dai et al., 2020). At a fundamental level, photochemical reactions not only facilitate the degradation of harmful compounds but also improve the performance of traditional treatment processes, ushering the doors to novel applications in environmental remediation. Despite their strengths, the efficiency of photochemical processes can be influenced by several factors, including the stability of the impurities as well as the existence of multiple environmental factors. For example, the conversion rate of some pollutants can be depending on their adsorption characteristics, photostability, and the generation of reactive intermediates (Dai et al., 2020). In addition, difficulty is experienced with some compounds that are photochemically stable, thus requiring a thorough understanding of the photochemical environment to maximize the conditions for treatment. More importantly, incorporating other modeling techniques, including Model Selection and Model Averaging, can be used to further improve the understanding of the manner in which these processes operate in practical applications. In conclusion, defining and optimizing photochemical processes is fundamental for developing effective water treatment technologies capable of addressing contemporary contamination challenges (Figure 5.2).

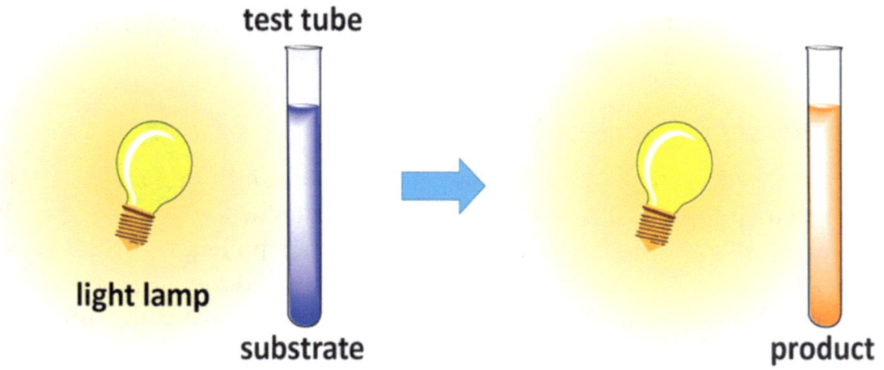

Figure 5.2. *Photochemical processes.*

Source: https://www.researchgate.net/profile/Heejin-Kim-19/publication/321311909/figure/fig4/AS:636977158299649@1528878597250/A-simple-system-for-photochemical-reactions.png.

Importance of Water Treatment

The significance of water treatment can hardly be overstated, particularly in terms of environmental health and public safety. With the world increasingly being threatened by contaminated sources of water, effective treatment systems become essential for safeguarding human health and the environment. Waterborne pathogen- and chemical contaminant-induced diseases continue to pose a serious threat to populations, and there is hence an urgent need to embrace cutting-edge water treatment technologies. The advent of photochemical water treatment brings new solutions to enhance the elimination of contaminants, particularly persistent ones. The process not only targets conventional contaminants but also new emerging threats, which exhibits the evolving role of water treatment in public health management. Recent studies highlight the efficiency of engineered nanomaterials (ENMs) in photochemical water treatment processes, particularly in breaking down toxic precursors such as perfluoroalkyl acids (PFAAs). While production of PFAAs ceased due to their persistence in the environment, precursors remain widespread, and novel ways of mitigating their impact have been called for. Research has revealed the photochemical transformation rates of precursors can be significantly enhanced upon exposure to sundry ENMs under sunlight illumination. The ability of ENMs to generate reactive oxygen species is a dominant factor in the process of degradation, leading to the transformation of dangerous precursors into potentially more dangerous by-products. This necessitates continued exploration of photochemical approaches in water treatment to effectively degrade and eliminate persistent environmental contaminants (Dai et al., 2020). Furthermore, the introduction of laser technology into photochemical treatments provides exciting routes for future water treatment technologies. By tapping into the ability of lasers to generate singlet oxygen and other reactive species, scientists are revealing new directions to addressing challenging water quality concerns. This technology not only allows selective treatment of selected pollutants but is also capable of boosting the overall efficiency of water treatment systems. Furthermore, the application of lasers in the generation of singlet oxygen shows their potential for the annihilation of dangerous agents with fewer risks from the traditional chemical treatment. As can be seen from the research, the ability to overcome shortcomings such as low light penetration further increases the applicability of laser technologies in the methods of water treatment, developing further towards water quality and safety (Bragina et al., 2017).

Overview of Photochemical Applications in Water Treatment

Photochemical water treatment processes rely on photochemical reactions for addressing the increasing demand for clean and safe water. Fundamentally, the technique exploits light, usually ultraviolet (UV) or visible light, to activate reaction that can dissolve organic pollutants, destroy pathogens, and reduce toxic chemical concentrations. The reactions incorporate photocatalysis, a process that uses light-activated catalysts to catalyze chemical reactions. For instance, titanium dioxide (TiO_2) is a renowned photocatalyst applied in water treatment due to its capability to break down organic contaminants into harmless chemicals. By the inclusion of light-based treatment methods, water treatment plants can enhance value addition on conventional water purification technology, thereby resulting in quality water and sustainability. The second most important feature of photochemical applications is that they also destroy newer pollutants such as pharmaceuticals and personal care products with high efficiency, which are not even removed by standard treatment technologies. Via the process of advanced oxidation processes (AOPs), the photochemical processes generate hydroxyl radicals, which are highly oxidizing species and can oxidize nearly any type of organic contaminant. Such AOPs have been discovered to significantly elevate the rates of the degradation of such persistent chemicals to the point of guaranteeing that they do not linger in water bodies to add to ecological pollution. The flexibility and effectiveness of photochemical treatment make a powerful choice for water treatment plants as they attempt to meet ever-strict regulation requirements while safeguarding public health and the environment. Furthermore, the latest studies highlight the promise of photochemical processes in the improvement of treatment for wastewater from agricultural origins that are prone to having pesticide residues as well as microbial contaminants. The study has demonstrated that the incorporation of photochemical processes not only reduces toxic phytochemicals but also reduces the use of toxic chemicals in traditional wastewater treatment plants. Delving into the resilience of genetically engineered species, i.e., MON810 maize, provides further understanding of how photochemical applications can be developed to accommodate the needs of diverse contaminants across varied environmental conditions (Fusaro et al., 2015). The integration of advanced scientific methods into paradigms for water treatment is an exciting avenue to pursue in the broader endeavor for sustainable and adaptive water management approaches.

5.1. PRINCIPLES OF PHOTOCHEMISTRY

Principles of photochemistry play an important role in optimizing its application in water treatment. Essentially, photochemistry is the science of the interaction between matter and light with particular emphasis on the way in which light energy initiates chemical reactions. In water treatment, the principle is utilized in the guise of ultraviolet (UV) light to drive photochemical reactions which help to degrade contaminants. Photolysis, or the light-induced cleavage of chemical bonds, is predominantly applied to degrade organic pollutants in water. Efficient contamination removal is achieved by techniques such as photocatalysis, where the photocatalysts, typically semiconductors like titanium dioxide, are excited by UV radiation and induce reactions that degrade harmful substances. Thus, an understanding of photochemical mechanisms can translate into innovations in treatment technologies more energy efficient and sustainable solutions to water resource purification that societies worldwide rely on. The environmental impacts of photochemical water treatment technologies are huge. With increased urbanization comes increased demand for safe and clean drinking water, pushing the need for efficient solutions to treatment. Comparative evaluation of various water treatment processes indicates the superiority of photochemical processes as compared to traditional processes. For instance, Life Cycle Assessment (LCA) methods indicate that newer photochemical treatments exhibit lower environmental impacts compared to traditional processes, particularly energy consumption and resource utilization.

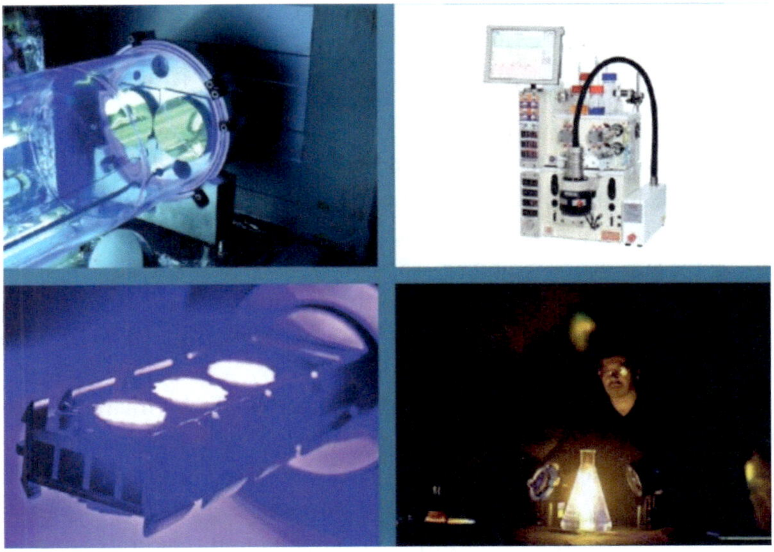

Figure 5.3. *Principles of photochemistry.*

Source: https://m.media-amazon.com/images/I/61fZqiXK0bL._SL1360_.jpg.

In the context of evidence that tap water is preferred over bottled water due to its lesser environmental footprint (Cadena et al., 2016), the integration of photochemical technologies not only enhances the quality of water but also helps towards sustainable causes to minimize carbon emissions and drain resources. The cross-disciplinary approach extends the scope of innovation in the practice of water management. In addition, solving the complexities underlying the development and application of photochemical technologies requires a systematic perception of various model choice techniques among researchers. Model selection and averaging allow for the ability to compare the effectiveness of various photochemical methods as well as providing insight towards enhancing their functionality. For example, Model Selection (MS) and Model Averaging (MA) might be the game-changers to enhance experiments on photochemical applications for water treatment (Abdullah et al., 2018).

With statistical metrics like the Corrected Akaike Information Criteria (AICc), researchers can gain reliable conclusions through model testing. With the investigation of control factors and interactions in photochemical processes, treatment systems can be rendered more dependable to achieve the stringent demands for environmental regulations while effectively alleviating water quality issues. Such harmony between photochemistry and advanced statistical assessment proves the pivotal position of interdisciplinary science for resolving emerging issues in water treatment (Figure 5.3).

5.1.1. Basic Concepts of Photochemical Reactions

Understanding basic principles of photochemical reactions is of utmost importance in exploiting their potential for application in the process of water treatment. At the heart of these reactions is the interaction between chemicals and light and, consequently, the reactants are converted into a number of chemical species upon absorbing photons. In water treatment, such reactions can be induced to degrade contaminants, inactivate pathogens in water, and produce reactive intermediates that destroy additional contaminants. The efficiency of photochemical reactions is also controlled by various factors including the wavelength of light utilized, the concentration of chemicals utilized, and the presence or absence of catalysts. Regulation of these variables enables practitioners to maximize the degradation of undesired compounds, enhancing the efficacy and safety of water treatment processes. Furthermore, application of engineered nanomaterials (ENMs) in photochemical reactions is a testimony to the development in the area. ENMs also act as catalysts that not only facilitate photodegradation but also enhance ROS generation upon illumination. This is particularly useful when the target is recalcitrant pollutants such as PFAAs, which are notoriously hard to degrade. Experiment has demonstrated that addition of ENMs can dramatically accelerate photochemical conversion of certain precursors, effectively eliminating toxic chemicals in treated water. However, the implications are complex; some ENMs promote degradation while others hinder degradation in other substances, proving the importance of thorough understanding of these processes (Dai et al., 2020). The strategic use and application of ENMs during water treatment can lead to improved results if adequately customized. The solar disinfection (SODIS) process also employs natural sun power to inactivate pathogenic microorganisms in contaminated water. Such treatment is basically linked with photochemical reactions wherein sunlight initiates several oxidative processes that make microbial cells weak. The susceptibility of microorganisms, such as bacteria and viruses, to light of different wavelengths determines the overall efficacy of SODIS. Natural organic matter (NOM) within water can affect these processes by scavenging harmful UV radiation or enhancing the formation of reactive intermediates that cause microbial inactivation (Boehm et al., 2018). In natural and engineered environments, obtaining detailed photochemical reaction mechanics besides illuminating efficient methods of water treatment also identifies areas of future research and innovation in water quality improvement.

5.1.2. Role of Light in Chemical Processes

Absorption of light is also pivotal in chemical reactions, particularly in photochemistry, where photochemical reactions are promoted by light energy. The process is critical in water treatment procedures, where the destruction of pollutants relies greatly on light-initiated reactions. For instance, photochemical reactions cause the breakdown of complex organic substances into simpler and less harmful products. This process is particularly relevant in purifying perfluoroalkyl acids (PFAAs), infamous long-lived environmental pollutants. The ability of engineered nanomaterials (ENMs) to initiate such reactions upon simulated sunlight irradiation is a clear demonstration of the promise that light can play in intensifying water treatment mechanisms. These rate increases in transformation emphasize the necessity of optimizing light exposure and understanding the interplay between chemical reactants, light, and photocatalytic material to achieve successful remediation in aqueous media. The interplay between light and chemical processes involved in water treatment also introduces complicating factors that need careful consideration. The transformation rates of perfluoroalkyl acid precursors (PrePFAAs) upon the introduction of particular ENMs demonstrate the potential that various materials have to influence photochemical reaction performance. According to research, while the introduction of ENMs can lead to the degradation of some PrePFAAs at a high rate, others get suppressed due to photo-stability and adsorption characteristics. These findings highlight the pivotal nature of ENM selection, as their catalytic activity can become detrimental or beneficial depending on the chemistry of the environment. As has been demonstrated in recent publications, the interaction between light and engineered materials not only affects the rate of degradation, but also the nature of more toxic by-products formed, raising questions about the overall process efficacy in real-world applications. Moreover, the understanding of the role played by reactive species generated in photochemistry can even further explain the dynamic character of photoinduced chemical reactions in water treatment. In light, various reactive oxygen species (ROS) are generated, which have a pivotal role to play in the degradation of pollutants like PFAAs. The production of ROS yield from ENMs is also highly responsible for the regulation of the efficiency of the degradation of such substances, with the direct relation of light, material properties, and chemical action. Thus, the mechanisms of such photochemical processes must be known in order to optimize water treatment systems. Adjusting the parameters of light treatment and material properties allows the stimulation of the degradation pathways of contaminants. While, as noted, the potential for the formation of harmful degradation products calls for a keen and balancing line through which water treatment operations do not create fresh environmental challenges inadvertently, echoing the ever-present need for whole studies on light's role in chemical reactions (Abdullah et al., 2018; Dai et al., 2020) (Figure 5.4).

Figure 5.4. *Role of light in chemical processes.*

Source: https://scx2.b-cdn.net/gfx/news/2017/withmoreligh.jpg.

5.1.3. Types of Photochemical Reactions Relevant to Water Treatment

Photochemical reactions play a key role in the inactivation and destruction of contaminants, including organic chemicals and microorganisms, in the process of water treatment. Direct photolysis of pathogens with ultraviolet (UV) light is one of the most studied technologies among them. UV irradiation is effective in inactivating protozoa, viruses, and bacteria by degrading their nucleic acids and disrupting cell processes. Experimental evidence indicates that the sensitivity of microorganisms to UV radiation varies greatly depending on their cellular composition and environmental environments. For instance, viruses actually require UVB wavelengths in order to be properly inactivated, while bacteria are sensitive to wavelengths that are only a little longer. It is imperative to comprehend the complex processes of sunlight-driven inactivation, as recently reported, so that design principles in engineered water treatment systems can be optimally maximized to efficaciously utilize such natural photochemical processes (Boehm et al., 2018). Another significant domain of photochemical reactions for water treatment includes the degradation of long-term pollutants by advanced oxidation processes (AOPs). AOPs harness the reactive species, such as hydroxyl radicals, formed in photochemical processes to degrade organic pollutants effectively. As an example, photoactivation of ENMs can hasten degradation of PFAAs precursors, which remain a challenge despite phased-down manufacturing. Follow-up reactions can proceed to convert these precursors to more stable and harmful byproducts, whose effects have to be controlled in order to ensure water quality. ENM research has revealed that ENMs have the ability to significantly accelerate the rates of transformation of certain contaminants, even though their action depends on factors such as the stability of the compounds and surface adsorption properties (Dai et al., 2020). The ongoing development of AOPs illustrates the feasibility of photochemical transformations in increasing the armory for the degradation of complex contaminants in water treatment operations. Finally, the integration of photochemical treatment processes into natural and engineered systems reflects the need for a profound understanding of the interaction between contaminants, water quality parameters, and photonic energy. Natural organic matter (NOM) in water both retards and accelerates photochemical reactions. While NOM is capable of absorbing UV light and attenuating the radiation needed to efficiently inactivate microorganisms, it also possesses the capability of acting as a sensitizer through the creation of reactive intermediates that facilitate enhanced breakdown of pathogens. The mixed functionality makes it challenging to anticipate inactivation rates and underscores the necessity of constructing robust models to better understand such phenomena. The interaction of photochemistry and the environment offer a viable platform for guaranteeing maximum water treatment processes, thereby promoting public health effects as well as environmental sustainability (Boehm et al., 2018).

5.2. PHOTOCATALYSIS IN WATER TREATMENT

Photocatalysis treatment of water relies on the ability of photocatalysts, primarily titanium dioxide (TiO_2), to activate the degradation of pollutants under irradiation with light. The mechanism is by the formation of reactive oxygen species that effectively oxidize the organic pollutants, leading to their mineralization into harmless byproducts such as carbon dioxide and water. Photocatalytic systems are particularly valuable as they are very effective at breaking down persistent organic pollutants that are otherwise resistant to conventional treatment. Advances in the design of photocatalysts, such as maximizing the morphology and crystallinity of TiO_2 thin films, enhance their efficiency in water purification further. As noted, variations in synthesis pathways, i.e., sol-gel or hydrothermal processes, significantly influence the photocatalytic activity, underlining the significance of careful selection of preparation processes for maximum treatment efficiency (Abdullah et al., 2018).

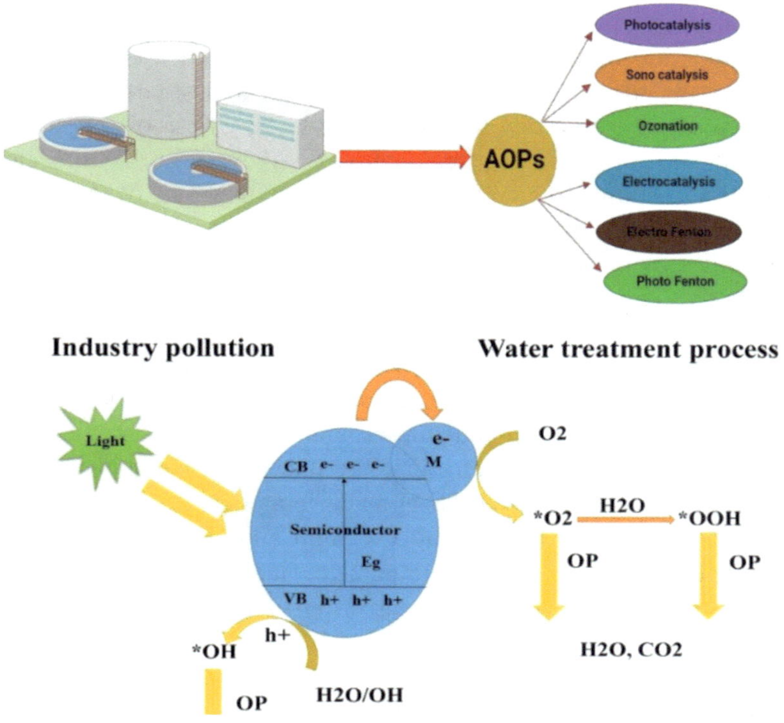

Figure 5.5. *Photocatalysis in water treatment.*

Source: https://www.researchgate.net/publication/373288366/figure/fig1/AS:1143 1281183113697@1692712889693/Shows-the-water-treatment-method-and-photo-catalysis-schematic-diagram-of-photo-redox.jpg.

Integration of photocatalysis into water treatment is a potential solution to the mounting risk of emerging pollutants, e.g., pharmaceuticals and personal care products. These micropollutants are now widespread in water bodies, and both environmental and human health can be compromised. Photocatalytic water treatment systems can address these issues through the use of sunlight-activated photocatalyst properties, enabling *in-situ* contamination of contaminated water sources. Studies have shown that photocatalytic degradation rates can be significantly affected by parameters such as light intensity, catalyst dosage, and the presence of other chemical species in the water. Hence, the optimization of these parameters is crucial in making photocatalysis applicable in real-life scenarios, which in itself demands a thorough knowledge of heterogeneous photocatalytic processes (Abdullah et al., 2018). Also, the scalability of the involved technologies in photocatalytic water treatment remains a central aspect of their applicability. While laboratory-scale experiments show encouraging results, upscaling the same to full-scale applications involve surmounting a range of challenges, such as reactor design and operating stability. Immobilization of photocatalysts on materials such as carbon or glass facilitates the manufacture of efficient photoreactors which can be integrated into existing water treatment facilities. New manufacturing techniques such as material printing and chemical vapor deposition improve the performance of photocatalysts while facilitating easier use of these technologies in large-scale applications. Overall, by enhancing photocatalytic processes and making them of practical use, we can bring about significant improvement in water treatment systems and resolve the grave issue of water pollution (Králová et al., 2012) (Figure 5.5).

5.2.1. Mechanism of Photocatalysis

Photocatalysis is an essential part of the majority of photochemical reactions, primarily in water purification. Photocatalysis refers to the acceleration of a chemical reaction with the aid of a light-sensitized catalyst. When ultraviolet (UV) or visible light photons strike a photocatalyst material, such as titanium dioxide (TiO_2), it forms electron-hole pairs. These charge carriers must be employed to trigger redox reactions capable of effectively degrading organic pollutants in water. The process's efficacy is reliant upon the material to photosensitize and consequent electron transfer to undertake reductive and oxidative reactions. An understanding of these fundamental processes allows optimization of photocatalytic systems to enhance pollutant degradation, which is crucial to develop effective water treatment technologies. In addition, the photodegradation reaction facilitated by photocatalysis enables one to degrade numerous water contaminants including pesticides and pharmaceuticals. For instance, the *in-situ* study of carbofuran photodegradation on a TiO_2 film catalyst demonstrates how irradiation with light induces the formation of reactive species that react with organic contaminants. Through the use of advanced analytical techniques, such as ATR-FTIR and multivariate curve resolution, scientists have been able to define the products and kinetics of such reactions. Simulation of these mechanisms, as in studies employing the Hard-Soft Multivariate Curve Resolution-Alternating Least Squares (HS-MCR-ALS) method, illustrates the complexity of photocatalytic systems. This complexity could be used to develop specific photocatalytic applications that lead to improvement in polluted water source treatment, eventually leading to improved public health status and environmental quality. Moreover, the application of photocatalysis in water treatment is not merely pollutant degradation but also water disinfection issues. Photocatalytic activities have the capability to inactivate disease-causing microorganisms efficiently via the formation of reactive oxygen species (ROS) upon light exposure. These ROS play significant roles in the inactivation of bacteria and viruses in water and hence represent a valuable technology for disinfection of water without the addition of chemicals. Ongoing advances in photocatalysts and processes facilitate possible scalability of such technologies to real-world applications. Since the research further explains the process of photocatalysis, it is evident that integrating such information in developing eco-friendly water treatment technologies is capable of efficiently curbing pollution in the environment as well as safeguarding water resources for the future (Figure 5.6).

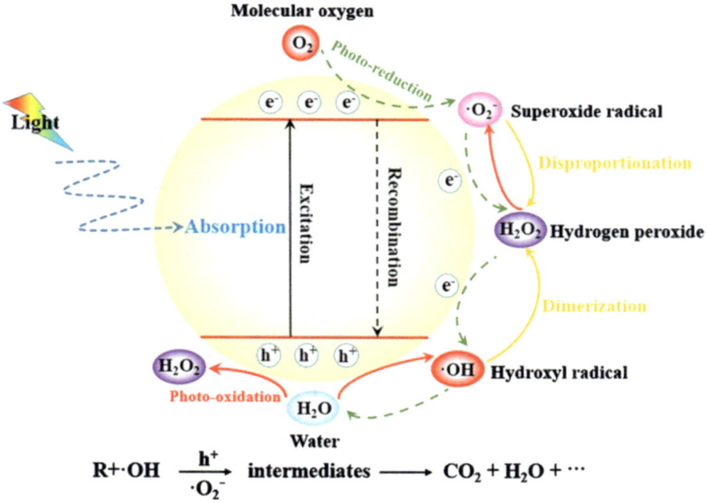

Figure 5.6. *Mechanism of photocatalysis.*

Source: https://www.researchgate.net/profile/Salma-Izati-Sinar-Mashuri-2/publication/345064926/figure/fig3/AS:959137340993537@1605687570401/The-basic-mechanism-of-photocatalysis-23.png.

5.2.2. Common Photocatalysts Used in Water Treatment

Photocatalysis is a predominant technology in water treatment, particularly due to its potential for oxidizing organic pollutants. Titanium dioxide (TiO_2) is among the most popular photocatalysts due to its high oxidizing power, stability, and low toxicity. TiO_2 efficiency in water treatment is largely dependent on its ability to generate reactive oxidative species upon irradiation using ultraviolet light.

A few fabrication processes, such as sol-gel and hydrothermal synthesis, have been employed to strengthen the photocatalytic performance of TiO_2 films. For instance, crystallite phase and structure of TiO_2 will influence its photocatalytic activity with very significant importance, and thus it needs to be controlled very accurately over the preparation process. Moreover, experiments prove that TiO_2 modification using additives like PEG maximizes activity by improving surface qualities and hence enhancing interaction with pollutants (Králová et al., 2012). Apart from titanium dioxide, other materials have been tested as photocatalysts, such as zinc oxide (ZnO) and carbon-based materials. ZnO, similar to TiO_2, has numerous properties such as high photocatalytic activity and stability but a distinct absorption spectrum that makes it very efficient when it operates under UV light, which provides a different avenue for pollutant degradation. Furthermore, new allotropes of carbon have brought new avenues in water treatment processes. Carbon materials are applicable as supports for traditional photocatalysts or can act singly in high-advanced oxidation processes based on hydroxyl radical formation, a key degradation mechanism for complicated chemical compounds in wastewater.

This ability is of immense potential for the development of cost-effective and efficient water treatment systems (Faria et al., 2014). Continuous research into photocatalysts for water treatment provides promising avenues for mitigating the difficulties posed by chemical pollutants. With continued improvement in the field, the combination of photocatalysts and supporting oxidants such as ozone or hydrogen peroxide maximizes the total efficiency of the treatment process. Synergy enables lower operating conditions than in thermal treatments, with conditions that are energy-efficient and sustainable. Also, key is the research into the scalability of these technologies for real-world implementation in actual systems. Continued studies must also examine the photocatalytic mechanisms taking place at the molecular level to allow the development of more efficient catalyst materials. Thus, the addition of new materials and concepts into photocatalytic systems is the way forward for allowing the formulation of robust strategies for effective water treatment solutions in the future.

5.2.3. Advantages of Photocatalytic Methods

Photocatalytic processes have emerged as a preferred method of water treatment owing to their unequalled efficiency in the degradation of organic pollutants. By deploying the use of light-activatable semiconductor materials, the processes catalyze sophisticated oxidation processes that break down toxic substances into benign byproducts. The availability of utilizing solar power also adds to their appeal, making photocatalytic processes both environmentally friendly and cost-effective. As emphasized in recent literature, continuous-flow photocatalysis enables accelerated reaction rates and increased selectivity, enabling laboratory-to-industry scale-up effectively. Efficiency is particularly sought after in treatment of large volumes of wastewater, where conventional methods may not be effective in providing the desired pollutant removal rates (Sambiagio et al., 2019). The integration of photocatalytic systems into traditional wastewater treatment systems has the potential to reduce system operating costs significantly and improve overall system efficiency. Another significant advantage of photocatalytic technologies is their versatility and flexibility. They can be applied to a broad spectrum of applications, including the inactivation and removal of pathogens and breakdown of pharmaceuticals in wastewater, which is generally recalcitrant to traditional treatment methods. Photocatalysis can be optimized by selecting appropriate catalyst materials and by modifying light wavelength, thus being able to optimize each reaction to oxidize a range of contaminants. Evolution of the science has also seen hybrid systems where photocatalytic technology is combined with other treatment processes such as adsorption or biological processes, thus achieving higher efficiency. The intrinsic robustness of photocatalytic processes also supports efficient treatment under different environmental conditions, expanding their potential to low-resource or low-infrastructure areas (Jingguang et al., 2018). This flexibility positions photocatalysis as a feasible solution for addressing advanced water pollution challenges globally. The favorable safety profile of photocatalytic water treatment technologies further broadens their advantages. In contrast to chemical processes, which typically use dangerous reagents and create dangerous byproducts, photocatalytic processes typically use harmless catalysts and produce primarily harmless minerals and water as byproducts. Not only does this reduce the issue of environmental health, but it also facilitates safer treatment practice in treatment plants. Additionally, the capability of photocatalytic systems to remove recalcitrant organic contaminants mitigates the risk of secondary pollution, making them suitable for application in sensitive ecological areas. Improved reaction kinetics in photocatalytic reactions also leads to shorter reaction times, with the result being smaller treatment systems that can be installed in smaller spaces without affecting efficiency. The overlapping of these efficiency and safety aspects emphasizes the potential of photocatalytic processes as a vital aspect of emerging water treatment processes to offer cleaner water with reduced environmental impact (Sambiagio et al., 2019) (Figure 5.7).

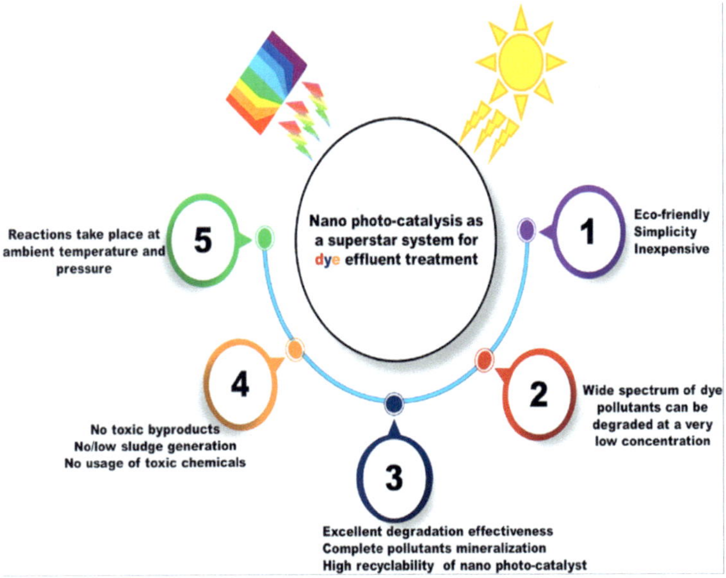

Figure 5.7. *Advantages of photocatalytic methods.*

Source: https://www.researchgate.net/profile/Christopher-Olawoyin-2/publication/370806426/figure/fig3/AS:11431281173738263@1689038788488/Major-advantages-of-photodegradation-for-wastewater-treatment.png.

5.3. ADVANCED OXIDATION PROCESSES (AOPS)

The development of Advanced Oxidation Processes (AOPs) is one of the most significant advances in water treatment technology, particularly the age-old problem of destroying toxic pollutants in wastewaters. AOPs utilize the generation of very energetic hydroxyl radicals, which are able to effectively oxidize a wide variety of organic contaminants that are very recalcitrant to conventional treatment methods. Of special interest in this case is growing global interest in the aspect of water scarcity, which has sparked research into sustainable solutions for effective wastewater treatment. As suggested in literature, countries such as China, Spain, and the USA are leading in AOP research, showcasing collaboration to come up with new approaches that can optimize water quality for various applications, including industrial, urban, and drinking water supplies (Agüera et al., 2019). This increased interest in AOPs emphasizes their potential not just for explicit pollution elimination but also in enabling sustainable practice in water management. The efficiency of AOPs is demonstrated through the Photo-Fenton process, a specific type of AOP, being reported to treat resistant wastewater, say from the textile industry. This process relies on photochemical activation of hydrogen peroxide in the presence of iron ions for the production of hydroxyl radicals, which are responsible for the degradation of complex dye molecules that cannot otherwise be biodegradable. Studies have indicated that under optimum conditions, the Photo-Fenton process has been found to be superb at COD and TOC removal.

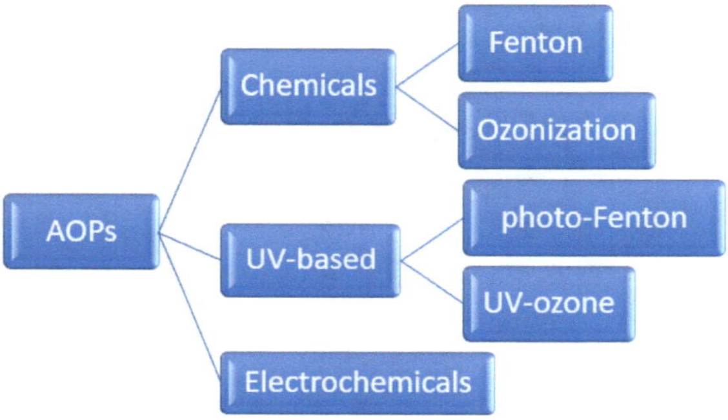

Figure 5.8. *Advanced oxidation processes (AOPs).*

Source: https://www.mdpi.com/processes/processes-10-01041/article_deploy/html/images/processes-10-01041-g001.png.

For instance, a study showed that in only 60 minutes of irradiation, the treatment had to a maximum of 100% color removal from wastewater with dyes, as well as significant reduction in TOC and COD levels (Ghaly et al., 2009). The findings illustrate the feasibility of using AOPs in wastewater treatment, ultimately promoting biodegradability and environmental compliance. Besides, AOPs can also help bridge the gap between chemical and biological treatment processes. Combining biotechnology with AOPs gives a synergistic process in which the initially treated effluents have significantly increased biodegradability and can be treated further by biological methods. By utilizing AOPs as a pretreatment process, wastewater conditions can be controlled to maximize the effectiveness of biological processes, which are typically limited by toxic substances. For instance, an increasing BOD_5/COD ratio following photo-Fenton treatment indicates a move toward increased biodegradability, allowing safely treated wastewater to be released into the environment. This new combination constitutes not only the versatility of AOPs but also their role in making holistic and sustainable wastewater treatment systems (Agüera et al., 2019; Ghaly et al., 2009) (Figure 5.8).

5.3.1. Definition and Significance of AOPs

Advanced Oxidation Processes (AOPs) are emerging water treatment technologies that are characterized by the generation of highly reactive hydroxyl radicals. These possess an unmatched capacity to destroy organic contaminants, even those that are resistant to conventional treatment. AOPs exploit a number of methods such as generation of ozone, hydrogen peroxide addition, and photocatalysis, which often utilizes ultraviolet light to enhance the efficiency of the reaction. These new processes not only target micropollutants but contribute to pathogen destruction, thereby presenting a mechanism of producing high-quality effluents. Their integration into the current systems of water treatment testifies to AOPs as being important in addressing the twin challenge of water quality and shortage in an era of increased environmental stresses. AOPs are more important than merely degrading the pollutants; they represent an integrated solution to addressing compound water quality issues. During a water scarcity scenario, there is a necessity to ensure an integrated watershed management, water saving in water utilization as a scarce resource and improve wastewater treatment plant efficiency for the assurance of treated effluent quality. Hence, tertiary wastewater treatments depending on advanced technology have been widely studied over the past 20 years (Sabrina et al., 2022). Not only do these treatments give better degrees of purification but also reduce the formation of toxic by-products, which collectively makes the environment more sustainable. The various types of AOPs available can be tailored to eliminate specific pollutants and are therefore essential for water treatment applications today. Besides, the technological innovations in AOPs make them applicable on a large scale. Systematic evaluations and extensive studies of different AOP techniques are imperative to provide effective tools to treat different kinds of wastewater pollutants. After successful implementation of pilot and full-scale projects, decision-makers are able to select appropriate AOPs based on the local environment and contaminant types. By using a technical feasibility consideration, operating costs, and environmental traces—factors often overlooked within traditional evaluation—future policy choices can be better informed (Sabrina et al., 2022). Finally, the value of AOPs lies in their efficacy and versatility that, when combined with robust evaluation protocols, can enhance water treatment processes significantly while maintaining sustainability.

5.3.2. Types of AOPs Utilizing Photochemical Methods

Advanced Oxidation Processes (AOPs) using photochemical processes are a novel water treatment technology, particularly useful in addressing the issue of persistent pollutants such as pesticides. Photochemical methods such as UV irradiation and photocatalysis rely on the formation of highly oxidative species, i.e., hydroxyl radicals, to break down organic contaminants in water effectively. AOPs became increasingly popular as alternatives to conventional treatment technologies, primarily due to the ability of AOPs to mineralize organic pollutants completely without the production of toxic byproducts. The treatments involved include direct photolysis, which is illumination with sunlight or UV light that leads to contamination degradation, and the use of photocatalysts, e.g., titanium dioxide, to enhance higher degradation efficiency. With the growing interest in pesticide contamination globally, the integration of AOPs into water treatment technologies is currently an essential path toward achieving improved water quality and safety for drinking water sources (Syafrudin et al., 2021). Furthermore, based on the flexibility of AOPs, combined systems involving the integration of photochemical reactions with other treatments can be developed for enhanced efficiency. For instance, coupling photocatalysis with biological treatment can achieve maximum removal of pollutants by capitalizing on the advantages of each treatment. Hybrid systems not only enhance degradation rates of the complicated pollutants but also reduce operation costs compared to conventional processes. A consideration of a variety of treatment processes emphasizes the need to select an efficient AOPs strategy according to the type of pesticide or contaminant in question. The integration of AOPs in treatment schemes is particularly compelling, as they offer a means to attain maximum removal efficacies at an intermediate level of secondary pollution, e.g., sludge generation seen in traditional practices (Saleh et al., 2020). The application of photochemical AOPs in decentralized water treatment plants is one instance of a novel solution to the present global problems of water quality and contamination. Such systems not only de-centralize treatment plants, improving access to clean water, but also enable on-site treatment of water contaminated with some substances, such as pesticides. Such application is very critical in agricultural or rural regions, where pesticide runoff is typical. Photochemical AOPs can solve the implications of agricultural activity on water bodies so that aquatic ecosystems and human health are preserved. As interest grows in economically viable and sustainable treatment procedures, AOPs founded on photochemical technologies will likely have a key role to play in shaping the next generation of water treatment paradigms, ultimately contributing to cleaner and safer water resources for future generations (Syafrudin et al., 2021).

5.3.3. Effectiveness of AOPs in Removing Contaminants

Advanced oxidation processes (AOPs) have emerged as the most prominent solution to the complex issue of eliminating contaminants, in the form of organic pollutants, from water bodies. The effectiveness of AOPs lies in their ability to generate reactive oxygen species (ROS), which can effectively degrade a vast number of persistent organic compounds that are generally recalcitrant to conventional wastewater treatment processes. As a result of the problem of drug residues, the demand for aquatic systems has increased, and AOPs have been observed to possess the potential to destroy pharmaceutical pollutants that have been found resistant and stable to conventional methods. Decomposition of most used drugs such as ibuprofen and tetracycline by AOPs is an evident suggestion of developing suitable mechanisms to offset the environmental impact caused by these contaminants. AOPs not only aid in the destruction of these compounds but also play a significant role in enhancing the overall quality of treated water (Bellardita et al., 2023). The Photo-Fenton process, being a specific AOP, illustrates the new application domains of photochemistry for wastewater treatment. This is a process that combines the catalytic nature of iron ions with hydrogen peroxide under ultraviolet light to form hydroxyl radicals that are able to decompose complex organic substances to significantly improve water quality. Remarkably, studies have established successful mineralization of wastewater from textiles, which typically contain biorecalcitrant and nonbiodegradable dyes. By optimizing parameters such as pH, irradiation time, and H_2O_2 and Fe^{2+} concentrations, significant reductions in COD and TOC have been achieved. From the experiments, it is seen that the photochemical process can bring more than 100% decolorization and enhanced biodegradability of treated wastewater, so that the effluents are suitable for further biological treatment (Ghaly et al., 2009). Moreover, AOPs provide a greener alternative to conventional wastewater treatment processes because they tend to work under moderate conditions but with great efficiency. That these processes are able to produce strong oxidants without producing toxic by-products also makes them more desirable, particularly in cases where other conventional treatments have not been successful. Application of AOPs in routine water treatment processes can reduce the environmental footprint considerably while keeping pace with the growing intensity of contamination from industrial activities and medication use. Furthermore, as more stringent regulatory atmospheres for effluent quality are put in place, application of AOPs can not only guarantee compliance but also bring about sustainable strategies towards water management. Thus, the investigation on the overall applications and efficiencies of AOPs remains significant to the evolution of water treatment technologies (Bellardita et al., 2023; Ghaly et al., 2009) (Figure 5.9).

Figure 5.9. *Effectiveness of AOPs in removing contaminants.*

Source: https://www.mdpi.com/materials/materials-15-08152/article_deploy/html/images/materials-15-08152-g004.png.

5.4. SOLAR WATER TREATMENT TECHNOLOGIES

Advanced solar water treatment technologies are gaining momentum as critical solutions to mitigate the global water crisis, particularly where there is limited access to clean drinking water. The technologies utilize the power of the sun to facilitate photochemical reactions that purify water, presenting a green solution against traditional disinfection methods. Solar disinfection (SODIS) is one of the promising approaches in which sunlight is used to destroy pathogens in water. But recent advances in photocatalysis, particularly with the use of semiconductor materials such as titanium dioxide (TiO_2) and zinc oxide (ZnO), have the potential to greatly enhance the efficiency and quality of this process. The integration of photocatalysis facilitates reduced time to disinfect and greater overall rates of removal of harmful microorganisms, addressing directly the urgent issues raised by the United Nations on access to clean drinking water (Byrne et al., 2014). Through synergism of solar disinfection and photocatalysis, new photo-reactor application maximizes water treatment efficiencies. These reactors contribute to increased photon flux onto the photocatalyst and thus the target reaction rates required for effective disinfection. Such important parameters as the illuminated volume/total volume ratio, catalyst loading, and flow rate play crucial roles in optimizing the effectiveness of these systems. New reactor designs have highlighted the importance of immobilizing photocatalysts since this process is more stable and reusable compared to the typical slurry systems. While early field results indicate encouraging trends, commercial viability remains challenging. Solving these issues as well as future reactor designs will be essential to commercialization of solar water treatment technologies and implementation in real-world scenarios (Agüera et al., 2019). Further, solar water treatment technologies comprise several advanced oxidation processes (AOPs) directed towards industrial and domestic wastewaters. AOPs have been recognized to comprehensively degrade organic pollutants like pharmaceutical residues and bacterial contaminants. It is clear from the research carried out that cooperative research by countries including China, Spain, and the USA characterizes the global AOPs scene, emphasizing the global prominence of this field. Using an examination of available scientific literature, certain sets of research practices can be identified with the objective of eliminating pathogens and industrial pollutants from various sources of water. The integration of solar energy with such advanced procedures not only facilitates sustainable water treatment procedures but also aligns well with the world's movement towards the achievement of sustainable development goals for clean water accessibility (Agüera et al., 2019) (Figure 5.10).

Figure 5.10. *Solar water treatment technologies.*

Source: https://pumps-africa.com/wp-content/uploads/2022/05/bZLl7vOopy4Jr-RaPcpkKDMxNPW_bQk5ALNocVql8tv5lMhq8NZsqYJkEu3pa4LIM-CVePCk-Ph3JgRXlmWMq2NuSkIB_-ecJggXX3xZaTpd97t7Y1DbmpeiWST9qMVPfc.jpg.

5.4.1. Overview of Solar-Driven Photochemical Processes

Sun-powered photochemical processes are an outstanding process in water treatment that harnesses the abundant energy of sunlight to catalyze reactions that purify water from pathogens. It is a process that largely relies on the contact between solar radiation and certain microorganisms in water, which significantly impacts their viability. The inactivation of the pathogens is brought about by a range of different mechanisms depending on the specific wavelengths of the sunlight, type of microorganism present, and quality of water itself. For instance, more recent work has confirmed that the effectiveness of solar inactivation can be enormously dependent upon environmental conditions and the type of contaminant and characterizes the photochemical dynamic complexity. With knowledge of such mechanisms, treatment systems can be engineered better by practitioners to enhance the benefits of solar energy to achieve greater public health impact through the use of a renewable resource (Boehm et al., 2018). The application of specific photocatalysts in solar-driven procedures enhances the effectiveness of the degradation process of contaminants. Zinc Oxide (ZnO), its semiconductor property, and band gap energy of approximately 3.2 eV make it an excellent example of an effective photocatalyst for such applications. When applied in thin film form onto glass substrates, ZnO is involved in oxidation reactions upon being irradiated by UV radiation, effectively transforming poisonous substances in water into inert products. Preparation of ZnO involves the creation of a gel between zinc acetate and the use of spray coating techniques, with subsequent sintering that makes the thin film rigid. Empirical data has established the capability of using ZnO with UV light to decompose up to 97.05% of compounds like methylene blue, demonstrating its practical utilization in solar photochemical process-enhanced water treatment (Hidayanto et al., 2012). The integration of solar photochemical processes for water treatment not only addresses the short-term issue of contamination but also offers a sustainable method of resource management. As technology advances, more interest lies in the potential of further optimizing these processes with new materials that maximize light capture and enhance reaction kinetics. For instance, natural organic matter in the environment can build up and influence the efficiency of solar inactivation, and thus methods must be found that are suitable to differences in water quality. Besides, future studies will play a key role in optimizing the potential of solar-powered photochemistry for water treatment through an examination of variable interactions to increase inactivation rates and designs. These dynamics will be crucial to know in order to deploy solar technologies in urban and rural water treatment plants to realize a more sustainable approach to public health (Boehm et al., 2018).

5.4.2. Case Studies of Solar Water Treatment Systems

The efficiency of solar water treatment facilities has gained prominence over the years, particularly in the form of case studies involving various methodologies and results. One primary instance is the application of solar disinfection (SODIS) facilities in rural communities that employ UV radiation to significantly reduce the levels of pathogens within drinking water. These systems rely on the natural capacity of sunlight to inactivate microorganisms as demonstrated by various rates of inactivation depending on solar spectral irradiance, water quality, and intrinsic pathogen sensitivity. Experiments have shown that under optimal conditions, solar disinfection can significantly decrease bacteria and viruses as described in recent reviews of the intricate mechanisms of sunlight-mediated inactivation. Such case studies not only emphasize the science behind photochemical applications but also introduce the pragmatic solutions to enhance potable water quality in conditions of limited resources. Another very well-stated case study is that which surrounds engineered nanomaterials (ENMs) integrated into solar water treatment systems, with their focus on accelerating contaminant destruction. By examining the photochemical degradation of perfluoroalkyl acids (PFAAs) with ENMs under artificial sunlight conditions, researchers have identified major interactions allowing environmental pollutants to degrade. Experiments revealed considerably increased degradation of certain PFAAs when coupled with certain ENMs, further highlighting the potential of such technologies for improving water treatment procedures. However, it is crucial to establish the implications of such interactions on adsorption, reactivity, and photo-stability under varying environmental conditions. Case studies like this illustrate the dual benefit of solar energy utilization, both for disinfection and degradation of refractory organic contaminants, thus illustrating a multi-faceted approach to water treatment. The integration of solar water treatment solutions into current infrastructure also reveals rich learning from many global case studies, particularly regarding responses to the challenges of water scarcity and contamination. These systems possess a distinct ability to adapt to conditions in their environment at the local level, making them particularly suitable for decentralized use in urban and rural communities. For instance, high-solar-irradiance region case studies reported considerable success in solar reactor usage to break down pollutants like industrial waste and pharmaceuticals. Mechanistic information obtained through such studies, particularly on the dynamics between sunlight, microbial communities, and water chemistry, forms the foundation of such systems' effectiveness to minimize water quality issues. Furthermore, one can learn from these case studies to implement in the design of even more efficient solar water treatment technologies that can be scaled up and made sustainable to address global water crises.

5.4.3. Benefits and Limitations of Solar Technologies

The incorporation of solar technologies in photochemical processes has significant benefits, particularly in water treatment. The main benefit is the sustainability aspect, as solar energy is renewable and cheap. Using solar energy to drive photochemical reactions can result in radical decreases in the carbon footprint that is typically linked with conventional sources of energy. Besides, technologies such as continuous-flow reactors, which are commonly applied in the chemical and pharmaceutical industries, can be employed to enhance the efficiency of solar-powered reaction. They allow for even irradiation, leading to reduced reaction time and discriminatory photochemical reactions and hence are suitable for bulk production of water treatment products. As pointed out by recent research, the applications of high-level solar technologies have registered improvements in yield and safety of reactions, affirming their abilities to contribute to sustainable use in water purification techniques (Sambiagio et al., 2019). Nonetheless, despite this advantage, several limitations are associated with the implementation of solar technologies to treat water. Firstly, the volatile nature of solar power is a disadvantage, particularly in regions characterized by varying weather patterns or limited sunshine. This discrepancy can hinder the consistency of the solar-powered processes, hence the need for auxiliary energy sources or means of storage. The initial cost of solar technologies is also typically high, which acts as a constraint to widespread application, particularly in developing nations where financing constrains. Additionally, there will be a need for long-lasting reactor designs that can optimize the use of the sun's energy while still having efficiency in water treatment processes. Thus, even as innovation in solar water treatment technologies promises much, it is essential to overcome these limitations in order to ensure effective use and durability in diverse environments (Singh et al., 2018). Additionally, developing green synthesis processes that come with the utilization of solar energy is also a new opportunity for enhancing the efficiency of water treatment. With the assistance of sunlight-powered photochemical reactions, researchers are studying eco-friendly methods for nanoparticle synthesis that can effectively detect and remove contaminants from water. Green synthesis of metal and metal oxide nanoparticles by natural extracts as reducers minimizes toxic byproducts and provides a less toxic method for solving water purification concerns. In addition, the integration of these technologies can also lead to improved biocompatibility of the resultant materials for their wider applications in environmental remediation. Although research efforts are still in the process of focusing on sustainable technologies, the advantages and limitations of solar technologies in water treatment will have a key role in shaping future research directions and industrial practice (Sambiagio et al., 2019; Singh et al., 2018) (Figure 5.11).

Figure 5.11. *Benefits and limitations of solar technologies.*

Source: https://www.solarsmiths.com/wp-content/uploads/2021/01/IMG-20210105-WA0014.jpg.

5.5. COMPARISON WITH TRADITIONAL WATER TREATMENT METHODS

The panorama in the treatment of water has radically changed with the introduction of photochemical applications, which present a formidable challenge to traditional methods such as chlorination and filtration. These traditional methods make use of chemical agents that, though efficient at disinfecting water, are prone to producing perilous byproducts that can pose adverse effects on consumers' health. Photochemical processes, particularly those ultraviolet (UV) light-based, present a safer and more environmentally friendly alternative. As highlighted in recent studies, UV light can efficiently counteract microbial load without the addition of other chemicals and thereby minimize the risk of toxicological harm. The advantages of UV technologies are not just within the safety bracket; they also produce operating efficiencies in treatment processes that are not very easily achievable with conventional systems. Relative to the efficacy of photochemical applications with traditional methods, consideration must not only be given to short-term outcomes but also operating costs and environmental impact. Traditional water treatment often involves the necessity for huge capital facilities and frequent chemical expenditures that can stress municipalities and water treatment plants. Photochemical technologies, particularly UV LEDs, have been proven to significantly decrease energy use while also enhancing the general reliability of water disinfection. Utilization of these technologies can lead to long-term cost savings due to lower maintenance and operational costs, and therefore they form an ideal solution for sustainable water treatment for communities. This paradigm is regarded to prioritize the integration of advanced techniques within existing treatment paradigms for improved water quality. Furthermore, the integration of photochemical reactions into water treatment is also emphasized by their scalability and the fact that they are easy to adapt. Traditional methods have been known to flounder in managing fluctuating contaminant loads prevalent in varied locations. Photochemical uses may be readily applied to rectify specific local conditions, such as the specific microbial regimes or kinds of pollutants found in different water sources. As a case point, it is evident from research the efficacy of UV treatment in fruit processing and its correlation with specific parameters such as color alteration in fresh fruit salads, evidencing the potential of light-based methods in broader environmental systems (Lante et al., 2016). Furthermore, the potential for photochemical reactions to augment traditional methods of agricultural and environmental monitoring, as indicated by advances in hyperspectral remote sensing to explore parameters like leaf area index and water content (Borzuchowski et al., 2010), reflects their use in integrated water resource management. In effect, the shift toward photochemical means of water treatment signals a step forward towards more efficient, effective, and sustainable approaches to managing water (Figure 5.12).

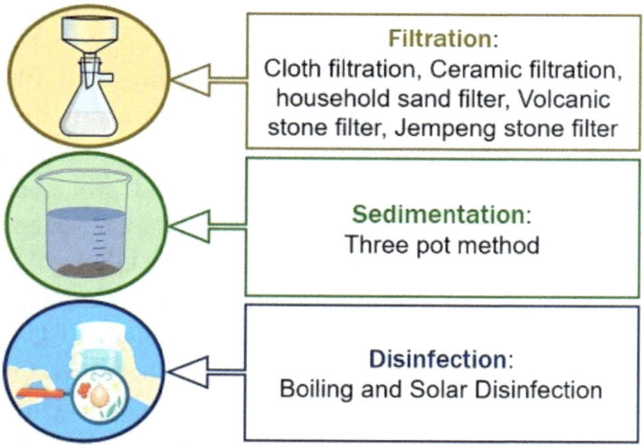

Figure 5.12. *Comparison with traditional water treatment methods.*

Source: https://biologyreader.com/wp-content/uploads/2022/07/traditional-methods-of-water-purification.jpg.

5.5.1. Efficiency of Photochemical Methods versus Conventional Methods

The efficiency of photochemical technologies in water treatment is a significant advancement over conventional methods, particularly in organic contaminants' destruction. The traditional chemical treatment typically employs the application of chlorine or other disinfectants that, though effective in pathogen elimination, produce harmful byproducts. On the other hand, photochemical treatments utilize ultraviolet (UV) light and advanced oxidation techniques, such as photocatalysis, to induce degradation of contaminants without the use of harmful chemicals. Studies have shown that photochemical treatment is capable of achieving degradation rates greater than in conventional technologies, particularly with resistant contaminants. This advantage comes not only from the immediate energy drawn from light but also from the generation of reactive species that can effectively oxidize organic molecules, thereby highlighting the superiority of photochemical processes in efficiency as well as in safety. The operational cost of photochemical processes can also be more cost-effective compared to traditional water treatment processes.

For instance, while it takes some initial expenditure on UV reactors and photocatalysts, the reduced chemical use and lowered energy inputs to maintain them lower the long-term costs dramatically. Comparative evaluation of water treatment in Barcelona showed that tap water treated by advanced systems contained a reduced environmental impact compared to bottled water, which can be extrapolated to conclude that improved treatment technologies such as photochemical processes are able to minimize energy input and costs in the long run.

These findings are in accordance with the growing evidence that the energy consumption of conventional methods, especially those reverse osmosis-based, is usually greater than their benefits, hence rendering photochemical methods a cleaner method of water treatment systems (Cadena et al., 2016).

In addition to economic benefits, photochemical methods have more flexibility to accommodate various qualities of water compared to conventional methods. Photochemical treatment can also be tailored to target individual contaminants by adjusting parameters such as light wavelength and reaction duration to create customized solutions based on water quality surveys. Such versatility increases general treatment efficacy, particularly in diverse environmental conditions where water sources may contain a range of contaminants. Standardized treatment methods are more likely to adhere to standardized procedures that do not react well to unorthodox contamination conditions. As signaled by advancements in photochemical studies, methods such as the use of different semiconductor materials in photocatalysis are more efficient for pollutant removal under different conditions, which bodes well for bright prospects in different water treatment processes across the globe (Fusaro et al., 2015).

5.5.2. Cost Analysis of Photochemical Applications

Cost analysis is a critical factor in the determination of the feasibility and use of photochemical technology in water treatment. With the increased need for effective water purification, stakeholders have to conduct cost considerations to ensure that these technologies are cost-saving. The initial cost of capital outlay, operating costs, and savings from the reduced consumption of chemicals must be examined. Moreover, the cost of system maintenance and energy consumption, particularly of photochemical processes relying heavily on ultraviolet (UV) light or other photocatalytic systems, would many times be the key factor in calculating the final cost. Knowing this, companies are able to make informed decisions regarding the adoption of photochemical technologies, particularly when compared with traditional methods, hence ensuring improved utilization of resources and sustainability in water treatment operations. The efficiency of photochemical processes is commonly complemented by their economic implications. For instance, some applications, like those that include UV disinfection and photocatalysis, are characterized by disparate operational costs in drawing them as treatment protocols. With the improvement of technology, the volume cost can decline per volume treated but initial installation costs may still be problematic to municipalities or industries with modest budgets. Literature indicates that the right cost-benefit analyses, including long-term costs avoided with reduced chemical use and increased public health gains, can underwrite earlier investments in such advanced treatment methods. Efficiency of photochemical applications in degrading contaminants not only alleviates environmental issues but saves healthcare costs downstream and hence presents a sound economic argument for adopting the same in water treatment systems. Additionally, initial research has indicated the viability of coupling photochemical processes with other treatment processes, further affecting cost dynamics. To illustrate, coupling photochemical processes with traditional filtration or biological processes can enhance the effectiveness of treatment while distributing operating costs across various processes. By evaluating the synergies among technologies, organizations can optimize the utilization of resources, resulting in lower expenditure. The flexibility of photochemical applications between various environments, for instance, agricultural terrain ventured in research comparing wet and dry regions, also indicates their cost-effectiveness (Borzuchowski et al., 2010; Fusaro et al., 2017). Thus, an understanding of the fine balance between treatment effectiveness, initial costs, and long-term economic benefit is crucial for stakeholders considering the adoption of photochemical applications in water treatment.

5.5.3. Environmental Impact Considerations

The application of photochemical processes in water treatment has substantial environmental impact factors that need careful examination. One of the major concerns is the generation of possible byproducts during the treatment process, which will ironically lead to further environmental damage. Because photochemical processes break down contaminants, it is worthwhile to understand the end products of substances for their environmental viability. The development of standards with a focus on ecological support systems, as seen in workshops, is an indication of the necessity for coordination among researchers across disciplines to present solutions to these issues in a productive manner (Carden et al., 2025). There must be a proactive strategy to prevent any adverse effects and ensure the sustainability of water treatment. Furthermore, the ecological aspects of using some microalgae in photochemical processes are another aspect of complexity.

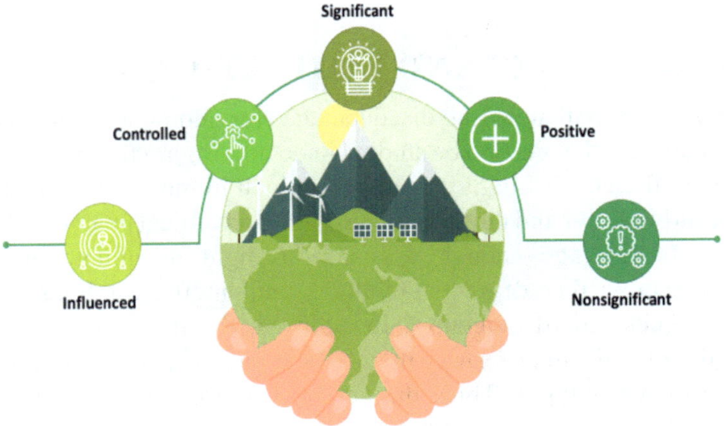

Figure 5.13. *Environmental impact considerations.*

Source: https://cdn.sketchbubble.com/pub/media/catalog/product/optimized1/d/7/d7d2dfc80402604aff614bd38dc94e4d91e9ff113657699a23be63b90cbe040a/environmental-aspects-slide1.png.

For instance, the dinoflagellate algae *Prorocentrum* has been a subject of research for the production of valuable chemicals such as okadaic acid, which holds immense value for various applications. However, low yield and slow cultivation rate of such microorganisms are serious scalability constraints in water treatment facilities, as highlighted by studies with conditions for growth and photochemical efficiency (Belarbi et al., 2014). This highlights the importance of balancing the trade-offs of employing such organisms in a bid to get the best treatment outcomes. Species and environmental factors are influential determinants that directly affect the effectiveness and sustainability of photochemical water treatment methods. Another important aspect of environmental factors in photochemical application is the energy consumption and carbon footprint of the processes. Elevated energy inputs will neutralize some of the environmental benefit that accrues due to pollutant breakdown, and thus it is significant to consider the overall energy efficiency of treatment plants. On top of that, energy optimization can lend to the overall sustainability goals in the field of water treatment since it can help reduce greenhouse gas emissions. Conducting complete lifecycle analyses can enable a greater understanding of how several different photochemical technologies differ not only in terms of performance but also in environmental sustainability and thus inform more effective decision-making about the application of these technologies (Figure 5.13).

5.6. CHALLENGES AND LIMITATIONS

There are various challenges and disadvantages for photochemical applications for water treatment that need to be studied in detail. Among these, one of the key issues is the efficiency of light penetration, which in most situations lowers in colored or turbid water and thus the scope of photochemical reactions decreases. Photocatalysts leverage enough exposure to light to initiate and sustain chemical reactions that degrade pollutants, and inefficient light absorption can lead to substandard outcomes. In addition, the dependence on specific wavelengths of light may restrict the applicability of particular technologies to a series of water types. Thus, the need for improved penetration of light and the development of more universal photocatalysts remains a core area in demand of academic attention to enhance the effectiveness of photochemical treatments under real-world conditions. The second problem is the potential for by-product formation when photochemical reactions take place. While the overall aim of such processes is to eliminate the pollutants, the processes can in the process inadvertently form toxic intermediates. For instance, photocatalytic processes may form reactive oxygen species that can efficiently degrade the contaminants but possibly bear toxicity risk if not adequately controlled. Identifying the complete pathways of reactions and establishing the potential toxic by-products are critical to ensuring water safety. All monitoring systems need to be integrated to assess by-product formation in real time to counteract unwanted release of harmful substances into treated water, which would ruin the purpose of photochemical applications. Finally, scalability is a challenge of intimidating proportions in using photochemical processes to treat water. Most of the present methodologies are laboratory proven where control and optimization of conditions can be achieved. However, implementation of these findings at the large-scale level is usually accompanied by logistical, economic, and technical difficulties. For example, pilot study systems can fail to perform efficiently upon enlargement owing to variable environmental conditions such as light availability, flow rates, and contaminant levels. Moreover, economic factors are also vital, as initial capital expenditure and running costs can deter large-scale application. Overcoming these limitations—via the development of low-cost technologies and strategies to ensure solid system performance in different environments—is essential for promoting practical implementation of photochemical water treatment methods (Bragina et al., 2017; Fusaro et al., 2015).

5.6.1. Technical Challenges in Implementing Photochemical Methods

While looking for effective photochemical technologies for water cleaning, there are several technical issues hindering their large-scale implementation. One of the key ones is the intricacy involved in the optimal light source design and its interaction with impurities. Impurities absorb at various wavelengths, for which the selection of an appropriate light source is necessary to achieve maximum efficiency. This specificity complicates photoreactor design because they might mismatch and cause partial destruction of pollutants and unsatisfactory treatment outcomes. The reactor design must also facilitate constant light penetration and flow regimes, presenting design complications that require advanced engineering solutions. These technical complications underscore the need for ongoing research into photochemical processes to make them more efficient and allow them to be incorporated into existing treatment systems (Boyd et al., 2025). Moreover, scalability is yet another major technical limitation in applying photochemical processes for water treatment. While small-scale laboratory experiments typically yield promising results, large-scale application tend to be often limited by energy consumption and operational efficiency. Conventional treatment technologies could typically be formulated fairly easily, but photochemical processes could require significant adjustments in terms of structure and operation. For example, scale-up calls for the development of new engineering solutions that are capable of handling variations in water quality and flow rates without deteriorating performance. It can deter investment in photochemical processes, which will be viewed as less resilient than traditional technologies. In this regard, enhancing the scalability of these processes remains paramount to their feasible application in large-scale water treatment plants (Toledo et al., 2016). Aside from the above challenges, regulatory compliance is a key technical challenge in the application of photochemical techniques. Technologies for water treatment have to be regulated to the highest levels to ensure both safety and effectiveness, and this can have a major impact on innovation rates. Photochemical reactions tend to produce by-products that need to be evaluated for toxicity and environmental concern, making it tougher to deal with regulations. In addition, the lack of regulation guidelines for photochemical applications may contribute to practitioner confusion when attempting to determine obligation to comply. This is combined with a relatively nascent state of many photochemical technologies in comparison to conventional processes that possess an established regulatory process. Addressing these issues of regulation is crucial to the advance of photochemical technologies for water treatment and realizing their potential to offer sustainable solutions for contemporary water quality issues (Boyd et al., 2025) (Figure 5.14).

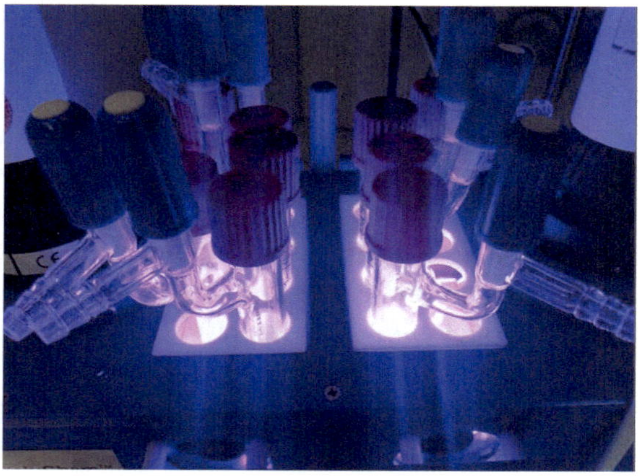

Figure 5.14. *Technical challenges in implementing photochemical methods.*

Source: https://www.scienceboard.net/user/images/content_images/sup_drug/2021_03_26_17_47_4222_2021_03_26_photochemistry_20210326171050.jpg.

5.6.2. Regulatory and Safety Concerns

The addition of photochemical applications to water treatment raises a number of regulatory and safety concerns that have to be properly addressed to ensure public and environmental safety. Regulatory frameworks play a critical role in managing the use of these new technologies. Governments and environmental bodies are usually inclined to set guidelines that offset risks resulting from chemical reactions and by-products associated with photochemical operations. The necessity of strict safety controls arises in the context of potential dangers posed by treated water residual chemicals, especially while employing state-of-the-art oxidation processes that can have toxic by-products. By creating and effecting broad-ranging regulatory policies, stakeholders are capable of balancing technological innovation with safety and thereby making the community embrace future water treatment technology. In looking back at the historical context of water treatment chemical management, evidence is seen that policy regulation has been effective in stimulating innovation, for example, limiting the application of environmentally damaging chemicals such as EDTA. Drawing from research into environmental policy effectiveness, voluntary agreements and initiative in communication among stakeholders have precipitated powerful changes in behavior among those industries relying on such chemicals (Conrad et al., 2000). These discussions instigated corporations to innovate using sustainable avenues and product choices, reiterating the importance of collaboration among regulators, scientists, and business strategists. The process must, nonetheless, be further refined because the interaction between regulatory regimes and innovation processes can be complex, often presenting problems that are more social than technically exacting in their demands. In addition, data related to the efficiency of various practices in dealing with waste provide clear evidence of the need for ongoing monitoring of the impact caused by human activities on water quality and safety (N/A, 2025). The regulatory authorities are also increasingly recognizing the need to monitor not only the chemical compounds that are used in treating water but also the subsequent environmental implications, including pollutant behaviors and degradation rates. Advanced observation techniques and data collection devices are essential to the proper assessment of the consequences of adopting photochemical technologies. Sustained data-driven assessments ensure adherence to safety policies, thereby facilitating evidence-based decision-making. Improved regulatory policies that result from such, along with innovative water treatment technologies, will offer a safer platform for the development and dynamic utilization of photochemical applications.

5.6.3. Future Research Directions in Photochemical Water Treatment

As the pursuit of sustainable and efficient means in water treatment intensifies, the future of photochemical water treatment calls on researchers to pursue new directions that leverage sunlight and advanced catalysts. The convergence of nanotechnology with photochemical reactions is promising, particularly in optimizing the performance of photocatalysts such as titanium dioxide (TiO_2). By manipulation of the surface properties and structure of such materials, researchers have the potential to achieve improved reaction rates and enhanced degradation of organic pollutants. Furthermore, the development of visible-light-active photocatalysts would greatly extend the applicability of photochemical water treatment for regions with intense sun irradiance, providing a promising application for many environmental conditions. Continued investment in this field of research will not only enhance degradation efficiency but will also play a critical role in fixing the world's water problem through the provision of cleaner sources of water. Besides advancements in catalyst technology, future research needs to address the importance of establishing the mechanisms behind photochemical reactions under conditions used in water treatment. Data acquired from *in situ* investigations, say carbofuran photodegradation on TiO_2 films, provides valuable insights regarding the reaction dynamics. The use of advanced analytical techniques such as ATR-FTIR and multivariate analysis techniques makes it possible to understand the reaction product evolution and pathways better.

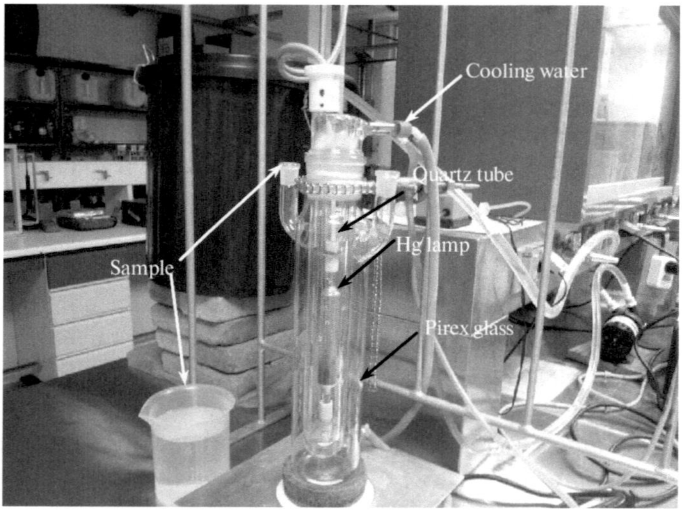

Figure 5.15. *Future research directions in photochemical water treatment.*

Source: https://www.researchgate.net/profile/Fernando-Hernandez-5/publication/335556733/figure/fig1/AS:849437677060096@1579533132361/Photochemical-system-used-for-wastewater-treatment.png.

These quantitative treatments hold the promise of solidifying the reliability of catalytic processes and unlocking more robust predictive models that can inform treatment methods. By prioritizing these research approaches, scientists will strengthen the theoretical foundation of photochemical water treatment and make contributions to optimized approaches for various water contaminants that occur within real-world applications (Atifi et al., 2013). Furthermore, interdepartmental studies combining expertise from environmental science, materials science, and engineering will be essential in establishing next-generation photochemical water treatment technologies. By studying the interactions between pollutants, catalysts, and environmental parameters, researchers can design tailor-made treatments that respond to specific local environments and patterns of contaminants. Joint endeavors using satellite technology in monitoring environmental contaminants and quantifying treatment efficacy can in the same vein provide inputs critical to local governments and industries that are trying to minimize water pollution. These types of integrated studies could play a crucial role in the development of treatment standards and driving policy initiatives towards water sustainability. As environmental challenges intensify, embracing a holistic research strategy will eventually result in better practical uses of photochemical techniques and their applications in addressing immediate water treatment needs (N/A, 2025) (Figure 5.15).

SUMMARY

When considering the sophisticated role of photochemical applications in water treatment, it is evident that such technologies have vast potential to enhance water quality. The integration of photochemical processes, such as advanced oxidation methods, will be capable of degrading a wide range of pollutants, from organic impurities to by-products of disinfection. Notably, studies indicate that the conventional techniques for water treatment often fall short, with natural organic matter (NOM) removal ranging from 4% to 60% (Tazkiaturrizki et al., 2025). This inadequacy underlines the necessity to discover innovative technologies capable of countering the long-term problems involved in water treatment plants. By applying techniques that take advantage of photochemical reaction, the removal capacities of toxic substances can be augmented to a considerable degree, providing a safer source of drinking water and correcting public health conditions. Moreover, the incorporation of hybrid systems consisting of granular activated carbon (GAC) and electrochemical membrane reactors (MER) provides a pointer towards a significant improvement in water treatment efficiency. This hybrid process successfully facilitates the removal of NOM, with removal efficiencies of up to 90% (Tazkiaturrizki et al., 2025). MER has also been shown to remove a considerable quantity of trihalomethanes (THMs), important disinfection by-products with public health implications, by removing between 40% and 62% of these pollutants when used in conjunction with GAC (Tazkiaturrizki et al., 2025). These findings reflect the revolution that can be unleashed by combining photochemical technologies with traditional practices and thereby aiming at an all-inclusive solution that not only eliminates current-day pollutants but also envisions issues for the future of water quality management. Secondly, developing portable water treatment systems offers a realistic approach towards photochemical innovations, particularly in unserved coastal communities. As illustrated by community service initiatives, these systems are designed for brackish water desalination and filtering out organic and inorganic content. Use of such technology empowers locals by providing them with clean and consumable water and fostering coping capacity in water-scarce environments. Focusing on user engagement and operation safety, these mobile systems enhance community capacity for sustainable water management. In short, the ongoing evolution of photochemical technologies in water treatment not only reflects what is currently available but also emphasizes the importance of efficacious, safe, and cost-effective solutions in meeting water security globally. Photochemical technology for water treatment leverages light power to facilitate reactions that degrade contaminants, providing a novel technology for improving water quality. Central to this process is the use of advanced oxidation processes (AOPs), which involve using ultraviolet (UV) light and photocatalysts to generate reactive species like hydroxyl radicals. These radicals are highly efficient in breaking down organic pollutants and pathogens and are thus highly beneficial in wastewater treatment and remediation of contaminated water sources. Most importantly, AOPs can break down pollutants that are normally recalcitrant towards conventional treatment technologies, thus addressing an important loophole in the current water treatment technologies. Moreover, the elasticity of photochemical processes provides real-time adjustability of treatment based on varying contaminant loads, leading to an improved and responsive water management system under changing conditions. The effectiveness of photochemical applications is also maximized through the utilization of appropriate photocatalysts, which form the core of their effectiveness. For instance, titanium dioxide (TiO_2) is one of the most researched photocatalysts due to its stability, non-toxicity, and high oxidizing power when excited under UV light. However, more recent studies have also explored new materials and composite systems that show promise for amplifying photocatalysis under visible light and thereby allowing photochemical approaches to be used in sunny regions of the globe with even greater benefits. This diversification is founded on the photocatalyst formulation optimization to enable sufficient absorption of light and induce effective electron-hole separation, aiming at achieving a maximum level of organic pollutant degradation. In addition, understanding the reaction kinetics and mechanisms is necessary for scaling up such technologies toward practical application, and thus it is an important component in developing effective water treatment approaches. Experimental evidence has revealed the possibility of integrating photochemical processes with traditional treatment systems towards creating hybrid technologies that optimize water cleansing mechanisms. The integration of photochemical treatments and biological processes, for example, has the possibility of creating a more comprehensive approach to manage a wider range of contaminants. This combined approach has the possibility of optimizing pathogen and resistant organic compound removal efficiencies while minimizing chemical usage and sludge production. Furthermore, studies on the operational conditions—temperature, pH, and light intensity—can provide valuable information regarding optimizing the performance of hybrid systems. This diversified approach not only aligns with the visions of sustainable development but also addresses issues generated by increased populations and industrialization, thus vindicating the intrinsic value of photochemical applications in providing clean and safe water to all (Belarbi et al., 2014; Cahoy et al., 2015).

REFERENCES

1. Abdullah, N., Mohd Padzil, S. A., Mohd Salleh, R., Pillay, V., et al. (2018). Model selection and model averaging on mortality of upper gastrointestinal bleed patients. https://core.ac.uk/download/18264202.pdf (accessed on 28th September 2025).
2. Agüera López, A. M., Esteban García, A. B., Garrido Cárdenas, J. A., Manzano Agugliaro, F., et al. (2019). Wastewater treatment by advanced oxidation process and their worldwide research trends. https://core.ac.uk/download/286590742.pdf (accessed on 28th September 2025).
3. Angrill Toledo, S., Gabarrell Durany, X., Josa Garcia-Tornel, A., Petit Boix, A., et al. (2016). Environmental performance of rainwater harvesting strategies in Mediterranean buildings. https://core.ac.uk/download/81571527.pdf (accessed on 28th September 2025).
4. Atifi, A., Czarnecki, K., Mountacer, H., Ryan, J., et al. (2013). *In situ* study of the photodegradation of carbofuran deposited on TiO_2 film under UV light, using ATR-FTIR coupled to HS-MCR-ALS. https://core.ac.uk/download/213076876.pdf (accessed on 28th September 2025).
5. Belarbi, E. H., del Cerón García, M. C., Gallardo Rodríguez, J. J., García Camacho, F., et al. (2014). Simultaneous effect of temperature and irradiance on growth and okadaic acid production from the marine dinoflagellate *Prorocentrum belizeanum*. https://core.ac.uk/download/286590302.pdf (accessed on 28th September 2025).
6. Bellardita, M., Kanwal, T., Loddo, V., Palmisano, L., & Umair, M. (2023). Review on recent advances in the removal of organic drugs by advanced oxidation processes. https://core.ac.uk/download/599092773.pdf (accessed on 28th September 2025).
7. Boehm, A., Davies-Colley, R., Dodd, M., Kohn, T., et al. (2018). Sunlight-mediated inactivation of health-relevant microorganisms in water: A review of mechanisms and modeling approaches. https://core.ac.uk/download/323080860.pdf (accessed on 28th September 2025).
8. Borzuchowski, J., & Schulz, K. (2010). Retrieval of leaf area index (LAI) and soil water content (WC) using hyperspectral remote sensing under controlled glass house conditions for spring barley and sugar beet. https://core.ac.uk/download/16431918.pdf (accessed on 28th September 2025).
9. Boyd, J. (2025). Searching for the profit in pollution prevention: Case studies in the corporate evaluation of environmental opportunities. https://core.ac.uk/download/pdf/9308183.pdf (accessed on 28th September 2025).

10. Bragina, O. A., Gekaluyk, A. S., Goltsov, A., Rafailov, E., et al. (2017). Laser-induced generation of singlet oxygen and its role in the cerebrovascular physiology. https://core.ac.uk/download/228178032.pdf (accessed on 28th September 2025).

11. Byrne, A. J., Dionysiou, D. D., Dunlop, P. S. M., Ibanez, M., et al. (2014). Solar photocatalysis for water disinfection: Materials and reactor design. https://core.ac.uk/download/pdf/60775675.pdf (accessed on 28th September 2025).

12. Cadena, E., Ferrer Martí, I., Garfi, M., Sanchez Ramos, A., et al. (2016). Life cycle assessment of drinking water: Comparing conventional water treatment, reverse osmosis and mineral water in glass and plastic bottles. https://core.ac.uk/download/81580447.pdf (accessed on 28th September 2025).

13. Cahoy, K., Fortney, J. J., Kempton, E., Lewis, N., et al. (2015). Thermal emission and albedo spectra of super Earths with flat transmission spectra. http://arxiv.org/abs/1511.01492 (accessed on 28th September 2025).

14. Carden, J. L., & Mason, R. M. (2025). Guiding the development of a controlled ecological life support system. https://core.ac.uk/download/pdf/42867040.pdf (accessed on 28th September 2025).

15. Chen, J. G., Crooks, R. M., Seefeldt, L. C., Bren, K. L., Bullock, R. M., Darensbourg, M. Y., Holland, P. L., et al. (2018). Beyond fossil fuel–driven nitrogen transformations. *Science, 360*. https://doi.org/10.1126/science.aar6611 (accessed on 28th September 2025).

16. Conrad, J. (2000). Technical innovations for reducing use and emission of EDTA. https://core.ac.uk/download/199417630.pdf (accessed on 28th September 2025).

17. Dai, Y., Guo, X., Hoffmann, M. R., Wang, D., et al. (2020). Photochemical transformation of perfluoroalkyl acid precursors in water using engineered nanomaterials. https://core.ac.uk/download/323085206.pdf (accessed on 28th September 2025).

18. Darwin, R. B., & Putra, B. S. (2025). Dissemination of portable water treatment for providing clean and fresh water for people living in the coastal area of Meuraxa District, Banda Aceh. *Jurnal Pengabdian Masyarakat*. https://www.semanticscholar.org/paper/e37f59d446a426c7129fbe850140d4107978a8da (accessed on 28th September 2025).

19. de Boer, S., González-Rodríguez, J., Conde, J. J., & Moreira, M. T. (2022). Benchmarking tertiary water treatments for the removal of micropollutants and pathogens based on operational and sustainability criteria. *Journal of*

Water Process Engineering, 46, 102587–102587. https://doi.org/10.1016/j.jwpe.2022.102587 (accessed on 28th September 2025).

20. Faria, J., Figueiredo, J., Gomes, H., Morales-Torres, S., et al. (2014). Carbon based materials for photocatalytic solar applications in water treatment. https://core.ac.uk/download/153412669.pdf (accessed on 28th September 2025).
21. Fusaro, L., Gullì, M., Manes, F., Marmiroli, N., et al. (2015). Comparison of drought stress response and gene expression between a GM maize variety and a near-isogenic non-GM variety. https://core.ac.uk/download/54511452.pdf (accessed on 28th September 2025).
22. Fusaro, L., Manes, F., & Salvatori, E. (2017). Effects of the antiozonant ethylenediurea (EDU) on *Fraxinus ornus* L.: The role of drought. https://core.ac.uk/download/98349675.pdf (accessed on 28th September 2025).
23. Ghaly, M., Haseneder, R., & Härtel, G. (2009). Integrated coupled systems as biodegradability enhancement of textile wastewater by photo-Fenton process. https://core.ac.uk/download/39152477.pdf (accessed on 28th September 2025).
24. Hidayanto, E., Nurhasanah, I., & Sutanto, H. (2012). Deposition of ZnO thin films by spray coating technique for photocatalytic and photochemical degradation of methylene blue (MB). https://core.ac.uk/download/76939057.pdf (accessed on 28th September 2025).
25. Králová, M. (2012). Self-cleaning properties of thin printed layers of titanium dioxide. https://core.ac.uk/download/30310816.pdf (accessed on 28th September 2025).
26. Lante, A., Nicoletto, M., & Tinello, F. (2016). UV-A light treatment for controlling enzymatic browning of fresh-cut fruits. https://core.ac.uk/download/53526712.pdf (accessed on 28th September 2025).
27. N/A. (2025). Environmental quality. https://core.ac.uk/download/pdf/42884433.pdf (accessed on 28th September 2025).
28. Reilly, K., Ellis, L. J. A., Davoudi, H. H., Supian, S., Maia, M. T., Silva, G. H., Guo, Z., et al. (2023). *Daphnia* as a model organism to probe biological responses to nanomaterials—from individual to population effects via adverse outcome pathways. *Frontiers in Toxicology, 5*. https://doi.org/10.3389/ftox.2023.1178482 (accessed on 28th September 2025).
29. Saleh, I., Zouari, N., & Al-Ghouti, M. A. (2020). Removal of pesticides from water and wastewater: Chemical, physical and biological treatment approaches. *Environmental Technology & Innovation, 19*, 101026–101026. https://doi.org/10.1016/j.eti.2020.101026 (accessed on 28th September 2025).

30. Sambiagio, C., & Noël, T. (2019). Flow photochemistry: Shine some light on those tubes! *Trends in Chemistry, 2*, 92–106. https://doi.org/10.1016/j.trechm.2019.09.003 (accessed on 28th September 2025).

31. Siddiqua, A., Hahladakis, J. N., & Al-Attiya, W. A. K. A. (2022). An overview of the environmental pollution and health effects associated with waste landfilling and open dumping. *Environmental Science and Pollution Research, 29*, 58514–58536. https://doi.org/10.1007/s11356-022-21578-z (accessed on 28th September 2025).

32. Singh, J., Dutta, T., Kim, K. H., Rawat, M., Samddar, P., & Kumar, P. (2018). 'Green' synthesis of metals and their oxide nanoparticles: Applications for environmental remediation. *Journal of Nanobiotechnology, 16*. https://doi.org/10.1186/s12951-018-0408-4 (accessed on 28th September 2025).

33. Syafrudin, M., Kristanti, R. A., Yuniarto, A., Hadibarata, T., Rhee, J., Al-onazi, W. A., Algarni, T. S., et al. (2021). Pesticides in drinking water—A review. *International Journal of Environmental Research and Public Health, 18*, 468–468. https://doi.org/10.3390/ijerph18020468 (accessed on 28th September 2025).

34. Tazkiaturrizki, T., Hartono, D., Adityosulindro, S., & Wahyuningsih, N. P. S. (2025). Integration of adsorption process and membrane electrochemical reactor to remove trihalomethanes in drinking water treatment. *Indonesian Journal of Urban and Environmental Technology*. https://www.semanticscholar.org/paper/d717f2b51fb40fd7bbf8ed32c94748f7af3fbcbe (accessed on 28th September 2025).

CHAPTER 6

PHOTOCHEMICAL METHODS IN AIR POLLUTION CONTROL

Contents

Introduction 221

6.1. Understanding Photochemical Reactions 227

6.2. Mechanisms of Photochemical Air Pollution Control 232

6.3. Photochemical Methods in Industrial Applications 237

6.4. Advantages of Photochemical Methods 242

6.5. Challenges and Limitations 246

6.6. Future Directions in Photochemical Research 251

Summary 257

References 259

INTRODUCTION

The use of photochemical methods in air pollution control is a milestone interface between environmental science and public health. With increased urbanization and ongoing industrialization, the need to fight air pollution, particularly photochemical smog, becomes increasingly more pressing. Photochemical smog is the result of the complex interactions between the sun and other atmospheric materials like volatile organic compounds (VOCs) and nitrogen oxides (NO_x). These reactions lead to the formation of ground-level ozone, a significant health risk, as well as respiratory disease and cardiovascular illness. The understanding of the processes is key to the identification of effective control strategies. Researchers like Peter K. Mueller have underscored the importance of precise measurement and analysis in identifying air pollutant sources and their behavior in the air environment, highlighting the importance of sound scientific inquiry in air quality management (James et al., 2025). Research in photochemical methods not only enhances our understanding of the air environment but also guides the implementation of effective regulatory policy.

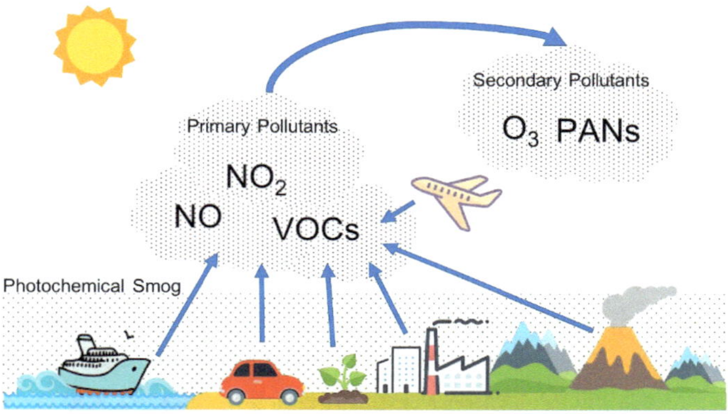

Figure 6.1. *Photochemical methods in air pollution control.*

Source: *https://www.worldatlas.com/r/w2000-h1125-q90/upload/52/be/75/photo-chemical-smog-formation.png.*

With the help of advanced techniques, scientists are able to build sophisticated models that can simulate air quality conditions and predict the effects of various pollutants. The models enable policy makers to make informed decisions based on reality, thus making room for innovative solutions to emissions mitigation. Further, recent studies have made new technologies, such as catalytic converters and photochemical oxidants, which can eradicate noxious emissions from automobiles and industries. Their inclusion within regulatory frameworks has been instrumental in reducing pollutants. For instance, studies have shown that areas implementing strict air quality standards are likely to note significant reduction of photochemical smog, and it directly correlates with policy efficacy and the improvement in the health of the population (James et al., 2025). Further, an application of the photochemical methodology promotes a more environmental approach to managing nature. Using the understanding of dynamic interactions between atmospheric chemical compounds, researchers are now capable of creating targeted interventions that not only mitigate air pollution but also enhance the quality of urban living. For instance, initiatives for mitigating VOC emission from industries can lead to improved air quality outcomes. In addition, community awareness about the causes and consequences of air pollution enables better public involvement and promotion of clean air campaigns. To this end, education and community activism become necessary to complement the effectiveness of photochemical methods. By the implementation of these strategies, society can work towards sustainability in order to mitigate the health impacts of depleting air quality and increase environmental resilience (Memoriam et al., 2023) (Figure 6.1).

Definition of Air Pollution

Air pollution is a serious ecological concern, which is the presence of harmful substances in the air that can inflict damage to human health, the environment, and the climate. They can be particulate or gaseous and can be produced through as diverse sources as industrial operations, motor vehicle emissions, agricultural operations, and natural processes. The complexity and diversity of air pollution are underscored by its various components such as from soot and dust to VOCs, NO_x, and SO_2. Each pollutant has varying health implications and effects on environmental deterioration, and it is thus necessary that their isolated and combined effects on the environment are assessed. A strong understanding of such factors goes towards the formulation of effective policies and technologies for the control and reduction of air pollution. Photochemical air pollution is in fact linked with reactions occurring in the atmosphere while being exposed to sunlight and results in the formation of secondary pollutants like ozone. Ground-level ozone is one of the main constituents of smog and is a major health hazard, causing respiratory issues and other health issues. Ground-level ozone formation takes place through intricate photochemical interactions of VOCs and NO_x, which can be emitted from many sources such as vehicles and industrial activities. Knowledge of the turbulent dynamics of these pollutants is necessary for designing effective control schemes. One of the techniques is by using knowledge acquired from the study of photochemical ozone formation potentials (POCPs), which provide the relative contribution of different VOCs to photochemical pollution formation. Such research is important for establishing regional variation in air quality and assisting in designing suitable emissions reducing strategies (Lama et al., 2013). In attempting to decrease air pollution, particularly photochemical air pollutants, the enhancement of agricultural practices has been a successful path. Agriculture per se, as indicated, serves both as a consumer and supplier of energy, affecting air quality in terms of its inputs of resources. Rational energy use in farming practices can result in greenhouses gas emission and other pollutant reduction and thereby contribute to sustainable production. Besides, using life cycle assessment techniques can provide a foundation for evaluating the environmental impacts of agricultural inputs throughout their life cycles and determining areas where emissions can be lowered. Once there is an understanding of the relationship between farm practices and air pollution, stakeholders can make well-integrated plans that not only enhance productivity but also target the enhancement of air quality. Hence, agriculture's dual purposes showcase the potential for innovation in the fight against air pollution while fulfilling the ever-increasing swelling needs imposed upon food production (مجید خانعلی et al., 2016).

Importance of Air Quality Management

Air quality management is of primary importance in the current environmental context since it directly affects the public health, ecological health, and climate regimes worldwide. The rising tendency towards air contaminants such as fine particulate matter ($PM_{2.5}$) and volatile organic compounds (VOCs) has established health risks, which have led to respiratory diseases, cardiovascular disorders, and increased mortality rates. Effective air quality management not only minimizes such health issues but also helps in sustaining cities. Photochemical pathways are pivotal to this management strategy, enabling the degradation of harmful substances with sophisticated processes like photocatalytic oxidation. By the enhancement of the efficiency of pollution control technologies, such methods aim to deliver cleaner air and thus improve the quality of life for populations that are exposed to airborne toxins. Additionally, the integration of emerging technologies into air quality management systems is required for regulatory compliance as well as achieving environmental goals. Satellite data and other surveillance techniques have revolutionized air quality knowledge such that it is now possible to discern trends and sources of pollution over time. For example, recent studies using Sentinel-5P satellite data illustrated the complex interactions between shipping and pollutants like nitrogen dioxide (NO_2) and sulfur dioxide (SO_2) in coastal areas. Such data is invaluable for the development of efficient intervention policies, including emission limits and anti-pollution controls. The ability to quantify the spatial scope of pollutants is essential in developing efficient strategies in reversing air quality deterioration, highlighting the fact that efficient management is an outcome of comprehensive data-driven policies that adapt to evolving environmental challenges (Yan et al., 2025). With respect to the use of photochemical technologies, one has to take into account the existing shortcomings of air quality control, particularly in terms of the operational efficiency and economic viability of these technologies.

According to literature, techniques like ultraviolet (UV) enhanced pyrolysis are promising for pollutant decomposition, although energy consumption as well as secondary pollutant formation remain issues. These problems are overcome by streamlining reactor designs and developing cost-effective sources of UV for mass use. Mutual progress in new catalytic materials also results in higher performance and reduced cost of operation. As the need for cleaner air continues to escalate globally, continued research and technological improvements will play a key role in enhancing air quality management measures, ultimately contributing to healthier communities and more robust ecosystems (Dhanuk et al., 2025) (Figure 6.2).

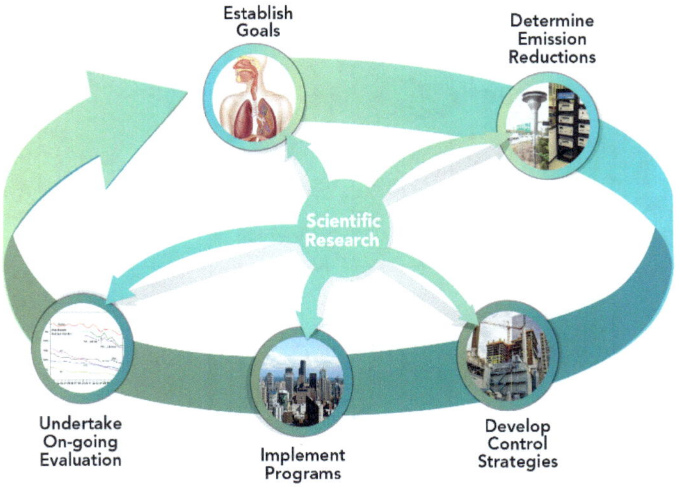

Figure 6.2. *Importance of air quality management.*

Source: https://19january2017snapshot.epa.gov/sites/production/files/2016-04/aqm_draft5.jpg.

Overview of Photochemical Methods

A thorough understanding of photochemical methods is required in order to fight the complexities of air pollution abatement. These methods harness the energy from light to power chemical reactions that will break down pollutants. Of particular interest are those methods involving the photolysis of obnoxious materials, in which light energy causes molecules to break into less harmful components. This is particularly important in urban areas where pollutants like volatile organic compounds (VOCs) lead to the formation of smog. Photochemical reactions are effective based on light intensity, the availability of catalysts, and the concentration of the pollutants themselves. Therefore, progress in photochemical technology is essential for the formulation of efficient air purification plans, highlighting the significance of continued research in environmental science. The role of photochemical methods in the general context of environmental regulation signifies their relevance in policy processes.

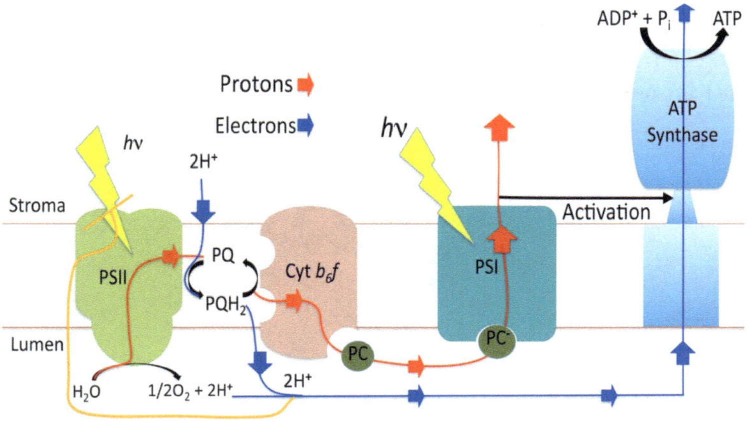

Figure 6.3. *Overview of photochemical methods.*

Source: https://www.pnas.org/cms/10.1073/pnas.1211017109/asset/5d9797f3-1db3-423b-8307-f344dc311915/assets/graphic/pnas.1211017109fig1.jpeg.

Through the history of air pollution control, photochemical reactions inform our knowledge regarding standards of air quality and public health concerns. Specifically, events in the past, such as the event of photochemical smog observed in Los Angeles, have led to laws aimed at preventing atmospheric pollutants that trigger such a reaction. Decade-by-decade systematic record of efforts in air pollution control, especially in U.S. legislation such as the Clean Air Act, reflects the growing concern of society towards the hazards created through contamination of the air. Through integration of results from earlier studies, available methods can be formulated to better maximize their real-life applications in an attempt to keep pace with evolving regulation needs and health requirements (Wang, 2024). Additionally, the innovative use of photochemical methods renders the development of measures specifically aimed at condensable particulate matter (CPM), particularly its organic component, whose environmental concerns are severe. CPM has the capability to increase the level of pollution and induce climate change, and hence intensive remediation needs to be called for. Photochemical degradation techniques, especially when combined with catalytic oxidation processes, have proved capable of reducing the air concentration of such pollutants. Integration of these techniques facilitates the simultaneous reduction of multiple air pollutants, keeping pace with the growing necessity for control measures on multiple dimensions. These advancements represent an important convergence of chemistry and environmental technology, paving the way for sustainable air pollution control (Peng et al., 2024) (Figure 6.3).

6.1. UNDERSTANDING PHOTOCHEMICAL REACTIONS

The understanding of photochemical reactions is vital to resolve the problem of air pollution, particularly in urban areas that are full of photochemical smog. In essence, photochemistry is founded on interaction between chemicals and light, leading to the formation of new chemical substances. This is best explained by observing VOC's interaction with sunlight. For example, in regions like the Pearl River Delta, photochemical trajectory models have demonstrated how VOCs act under sunlight to create ozone. The tropospheric ozone structure is a central component of air quality dynamics that performs the double function of an urban pollutant and a natural atmospheric process, emphasizing the duality of its role in urban air pollution (Blake et al., 2010). By deconstructing discrete components of photochemical reactions, we can identify the pathways through which pollutants are formed and develop. The partitioning of VOCs on the basis of reactivity is particularly germane; alkenes and aromatics possess considerably higher reactive potential than alkanes. Such data are invaluable when designing particular pollution control methods because optimum strategies must consider the dominant oxygenated organic compounds that contribute to ozone formation. Moreover, longitudinal investigations, such as under the Megacity Air Pollution Study-Seoul (MAPS-Seoul), further identify the photochemical ozone formation processes. During some seasonal months, the studies establish that the relative contributions of VOCs and nitrogen oxides (NO_x) vary, which influences the overall effectiveness of removal processes for ozone (Ahn et al., 2018). Photochemical reactions are also about comprehending the climatic and environmental factors that influence them. The speed of these reactions is governed by conditions like temperature, humidity, and even surrounding geography that affects the movement of air masses. For instance, stagnant meteorological conditions can significantly alter the photochemical environment, which makes it possible for pollutants to accumulate during summer months. Such a complicated relationship necessitates advanced modeling methods, like photochemical models, which enable one to predict air quality outcomes in terms of atmospheric chemistry. These models integrate the behavior of numerous distinct pollutants, giving policymakers important information on how to develop effective, photochemical-based methods of mitigating the adverse effects of air pollution (Blake et al., 2010).

6.1.1. Definition and Principles of Photochemistry

Photochemistry, in short, is the study of the chemical transformations brought about by light interaction. It is founded on the processes of quantum mechanics, which explain how photons, the constituent units of light, may impart energy to molecules and induce changes in chemistry. Molecules, by absorbing specific wavelengths of light, get excited into a position where bonds are broken or new ones created. This principle is of critical importance to environmental science and specifically to the context of air pollution control, where photochemical reactions are used to break down harmful pollutants in the air. Through an understanding of these processes, it is possible to design photochemical methods for reducing air quality problems, highlighting the importance of total knowledge of the molecular interactions initiated through light and their resultant consequences. The principles of photochemistry are based on several basic concepts, including that of excitation and the nature of photochemical reactions. When a molecule absorbs a photon, it may undergo electronic changes, often shifting from a ground state to one of excitation. It commonly results in molecular structure changes, sometimes generating free radicals or other active species. These reactions may either lead to the degradation of air pollutants or, conversely, the production of secondary pollutants that exacerbate environmental concerns. For instance, the photodegradation of volatile organic compounds (VOCs) through sunlight-catalyzed reactions can convert the precursors into secondary pollutants such as ozone, which are detrimental to air quality. Thus, understanding the interaction between excitation energy and subsequent chemical processes is central to modeling and predicting the environmental impact of a variety of pollutants. Moreover, the application of photochemical concepts in the control of air pollution has gained immense importance in recent times. Techniques such as photocatalysis are able to exploit these principles efficiently using light-activated catalysts to photoinduce and speed up the degradation of the pollutant. This strategy demonstrates possible advantages over traditional chemical treatment, such as less energy requirement and innocuous byproduct formation. Advances in materials science, together with improved reaction mechanism understanding, have contributed to more efficient photocatalytic material. For example, titanium dioxide (TiO_2) has emerged as a prominent photocatalyst due to its activity and stability under UV illumination. As established in studies, the organic aerosol transformation chemistry and environmental conditions controlling such reactions over time need to be taken into account in the design of photochemical technologies for air pollution control, stressing the supreme importance of embracing a high-level understanding of photochemistry in ecological planning (Feng et al., 2023; Pagola, 2023) (Figure 6.4).

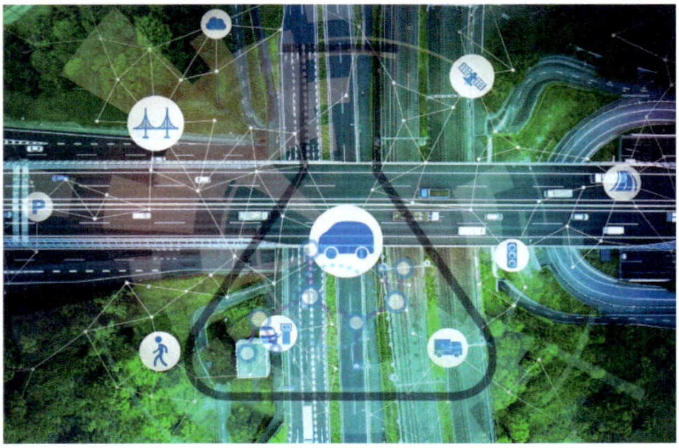

Figure 6.4. *Principles of photochemistry.*

Source: https://blog.phoseon.com/hubfs/photochemistryapplication.png.

6.1.2. Types of Photochemical Reactions Relevant to Air Pollution

The intricacy of air pollution is intrinsically tied to the types of photochemical reactions in the air, including the transformation of volatile organic compounds (VOCs) into secondary pollutants. Biogenic volatile organic compounds (BVOCs) emitted by vegetation are a key component in this process. Upon exposure to solar radiation, photochemical oxidation is possible, leading to the formation of ozone and secondary organic aerosols, which are detrimental to air quality and human health. Understanding mechanisms and variability of BVOC emissions is crucial in the context of effective air pollution management. Currently available measurement methods for BVOC emissions, while sophisticated, require refinement to achieve an optimal balance between simplicity, cost, and precision. These additions are essential to design better predictive models concerning VOc behavior and photochemical reactions, which are essential in implementing strategic controls to combat secondary air pollution effectively (Luo et al., 2025). Another important category of photochemical reactions of relevance in the context of air pollution is the formation of ozone through precursors such as nitrogen oxides (NO_x) and hydrocarbons. In urban areas, strong sunlight facilitates these processes, causing ground-level ozone to be formed, a principal component of smog. The photolytic breakdown of NO_x is one step in a series of processes by which these nitrogen compounds are oxidized to ozone when they react with VOCs. Seasonal trends have been observed, particularly in urban areas like Shanghai, where the warmer months of the year are associated with high levels of ozone due to greater photochemical activity. Therefore, understanding the dynamics of these reactions is crucial in designing focused air quality management policies, particularly during maximum ozone formation periods (Johnson, 2025). These include regulatory measures to limit emissions of vehicle and industrial emissions at such critical times. The interplay of different photochemical reactions highlights the importance of integrated air quality monitoring and modeling in addressing the distinctive challenges posed by air pollution. Methods that use measurement and modeling will enhance the explanation of how several factors such as meteorological situations and human behavior impact photochemical processes. For instance, interdisciplinary methods using advanced technologies can lead to high-resolution models having the capability of making predictions on the impacts of individual emissions on air quality. Besides, novel VOCs and NO_x emissions monitoring and controlling techniques would significantly contribute to the mitigation of ground-level ozone and particulate matter formation. Science-backed and proactive steps should be initiated, particularly in very sensitive cities, to minimize single and combined impacts of these photochemical reactions on public health and the environment (Luo et al., 2025; Johnson, 2025).

6.1.3. Role of Sunlight in Photochemical Processes

The photochemical action of sunlight is at the focal point of air pollution control. Sunlight provides the energy needed to initiate a series of atmospheric reactions, primarily between volatile organic compounds (VOCs) and nitrogen oxides (NO_x). When solar radiation interacts with such pollutants, it initiates the formation of secondary pollutants such as ozone (O_3), both a byproduct and source of poor air quality. The rate and effectiveness of these processes can vary with intensity and wavelength of sunlight, highlighting the need to understand the photochemical dynamics in different environmental conditions. As such, the utilization of sunlight by way of photochemical processes can result in innovative methods of mitigating air pollutants, improving air quality and public health effects. In addition, the effect of sunlight goes beyond the mere act of initiating reactions; it also plays a significant role in the degradative processes of harmful substances. Photolysis, which is the degradation of chemical bonds by light energy, is required for the abatement of air pollutants such as particulate matter and VOCs. Temperature inversions in the urban setting are a contributing factor towards air pollution, leading to an increase in ozone content.

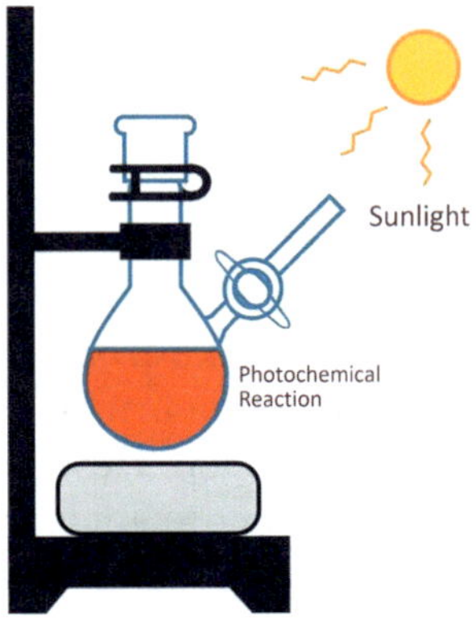

Figure 6.5. *Role of sunlight in photochemical processes.*

Source: https://chemistrytalk.org/wp-content/uploads/2024/09/image-2.png.

Under circumstances of high sun exposure, photochemical reactions are accountable for the rapid formation of ozone, which becomes a health threat to the populace. For instance, research in the Beijing–Tianjin–Hebei region demonstrated a strong correlation of increased ambient temperatures and high levels of ozone. It is important to note the influence of atmospheric conditions on photochemical reactions, particularly in relation to sunlight, in order to develop effective strategies to mitigate ozone pollution (Li et al., 2022).

Apart from the control of air quality, the role of sunlight in photochemical reactions is vital in city planning and implementing sustainable urban policies. Urban areas can be well planned to be exposed to the sun as much as possible, hence maximizing the efficiency of photochemical reactions that remove air pollution. Technological advancements such as green roofs, urban vegetation, and photochemical reactive surfaces that have the ability to absorb sunlight to break down pollutants can transform urban ecosystems into cleaner, healthier spaces.

Furthermore, the inclusion of publicly available spaces for the purpose of prioritizing sunlight accessibility can also push the community to actively participate in air quality campaigns. As urban areas keep on developing, the utilization of sunlight's ability in photochemical reactions will be critical towards realizing sustainable air quality objectives, proving that environmental stewardship and city growth can complement each other in perfect harmony (Priyadarshini, 2023) (Figure 6.5).

6.2. MECHANISMS OF PHOTOCHEMICAL AIR POLLUTION CONTROL

The mechanisms of photochemical control of air pollution depend on the understanding of the complicated reactions that occur when air contaminants are exposed to sun. The essence of this process is biogenic volatile organic compounds (BVOCs), which are produced by plants and contribute significantly to the formation of secondary pollutants like ozone. BVOCs undergo complicated photochemical reactions based on environmental parameters like temperature, humidity, and intensities of solar radiation. However, accurate description and quantitation of BVOC emissions also pose a major challenge in successful air quality management. Current measuring methods are often not precise or cost-effective enough, which necessitates the development of advanced measurement techniques and high-resolution atmospheric chemistry models. Such innovation is critical for accurate assessment of the contribution of BVOCs to air pollution processes and developing effective mitigation options that can help to constrain secondary pollution and enhance the quality of urban air (Luo et al., 2025). In urban areas, the integration of advanced technologies can improve the effectiveness of photochemical air pollution management methods. The new use of artificial intelligence (AI) in pollution management is very promising, as it can facilitate predictive modeling and advanced visualization techniques. Through the use of AI, researchers can analyze large volumes of data to predict patterns of pollution and identify the most effective intervention strategies. Real-time monitoring and response to fluctuating air conditions are made possible through this new approach, optimizing waste management and air pollution control. Moreover, citizen participation through this process empowered by decision-making tools that utilize data makes the process a collaborative effort towards the creation of healthier cities. Therefore, the use of AI in conjunction with conventional photochemical techniques can particularly maximize urban resilience to air pollution as well as spur the creation of sustainable, intelligent cities that value air quality (Chandran et al., 2024). In addition, strategic landscaping with low-BVOC plant species in urban environments is an essential element of a comprehensive strategy for photochemical air pollution abatement. By establishing and sustaining such cover in polluted areas, cities will cut down on the harmful effects of secondary pollutants like ozone. This form of biological control complements strict anthropogenic volatile organic compounds (AVOC) regulation, providing a balanced approach that harmonizes urban development with ecosystem integrity. Strict ecological analysis is required to ascertain that selected species not only reduce VOC emissions but also contribute positively to the urban ecosystem. The inclusion of green infrastructure, such as green roofs and urban forests, can add vigor to this strategy, fostering biodiversity in addition to keeping air cleaner simultaneously. Lastly, the combination of scientific investigation and urban planning will be essential for keeping photochemical air pollution under effective control (Luo et al., 2025) (Figure 6.6).

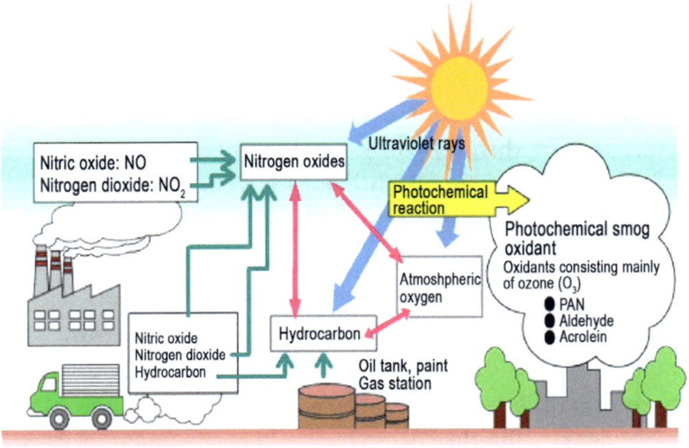

Figure 6.6. *Mechanisms of photochemical air pollution control.*

Source: https://theshawclassroom.weebly.com/uploads/1/5/3/6/15361694/612413222_orig.jpg.

6.2.1. Photocatalysis and Its Applications

Photocatalysis has emerged as a fundamental technology in air pollution control due to the ability of photocatalysts to use light energy to power chemical reactions. In ultraviolet or visible light, materials of photocatalysts, typically semiconducting materials like titanium dioxide, are energized to power the reactions capable of degrading poisonous pollutants into non-toxic compounds. The most promising green applications of photocatalysis are, as it offers a clean means of addressing air pollution issues without the need for massive chemical additives. These complements growing interest in green technology that does not disrupt the environment but effectively fights pollution. The efficiency of photocatalytic systems largely relies on the generation of advanced materials that are able to optimize light absorption and generation of reactive species, thereby attaining higher overall reaction rates and efficiencies in pollutant degradation (Yue et al., 2021). Besides simple air purification, photocatalysis also has other applications such as water treatment and self-cleaning materials. The capacity to decompose volatile organic compounds (VOCs), which are among the largest contributors of indoor air pollutants, points to its potential for application in enhancing urban city air quality where the compounds have a high prevalence. The study has also proved that the integration of photocatalytic reactions in building material can design self-cleaning surfaces that reduce the amount of pollutant accumulation over time (Singh et al., 2018). These developments not only promote a healthier environment but also building longevity and beauty. As the increasing demand for cleaner air continues, photocatalytic applications will be a determining role in the development of innovative solutions in addressing pollution. The broader context of photocatalysis in preventing air pollution can be shown through its complementarity with renewable energy sources, especially solar energy. Such synergy indicates the prospect of a green future, with solar-powered photocatalytic systems poised to be efficient air cleaners with minimal reliance on traditional energy-dependent approaches. Additionally, ongoing research into biocompatible and eco-friendly synthesis routes to photocatalytic materials, as indicated by green synthesis mechanisms, leads to more environmentally friendly fabrication processes (Yue et al., 2021). Developments in this area not only improve the performance of photocatalysts but also add towards worldwide efforts towards more sustainable manufacturing methods. For this reason, photocatalysis is a change that addresses not only mitigating air pollution but also heading towards a cleaner and greener world.

6.2.2. Photochemical Oxidation Processes

The photochemical oxidation reactions have significant implications for controlling air pollution, because they occur through a series of complex light-induced chemical processes. These processes are crucial mechanisms by which pollutants are oxidized to become fewer toxic substances, somewhat diminishing their impacts on human health and the environment. Photochemical reactions occur primarily in the atmosphere where sun's radiation interacts with volatile organic compounds and nitrogen oxides, among other gas and particulate matter pollutants. The sun's energy input leads to the formation of reactive radicals that oxidize and break down the pollutants. This conversion not only eliminates harmful compounds by depolarization but is also involved in the formation of secondary pollutants, such as ozone, with adverse implications if not well regulated within urban areas. The dynamic character of photochemical oxidative processes suggests the importance of identifying the environmental conditions of occurrence of the reaction. All of these issues can critically influence the efficiency of these processes.

For instance, the Los Angeles photochemical smog episode demonstrates how local conditions can cause an enhancement in secondary pollutant formation, requiring stringent regulatory control. As air pollution control research evolved over the years, scholars have become aware that photochemical processes serve two purposes: while they can break down certain pollutants, they also have the ability to form new, and possibly harmful, substances. Hence, specific frameworks such as those outlined within the Clean Air Act have become essential for managing the complex photochemical reaction equilibrium within urban areas so that air standards are maintained for public health and safety purposes. Photochemical oxidation process-based engineering solutions have emerged as possible mechanisms in combating air pollution in recent advances. Advances in sensor technology have made possible real-time observation of releases of pollutants to enable intervention to limit release of toxic materials from various sources.

For example, photochemical oxidation equipment can be integrated into energy conversion processes for capturing and detoxing releases before discharge into the atmosphere. This approach not only demonstrates the innovative application of photochemical methodologies but also underscores the relevance of adaptive governance methods that are shaped by varying pollution conditions.

As characterized in available scholarly research studies, e.g., those discussed in the context of combustion process-controlling systems, these advancements signify a major leap in achieving sustainable urban air quality regulation, hence highlighting the central role of photochemical processes in the broader context of air pollution control (Armenta, 2025; Wang, 2024) (Figure 6.7).

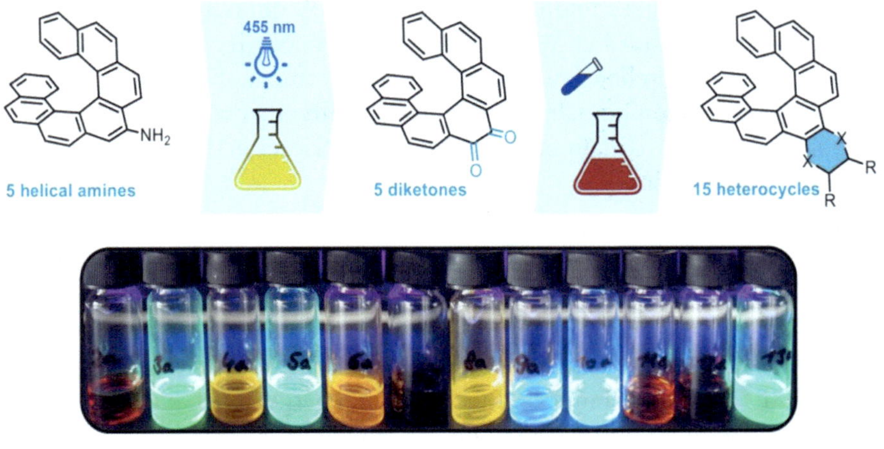

Figure 6.7. *Photochemical oxidation processes.*

Source: https://www.icpf.cas.cz/wp-content/uploads/2020/03/Fotochemicka_oxidace-1024x492.jpg.

6.2.3. Role of Reactive Oxygen Species in Pollutant Degradation

Air pollutant oxidation through photochemical reactions predominantly relies on the generation and activity of reactive oxygen species (ROS). The extremely reactive molecules like singlet oxygen, hydroxyl radicals, and superoxide anions form the backbone of the degradation of a wide variety of pollutants like volatile organic compounds (VOCs) and particulate matter. With mechanisms such as photocatalytic oxidation, ROS can initiate complex chemical reactions that degrade harmful pollutants into less toxic materials. Not only is the air quality enhanced, but also potential danger to health through long-term exposure to these toxic pollutants is lessened. Photocatalysts use visible or UV light to generate ROS from oxygen or water and thus effectively power the degradation process. An understanding of the way ROS interact with other air pollutants is important as it can inform the design of more efficient photochemical systems for air pollution abatement. How well reactive oxygen species can break down pollutants is evident in various studies, as much emphasis should be placed on further studying their potential.

The application of advanced oxidation processes (AOPs) has proved ROS can effectively oxidize organic molecules, leading to total mineralization in some cases. Some contaminants, like harmful substances released during industrial processes, have been shown to break down effectively using AOPs that generate ROS. These findings suggest that ROS not only play central roles in pollutant degradation but also underscore the necessity of integrating such processes into comprehensive air quality management. For instance, hybrid systems that combine photochemical methods with other technologies, such as pyrolysis or thermal oxidation, can enhance ROS production, thereby overall pollutant removal efficacy.

Such hybrid systems present new avenues for addressing air pollution issues more efficiently (Dhanuk et al., 2025). Although the promise for pollutant degradation through reactive oxygen species is great, there are relevant issues to be addressed. By-product formation during ROS-mediated degradation has in some instances created secondary pollution scenarios, raising concerns over the overall safety of such techniques. Furthermore, energy input and photocatalyst stability to facilitate the production of effective ROS remain important factors influencing the practicality of these technologies. As an example, it has been proven that certain photocatalysts have been discovered to be prone to deactivation when used over the long term due to the accumulation of contaminants on their surfaces, lowering their efficiency (Wardana et al., 2025). Thus, sustained research aimed at enhancing catalytic materials and identifying suitable environmental conditions will be essential towards optimizing the benefits of ROS in air pollution control. Mitigation of such challenges will be essential towards making photochemical processes with ROS contribute their share to sustainable cleaner air initiatives.

6.3. PHOTOCHEMICAL METHODS IN INDUSTRIAL APPLICATIONS

The creation of photochemical methods for commercial application highlights a key shift in air pollution control technology. Light energy is employed by photochemical processes to initiate chemical reactions that are highly effective in breaking down poisonous pollutants. This novel method not only profoundly enhances the efficiency of VOC and other poisonous emission degradation but also can be readily integrated into existing industrial systems. Employment of solar power in these processes suggests a mechanism that is green to reduce reliance on conventional sources of energy. Moreover, with industries facing tougher environmental regulations on emissions, utilization of photochemical processes suggests a promising direction towards compliance at no exorbitant operating cost. This suggests that industries can have greater care for the environment while still being economically sustainable too, hence fostering business culture towards sustainability and responsibility. After looking at the practical applications of photochemical processes, it is evident that their use has a monumental significance in air quality control. For instance, photochemical technology has been used in effluent treatment and air pollution removal at petrochemical industries with great effectiveness. Industries can also convert lethal chemicals into a less harmful form using photochemical reactions, significantly minimizing their impact on the environment. Scientists have proposed using satellite systems for monitoring the effectiveness of such measures so that data collection records fluctuations in air quality caused by industrial processes (N/A, 2025). The blending of photochemical technology with advanced monitoring systems enables regulatory bodies and industries to promote accountability and transparency towards pollution control, thus enhancing public confidence in industrial processes. In addition to this, the application of advanced computational techniques such as finite element modeling facilitates understanding of photochemical process dynamics. Pollutant transport and reaction mechanism simulation related to complex terrains is enabled by these techniques, thereby improving the design and operation of air pollution control strategies (Escobar et al., 2012). By modeling situations with composite atmospheric conditions and source types of pollutants, industries are able to predict the effectiveness of photochemical intervention and plan their strategy accordingly. This proactive strategy not only works towards reducing short-term pollution issues but also works towards improving long-term air quality. The use of photochemical techniques in industrial applications, thus, reflects a qualitative jump in solving air pollution and indicates the merits of innovation, adaptability, and increased environmental protection (Figure 6.8).

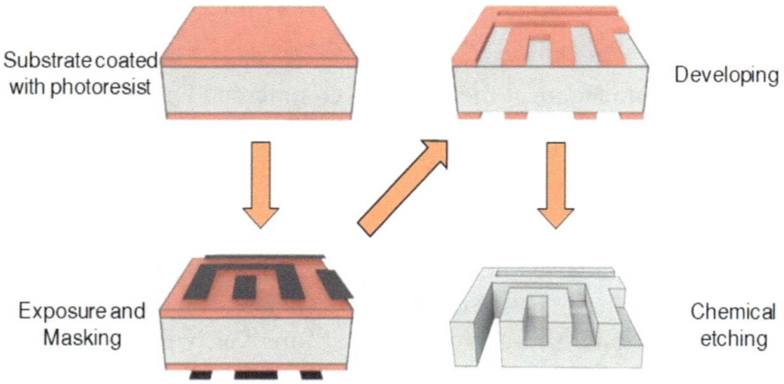

Figure 6.8. *Photochemical methods in industrial applications.*

Source: https://fotofab.com/wp-content/uploads/2021/04/basic-photochemical-etching-process-opt.jpg.

6.3.1. Use of Photochemical Methods in Waste Treatment

The utilization of photochemical methods in waste treatment exemplifies an innovative approach aimed at mitigating environmental pollution, particularly in urban settings where waste generation is rapidly increasing. These methods leverage ultraviolet (UV) light and advanced oxidation processes to decompose hazardous substances found in wastewater, effectively converting them into less harmful materials. For instance, photochemical oxidation can break down organic pollutants and heavy metals, rendering them less toxic and easier to handle. The importance of such an approach is underscored by the estimated urbanization rates hikes that raise the stakes on waste and air quality management efficiency, as argued in recent studies of the urgent necessity for sustainable city planning and waste treatment processes (Ren et al., 2020). Thus, photochemical treatment is not only a method of waste treatment but also contributes to broad objectives of air pollution control by inhibiting emissions from waste treatment operations. The suitability of photochemical waste treatment is in direct proportion to its ability to reduce specific contaminants that are prevalent in industrial effluent and urban runoff. For example, sophisticated photochemical destruction of volatile organic compounds (VOCs) has been of interest because it is effective. Utilizing UV irradiation, such methods allow the destruction of complex organic matter into simpler, harmless forms. Moreover, as illustrated in zeolites synthesized from coal fly ash studies, the materials can catalyze or adsorb the photochemical processes. This dual role by zeolites allows the simultaneous treatment of air and solid waste pollution and thus presents an integrated method of pollutant control. By focusing on synergistic application development and optimization, the effectiveness of photochemical processes towards waste treatment can significantly improve environmental effects and air quality (Ren et al., 2020). Furthermore, integration of photochemical processes into the existing infrastructure of waste management not only ensures environmental sustainability but also offers economic advantages. As urban areas have mounting waste treatment costs due to higher population densities, the use of photochemical technologies is a less expensive alternative to conventional processes. By harnessing power from solar energy as a source for treatment, operating expenses can be greatly reduced, allowing resource-poor cities to access them at lower costs. Second, reduction of harmful emissions by enhanced waste treatment not only meets environmental regulations but enhances public health, a parameter that is emerging as a key factor in urban pollution control strategies. As cities emerge and aim for sustainable development goals, integration of advanced photochemical techniques is of critical interest in addressing the interrelated challenges of waste disposal and air quality (Wang et al., 2013).

6.3.2. Implementation in Power Plants and Manufacturing

Application of advanced technologies in power plants is crucial to prevent air pollution, particularly nitrogen oxides (NO_x) that are produced during combustion. Nitrogen oxides are credited with the production of ground-level ozone as well as acid rain, which are grave environmental threats. Because of stringent emission standards, the majority of coal-fired thermal power plants have adopted technologies like selective non-catalytic reduction (SNCR). For instance, ammonia reduction agent usage needs precise control to prevent unstable operation and achieve optimal NO_x reduction effectiveness. Specifically, the case of the Nikola Tesla A plant highlights how temperature mapping solutions can better optimize ammonia injection so that it only occurs in suitable thermal regions, thereby achieving NO_x below European standards' regulatory levels. This particular strategy is the most effective demonstration of how advanced technologies used in power production can effectively address pollution concerns (Vladimir et al., 2024). Production procedures are also critical in terms of air pollution control, especially as companies continue to aim for a positive environmental outcome. Developing new production methods that align with green practice can be a significant factor in emissions mitigation. One such promising technology is the use of solar tower power plants, which incorporate open volumetric receivers based on air as a heat transfer fluid. This not only improves thermal efficiency in energy conversion but also decreases the reliance on fossil fuel resources, thereby decreasing toxic emissions. The use of 3D printing to fabricate intricate porous geometries in such solar receivers allows for optimal design of absorbers according to specific manufacturing constraints.

Figure 6.9. *Implementation in power plants and manufacturing.*

Source: https://s3.amazonaws.com/s3-biz4intellia/images/benefits-of-energy-monitoring.jpg.

These developments thus represent the intersection of clean energy technologies and manufacturing methodologies that focus on air quality improvement without sacrificing operational efficiency (Broeske et al., 2024). Moreover, the interaction between power stations and advanced manufacturing technologies offers an avenue for environmentally friendly growth. The utilization of photochemical processes, such as those involving new materials and absorptive technologies, could supplement existing pollution control devices. For example, the incorporation of optimized absorbers in manufacturing plant processes could enhance the trapping of volatile organic compounds (VOCs) before emission to the environment. Through the combination of different photochemical methods, industries are not only able to comply with environmental standards but ultimately contribute towards the minimization of greenhouse gases. The integrated implementation process favors industry-to-industry collaboration, encouraging innovation towards the generation of cleaner energy as well as manufacturing solutions. As industries continue to evolve, technological integration and evolution of manufacturing and power generation technologies shall be the key to the achievement of a more sustainable world through the drastic reduction of the air pollution level (Vladimir et al., 2024; Broeske et al., 2024) (Figure 6.9).

6.3.3. Case Studies of Successful Industrial Applications

The use of photochemical technologies in air pollution management has witnessed considerable advancements that are exemplified through various successful industrial applications. An example is the use of advanced oxidation processes (AOPs) to treat volatile organic compounds (VOCs) emitted as a result of industrial activities. Industrial facilities located in geographies with strict air quality regulations have incorporated photocatalytic oxidation equipment, which uses ultraviolet (UV) light to activate photocatalysts that degrade organic pollutants. This approach, aside from its role in the enhanced VOC degradation, also minimizes secondary pollutants formation and thus promotes an industrial model that is sustainable. Case studies demonstrate that these technologies have been able to reduce VOC emissions significantly, thus improving environmental standards compliance. Another intriguing case study is found in the Bukhara province where innovative photochemical methods have been applied to address air pollution problems arising from industrial effluents. In this instance, comprehensive assessments located major sources of pollution, and the application of photochemical air purification technology aimed to minimize harm to public health and local ecosystems. With the addition of UV-based photochemical reactors and advanced systems of filtration, industries have successfully turned toxic pollutants into more benign byproducts. This is an example of the applicability of utilizing photochemical methods as a functional process for regions with unforeseen problems of pollution, ultimately resulting in a cleaner air environment that is beneficial to humans as well as the economy (Hotamova et al., 2024). One such critical initiative comprises the development of air pollution susceptibility maps based on the use of photochemical values to delineate areas most susceptible to industrial emissions. Scholars, together as a team, have conducted a study wherein multifactorial air pollution indicators were integrated into a comprehensive index, which served as the basis for formulation of intervention measures. The approach has been positively utilized in urban as well as peri-urban areas, wherein industrial emissions are dense in concentration. The mapping exercise allows ranking pollution control measures based on particular environmental exposure factors, culminating in more efficient monitoring and regulation. Based on this methodology, case studies emphasize its utility in catalyzing advancements in air quality management and policy decision-making aimed at curbing pollution-intensive activities (Aniceto et al., 2025).

6.4. ADVANTAGES OF PHOTOCHEMICAL METHODS

The advantages of photochemical methods in controlling air pollution are numerous, particularly when considered in the context of their ability to eliminate air pollutants under ambient conditions. Of prime benefit is the fact that they harness sunlight to trigger chemical reactions, and hence they are inexpensive and eco-friendly. In cities, where air pollution is generally strong, the natural resources of sunlight can be used to initiate the breakdown of harmful chemicals, such as VOCs and BVOCs. Photochemical reactions allow the possibility of such pollutants to be decomposed into less harmful chemicals, ultimately improving air quality without the need for extensive energy input or expensive infrastructure. As pointed out in current research, addressing the issues of BVOC emissions can significantly maximize air pollution control and enable environmental sustainability, which is in harmony with current ecological and public health objectives (Luo et al., 2025). In addition, photochemical methods can also effectively be integrated with modern monitoring devices, which present real-time data to control pollution means. The development of wireless sensor networks (WSNs) and cloud computing has made possible dynamic monitoring, and it provides information-based insights on trends in air quality. With the deployment of low-cost sensors, critical data on polluters such as particulate matter (PM), carbon monoxide (CO), and nitrogen dioxide (NO_2) can be gathered and analyzed. The integration of this information with photochemical technology allows predictive analysis, such that intervention is targeted, and a significant decrease in the levels of pollution occurs. Moreover, the feedback loop provided by these technologies allows adaptive management approaches, such that cities can inform decision-making on urban development and pollution control policy, thus allowing a more interconnected approach to air quality management (Ranganathan et al., 2024). Last but not least, photochemical processes are highly versatile for treating many forms of pollutants and therefore can be employed in many different environments. Their use varies from large-scale industrial application to localized and community-based small-scale use. In their interaction with many contaminants in many environmental conditions, photochemical processes have the capability to significantly enhance the effectiveness of air pollution control systems. In addition, their power to address multiple pollutants simultaneously presents an integrated solution to the resolution of advanced air quality issues. While cities continue to grapple with both the double-edged sword of urban air pollution and climate change, the adoption of photochemical methods can be an enlightened way of cultivating healthier cities. Thus, by leveraging the inherent advantages of such methods, stakeholders can work towards healthy solutions that are at once good for the public health as well as environmental integrity.

6.4.1. Environmental Benefits

The use of photochemical methods in air pollution control is environmentally significant with respect to more than mere reductions in emissions. The methods, which harness solar energy to power chemical reactions, provide an innovative alternative to traditional pollution control methods. The methods can effectively degrade hazardous pollutants like VOCs through advanced oxidation processes. In addition, utilization of photochemical systems could facilitate on-site air purification, and thereby further enhance community well-being while preserve biodiversity. Through embracing strategic application of these techniques, cities will be able to mitigate smog and improve the quality of air as a whole, resulting in a healthier city overall. Such a measure is supplementary to broader sustainability initiatives that try to minimize carbon signatures and promote ecological sustainability, thus placing photochemical strategies as an important contribution towards present-day air pollution mitigation measures. Moreover, employing photochemical air pollution control systems can yield considerable co-benefits of climate change mitigation. Since cities are confronting the twin challenges of air pollution and greenhouse gas emissions increasingly, the incorporation of potent technologies is really essential. For instance, research highlights the fact that modest policy shifts towards pollution control could have the effect of spectacular emission reduction of not only CO_2 but also particulate matter (PM) and toxic nitrogen oxides (NO_x) (Chen et al., 2021). Such comprehensive measures prioritize environmental health and support fulfillment of world carbon neutrality targets. By optimizing energy efficiency and reducing reliance on fossil fuels, these plans realize a double advantage: they reduce harmful atmospheric pollutants as well as climate change while fostering sustainable urban growth. Examining the possibility of integrating renewable energy, photochemical methods can also facilitate accelerating the transition towards a carbon-free world. Technologies such as hydrogen production through electrolysis present exciting possibilities for energy conversion and storage, with opportunities to utilize excess renewable energy efficiently (Yue et al., 2021). Combining hydrogen production with air pollution abatement can also change the game of converting waste products to valuable resources in the long term and closing the loop of sustainability. Through the demonstration of interdependence in energy systems, such photochemical applications not only provide a window of opportunity to purify air quality but also create scope for revolutionary energy solutions. In this way, they embody the circular economy and resilience spirit, demonstrating that effective pollution control measures can at once contribute towards environmental health as well as drive an overall agenda of sustainable energy transition (Figure 6.10).

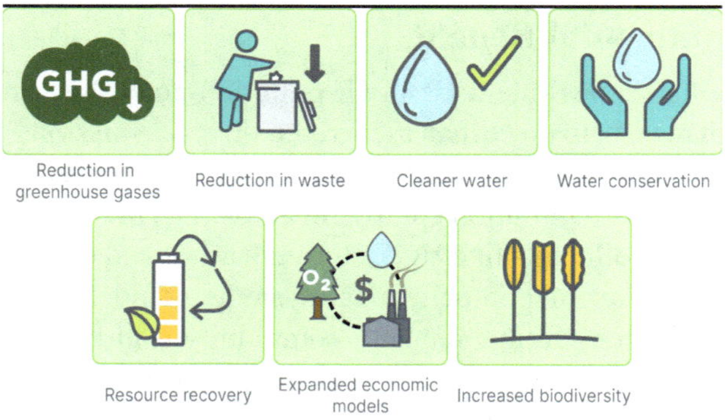

Figure 6.10. *Environmental benefits.*

Source: https://www.stjohns.ca/en/news/resources/News-Releases/IMG_CEM-week-2_Enviro-Benefits.png.

6.4.2. Cost-Effectiveness Compared to Traditional Methods

Cost-effectiveness analysis is a key element in analysis of photochemical methods of air pollution control against traditional methods. Traditional methods, such as thermal incineration and chemical scrubbing, often come with high operating and maintenance costs due to complexity and energy demands. Photochemical methods, on the other hand, are based on light-activated processes that dismantle pollutants, hence more energy-conservative and possessing lower operating costs. Studies indicate that newer photochemical systems can minimize energy consumption during treatment and ultimately be associated with significant long-term cost savings. Furthermore, scalability of the process assists in integration in various environments, which is an additional factor from a cost perspective. As increasing numbers of municipalities seek more sustainable ways of mitigating air pollution, the cost-effectiveness of photochemical processes assists in its acceptance and becoming beneficial in pollution abatement efforts. In addition, the working life of photochemical systems makes them overall cost-effective. In contrast to traditional technologies that involve regular replacement of scrubbers or filters, photochemical technologies often possess longer working lives, which enable lower long-term apparatus costs. Maintenance costs of photochemical systems are mostly significantly lower than traditional technologies. For instance, systems utilizing advanced semiconductor materials can minimize wear and tear but maximize performance efficiency, which eventually results in lesser downtimes and replacement costs. Such attributes render photochemical processes not only environment-friendly but economically beneficial as well. The growing need for smart materials and advanced control programs even further increases the efficiency of such systems, which allow real-time optimization of photochemical treatments for air pollution control, thereby maintaining efficiency at a lower cost. Further, the investment in capital to install photochemical methods may be justified by the resulting environmental compliance cost benefits and public health advantages. With increasing regulatory pressures on emissions and air quality standards, traditional methods of operation are likely to translate into enormous costs in terms of penalty fees and lawsuits. The application of photochemical methods, however, is capable of enabling compliance with such standards, thereby preventing additional cost incurred due to non-compliance. The control of air pollution also has a direct correlation with improved public health, which leads to reduced healthcare costs for pollution-related illness. The ability of photochemical methods to provide a twofold benefit of addressing environmental issue and being economically inexpensive positions them as a potential option compared to traditional strategies, promising much in transforming air quality management tactics successfully.

6.4.3. Potential for Scalability and Adaptability

The scalability and flexibility of photochemical technologies for air pollution abatement is most important in response to the increasing anthropogenic impact on the air. As urban populations expand and the attendant rises in pollution, traditional control technologies prove restrictive or impossible. Photochemical reactions, particularly photocatalysis with high-end materials like titanium dioxide, offer a scalable technology that can be diversified to respond to diverse urban settings. Their deployment in low-cost wireless sensor networks is a shining example of this versatility. They allow real-time monitoring of air quality and the localization of photochemical reactors in high-priority zones of contamination, representing a potentially powerful avenue to improve urban air quality that continues to be economically feasible (Boente et al., 2024). Flexibility is also highlighted by applying machine learning techniques to the optimization of photochemical processes in managing air pollution. By the use of data-driven approaches, scientists are able to chart the most effective configurations of photochemical systems according to local environmental conditions to enhance overall effectiveness of procedures for reducing pollution.

The utilization of a wide range of meteorological and traffic variables is also feasible, and this subsequently allows for a more holistic method of air quality management. Studies have found that the adaptability of methodologies in various urban environments enriches their effectiveness, leading to improved forecasts of air quality and a better overall understanding of pollutant behavior (Giovia et al., 2024). This dynamic interplay between advanced photochemical methodologies and real-time data analysis indicates the pertinence of these methodologies to scalable and situational applications. The scalability of photochemical technologies is also extended to their use in different geographic locations beyond city boundaries. For instance, the technologies can be adapted for usage in industries, agricultural production, and rural areas with challenging access for air quality monitoring. Developments in photocatalytic materials ensure that these technologies can be optimized according to specific pollutants or by local regulatory standards, thus being both effective and locally relevant. This versatility not only facilitates the use of these methods across various contexts but also offers scopes for interaction between policymakers, researchers, and industry leaders to create sustainable, long-term air quality management systems. With growing global focus on addressing air pollution, scalability and adaptability of photochemical methods come into the center stage to enable perceptible and sustainable air quality changes (Boente et al., 2024) (Figure 6.11).

Figure 6.11. *Potential for scalability and adaptability.*

Source: https://www.collidu.com/media/catalog/product/img/7/3/7387526f659dab74667a363bcbed4a9b450bd607cb546d2699121e889c087d63/scalability-plan-slide6.png.

6.5. CHALLENGES AND LIMITATIONS

Use of photochemical methods for air pollution control is also accompanied by many challenges and limitations that should be critically assessed. Among the main challenges is the efficiency of utilization of light in varied conditions of atmosphere. The effectiveness of photocatalytic reactions is very dependent on the quality and quantity of ample sunlight, which could be inconsistent in different regions and environments. Besides, the efficiency of photocatalysts such as titanium dioxide can also be greatly diminished when exposed to impurities that disrupt the absorption of light. Based on the literature, the ongoing research is also a necessity to optimize the performance of these materials in order to be able to continuously operate efficiently in different conditions. This inconsistency is indeed a major potential obstacle to the scalability and viability of employing such technologies in various regions geographically, where reliance on variable environmental aspects can exert profound impacts on the effectiveness of air purification techniques. Another necessary limitation is derived from the technological and economic difficulty of installing and advancing enhanced photocatalytic systems.

The current techno-economic environment proves that although there has been steady progress in reducing the cost aspect associated with photocatalysts, durability and efficiency-related issues are the foremost concerns. Current research shows that improving the lifetime as well as efficiency of photocatalytic materials requires huge expenditures on research and development. Besides that, the majority of the mentioned technologies are still not cost-effective compared to traditional methods of air pollution control. So, without strong financial incentives and enabling policies, there is little reason for industries to make a change over to using these more sustainable methods. The issue is not so much the technology as creating a market that offers incentive to make the change to cleaner air solutions through new photochemical methods. Finally, the integration of photochemical methods into existing schemes of pollution control represents a significant operations challenge. The efficacy of these plans can be greatly impacted by existing infrastructure and legislative frameworks, which can be unsuitable to facilitate emerging technologies. For instance, traditional pollutant control regimes are generally incompatible with the dynamic nature of photocatalytic processes and thus are likely to result in potential inefficiencies and areas of reduced effectiveness overall. In addition, regulatory agencies may lack familiarity with new photochemical techniques, leading to a lag in official support and accreditation. Lacking proper knowledge and policy implementation, there is always a possibility of underexpansion of advanced technologies like photochemical process-based solutions. Overcoming these incorporation and regulatory obstacles is essential to establishing an environment conducive to progress in photochemical means with the potential to alleviate air pollution problems on a larger scale.

6.5.1. Technical Challenges in Implementation

The application of photochemical processes in air pollution abatement is confronted by several technical issues that must be solved in order to ensure successful use. Among such issues is the variability in environmental conditions, which affects the efficiency of photochemical processes. Temperature, humidity, and the presence of interferences pollutants are some factors that may significantly influence the outcome of these processes. For instance, in regions like Riau Province, long-term air pollution issues arise due to forest and land fires, followed by climatic conditions that hinder the possibility of having the regular occurrence of photochemical processes. If the atmospheric conditions are not supporting such reactions, the desired benefit is lost, leading to insufficient reduction of the pollutants and violating public health effects (Astini et al., 2025). Thus, continuous monitoring and fine-tuning of such techniques are required to deal with such environmental issues. Incorporation of photochemical techniques into the present air pollution control systems is another extremely critical challenge. Effective air quality control requires coordination between various stakeholders like industry, government, and society. Regrettably, the lack of coordination results in disconnected strategy implementation that reduces the overall effectiveness of the implemented measures. For example, despite attempts to improve air quality within the urban setting, poor dust management and regulation enforcement can compromise the effectiveness of photochemical methods to function optimally. This fragmentation is seen through various case studies from around the world, where even if policies exist, there is not necessarily an implementation that reflects realization of the local settings or totality of causes of air pollution (Li, 2025). Therefore, the solution to avoiding these technical problems depends on enhancing inter-agency coordination and popular participation. In the final analysis, the technological requirements of developing photochemical techniques of air pollution control usually result in great hurdles needing to be cleared before they are effective. The need for expensive photoreactors and sensing equipment usually outstretches budgets, particularly in poorer regions. Further, the complexity of the systems demands special expertise, which may not be available where the air pollution is greatest. This lack of knowledge hinders the implementation of photochemical solutions, particularly in developing nations with limited budget and resources. Second, the technology must be adapted to meet local realities, implying continuous research and development to appropriately make the solutions workable. These are an eye-opener to the necessity of adequate training programs and increased investment in clean technologies to ensure that photochemical methods can be effectively utilized for sustainable air quality benefit (Astini et al., 2025; Li, 2025) (Figure 6.12).

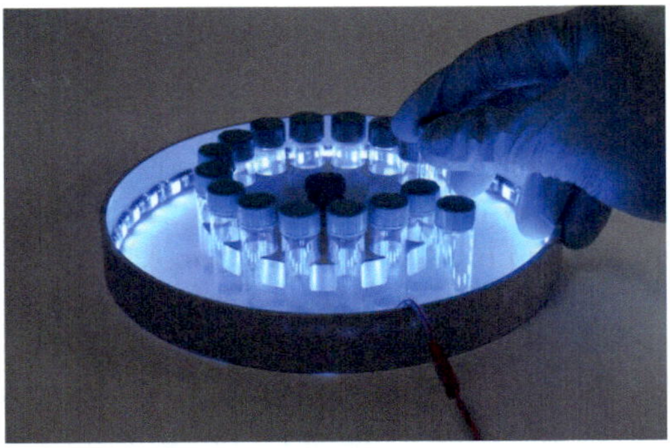

Figure 6.12. *Technical challenges in the implementation of photochemical processes.*

Source: https://www.sigmaaldrich.com/deepweb/assets/sigmaaldrich/marketing/global/images/technical-documents/articles/chemistry-and-synthesis/photocatalysis/photochemical-reactor/photochemical-reactor.jpg.

6.5.2. Limitations in Pollutant Types and Concentrations

The limits on types and amounts of pollutants have a significant influence on the effectiveness of photochemical methods in air pollution control. While photochemical processes are so adept at the destruction of a broad variety of pollutants, there are specific pollutants such as particulate matter (PM) and certain volatile organic compounds (VOCs) that pose unique challenges. For example, fine particulate matter ($PM_{2.5}$) and PM_{10} are prevalent in city centers, where traffic emissions are high. The complexity of PMs makeup also complicates treatment, since its smaller size can cause more lung penetration and a wider range of adverse health effects. This renders the use of photochemical approaches, which are better suited to treating gaseous emissions like nitrogen oxides (NO_x) and ozone (O_3), problematic. Thus, a fine balancing act among the groups of pollutants targeted in a control strategy is essential in an effort to maximize the efficacy of such advanced strategies. Besides, concentrations of some of the pollutants may significantly impact the kinetics of the photochemical reaction and thereby operational efficiency. For instance, it has been demonstrated through research that there are seasonal variations in the concentration of some of the pollutants that affect their correlation with health outcomes as evidenced in a recent study establishing linkages between increased concentrations of NO_2 and O_3 with increased emergency room admission during the winter season (Wang et al., 2025). In addition, the effectiveness of photochemical reactions changes with variations in pollutant concentration;

low concentrations may be too low to provide the energy required for creating efficient photochemical breakdown, while extremely high concentrations can produce byproducts that might be more hazardous than the initial pollutants. Thus, understanding these phenomena in the context of concentration is essential in designing effective air pollution abatement strategies. Besides, the spatial setting is crucial in determining the constraints in terms of types and levels of pollutants. Urban settings tend to be exposed to more pollutants due to human activities as well as emissions resulting from vehicle use on roads. For example, findings from research studies on urban traffic microenvironments show that occupant exposure to highly lethal concentrations of PMs results from vehicles, particularly non-AC buses and three-wheel taxis (Adikaram et al., 2025). These findings call for intervention targeting and geographically specific solutions to air quality control. Subsequently, the approaches would need to be context-specific to the local setting, highlighting the significance of knowing in detail both the categories of predominant pollutants as well as their concentrations to effectively optimize photochemical air pollution control measures.

6.5.3. Economic and Regulatory Barriers

The use of photochemical technologies in the mitigation of air pollution is fraught with a grave economic and regulation challenge that thwarts progress towards cleaner air. These inhibitions are often created by the gigantic capital expenditure required for new technology introduction. Most countries, particularly those on the economic doldrums, have a preoccupation with short-term economic advancement over long-term environmental improvement. Technologies such as photochemical oxidation, in this context, are deemed to be costly and unnecessary and hence underfinanced for research and development. To solve air quality issues meaningfully, governments must make sure there is enough money spent toward encouraging innovative technologies that would be capable of stopping pollution effectively. Without adequate financial support, initiatives towards enhancement of air quality through photochemical methods are likely to be elusive, hence remaining in the cycle of poor air quality and related public health problems that erode sustainability and economic security (Venkataraman et al., 2018). Apart from the economic factor, regulatory challenges significantly erode the efficiency of photochemical methods for air pollution reduction. Although regulations have a tendency to lag behind technological advancement, they form a lacuna that represses progress. For instance, administrative systems may overlook the challenges and benefits of using nanomaterials in air cleaning. Reports indicate that increased production and use of nanomaterials pose toxicity and ecological issues that need serious assessments (Jeevanandam et al., 2018). Without proper timely and responsive administrative systems, integrating such advanced solutions into application becomes extremely challenging. Policy hurdles could deter investment in promising technology, and therefore reconsideration of existing policies is the need of the hour. A proactive approach with the collaboration of scientists, policymakers, and business leaders will assist in developing an appropriate setting for embracing effective control measures for air pollution. Ultimately, overcoming these economic and policy hurdles involves a holistic strategy with public opinion, policy reforms, and research expenditures. Creating greater awareness among the stakeholders on the financial advantages of effective air pollution control can alter behavior, forcing the governments to formulate supportive policy. Incentive programs, such as tax relief or subsidies to firms utilizing cleaner technology, can push investments into photochemical methods. As countries face the challenge of reducing air pollution, creating robust regime of regulations that are responsive and attuned to technological developments will become crucial. It is only through a concerted effort that can the hurdles created by financial limitations and lag in regulatory affairs be overcome successfully to provide entry to improved public health outcomes and cleaner air (Venkataraman et al., 2018) (Figure 6.13).

Figure 6.13. *Economic and regulatory barriers.*

Source: https://fastercapital.co/i/Regulatory-Barriers--Navigating-the-Red-Tape--Regulatory-Barriers-as-an-Economic-Moat--Understanding-Regulatory-Barriers-as-an-Economic-Moat.webp.

6.6. FUTURE DIRECTIONS IN PHOTOCHEMICAL RESEARCH

Photochemical research is required to improve the efficiency of air pollution control measures. As international concern about the degradation of the environment continues to rise, new technologies are in great demand that can utilize renewable energy. Photochemical methods, which use light to generate chemical change, have tremendous potential for photochemical conversion of pollutants to harmless byproducts. The future directions for this field include the development of new photocatalysts that can perform under a wider range of solar wavelengths, making them more efficient and effective in various applications in need of environmental improvement. Apart from that, the integration of such advances with fuel cell technology has the promise to advance sustainable energy systems further along the path toward a carbon-free world, as emphasized in recent studies on the role of hydrogen technologies in the process (Yue et al., 2021).

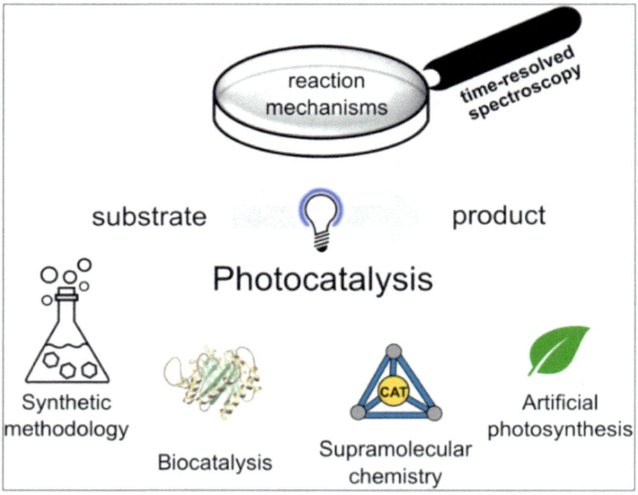

Figure 6.14. *Future directions in photochemical research.*

Source: https://pubs.acs.org/cms/10.1021/jacsau.4c00527/asset/images/jacsau.4c00527.social.jpeg_v03.

Another focus in photochemical advancement relates to the application of biomass-related materials, such as nanocellulose, for the control of air pollution. As the social attention towards replacing petroleum-based products with eco-friendly alternatives grows, nanocellulose becomes a promising option due to its facile replenishment and functional properties. The future of the green material can support the development of advanced photocatalytic systems that can degrade harmful air pollutants in an efficient manner. Its ability to interact with other chemical molecules can be utilized in the production of multifunctional materials that not only purify air but also contribute to energy storage and conversion. The importance of such advancements cannot be overstated as they align with global sustainability goals and the demand for environmentally friendly alternatives relative to conventional air pollution control technologies (Thomas et al., 2018). In addition, future research work must involve the integration of artificial intelligence and machine learning in photochemical processes. Through these technologies, scientists are able to attain optimal reaction conditions, predict catalytic activity, and enhance scalability in photochemical reactors. Such a digital revolution has the potential to bring more precise control over the pathways of reactions and product yields, which, in turn, is reflected in enhanced efficiency in air pollution abatement measures. Besides, AI-aided modeling will be able to find novel materials and formulations that maximize sunlight absorption and catalytic performance. With further research, the synergy between advanced computation tools and photochemistry will be central to developing more efficient approaches towards fighting air pollution and achieving sustainable environmental systems (Figure 6.14).

6.6.1. Innovations in Photocatalytic Materials

Photocatalytic material advancements have contributed significantly to controlling air pollution, particularly with the development of new semiconductors. Material like graphitic carbon nitride (g-C_3N_4) has caught the eye of scientists owing to its great stability and activity upon visible light irradiation. g-C_3N_4 preparation has attracted interest since it is green and inexpensive, hence a worthy candidate for scaled-up applications. Its new electronic properties enable effective charge separation, which enhances photocatalytic degradation of pollutants. In addition, the ability of g-C_3N_4 allows for its modification for the adjustment of its bandgap and surface properties in order to optimize its functioning under different environmental conditions. These innovations, therefore, provide a means of better utilization of photochemical processes to improve indoor and outdoor air quality in terms of enhanced pollutant elimination strategies, as highlighted by recent literature (Bhanderi et al., 2023). The advances in nanotechnology in the recent past have also played a key role in enhancing photocatalytic processes to manage air pollution. The development of nanoscale material such as metal-doped metal oxides and plasmonic nanostructures has increased the surface area and the reactivity of photocatalysts. Nanomaterials specifically can be engineered to have improved catalytic activity by adjusting their structure and size, which improves their effectiveness in breaking down air pollutants by a significant amount. For instance, incorporation of materials such as titanium dioxide (TiO_2) with noble metals can lead to enhanced light capture as well as the mobility of electrons, hence facilitating better charge separation. Such a characteristic is pivotal in facilitating the photochemical reactions that allow for the degradation of pollutants, hence cleaner air. Furthermore, machine learning and artificial intelligence utilization in photocatalyst material optimization ensures that the optimal-performing compounds are synthesized and engineered for specific environmental concerns (Mohamed et al., 2023). Cost-effectiveness and sustainability of photocatalyst materials are also enhanced by green synthesis, leading the research community to green manufacturing pathways. The conventional synthesis methods tend to employ risky chemicals, generating harmful byproducts that are harmful to both the environment and health. The green approach, on the other hand, relies on natural resources, including plant extracts, to generate photocatalysts with zero environmental footprint. Not only does this follow environmental sustainability objectives, but it also adopts the new emerging principle of zero discharge of waste in industrial processes. Techniques utilizing biogenic agents in green synthesis of photocatalytic material can enhance the performance of these compounds while reducing production costs simultaneously. Further, with rising international demand for efficient wastewater treatment technology, new green synthesis techniques will play a fundamental role in developing stable photocatalytic systems to support technologies to advance alongside rising environmental needs (Mohamed et al., 2023).

6.6.2. Integration with Other Air Pollution Control Technologies

In air pollution control, integration with other technologies is necessary for photochemical processes to maximize overall efficiency. Solar energy is typically utilized by photochemical reactions to facilitate the degradation of pollutant chemicals, such as volatile organic compounds (VOCs) and nitrogen oxides (NO_x). When integrated with other pollution technologies, such as adsorption or catalytic converters, photochemical processes can significantly improve the rate of pollutant removal. For instance, adsorption technologies can pre-concentrate targeted contaminants, making them easier to be treated through photochemistry. Synergistic application of such technologies not only leads to increased efficiency of operation but also reduces the cost and energy-intensive nature of air quality management. Consequently, understanding how to synergistically combine such technologies can serve as a basis for developing effective air pollution control strategies. The integration of photochemical technology with emerging monitoring systems also enhances the effectiveness of pollution control strategies. The application of sophisticated sensor networks enables real-time monitoring of key air quality indicators, thereby enabling an immediate response to pollution peaks. Monitoring systems are able to identify which pollutants are there and in what amounts, determining when to trigger the photochemical treatment process. Increased data collection and examination through such systems facilitate evidence-based decision-making, which in turn results in more efficient and focused interventions. The combination of photochemical methods with cutting-edge monitoring technology is an exemplary instance of an integrated approach to air quality management, which is made agile and adaptive to dynamic environmental needs. Integration is the essence of managing long-term compliance with regulation requirements and safeguarding public health. In addition, application of photochemical processes in clean energy technology can lead to new air pollution control improvements. Such processes can be integrated with renewable sources such as solar or wind energy in order to enhance the overall sustainability of the system. Solar-powered photochemical reactors, for example, can utilize the sun's energy efficiently to break down pollutants and reduce the dependence on fossil fuels. In addition, integrating these technologies with urban planning initiatives, as seen with the numerous green city projects, can promote sustainable development in addition to rehabilitating air quality issues. Vancouver and Copenhagen are a couple of cities with comprehensive environmental policies which aim towards the incorporation of pollution abatement technology and renewable energy strategies. These examples reinforce the place holistic integration methods can occupy in producing healthier urban environments and enabling global carbon neutrality targets ultimately to pave the way for a sustainable urban air quality future.

6.6.3. Potential for Policy Development and Funding

The future of policy development and funding in the field of photochemical solutions to controlling air pollution depends on national and international collaboration for controlling air quality issues. Programs like the Korea–United States Air Quality (KORUS-AQ) field study illustrate how global collaboration may advance knowledge regarding the dynamics of air pollution. This collaborative approach is significant in that it provides an integrated view of air quality conditions by merging different sources of information from aircraft, ground stations, and space platforms. The findings of such research have significant policy implications in that they clarify the causality of pollution and offer targeted suggestions for the design of regulation. Therefore, the need for urgent investment in this cooperative research is required to make them become sustained policy that is informed well by empirical evidence and advanced monitoring technologies (Crawford et al., 2021). Emerging sources of funding need to be identified to allow photochemical techniques to advance in air pollution control. Due to the complexity of air quality management, funding is needed when it comes to studying nanomaterials and how they help influence atmospheric processes. Accidental nanomaterials and engineered nanomaterials, now more prevalent due to industrialization, are unique challenges and opportunities for air quality interventions. Emphasis on their unique characteristics can be the solutions to new alternatives that improve the air quality.

Figure 6.15. *Potential for policy development and funding.*

Source: https://www.collidu.com/media/catalog/product/img/8/f/8f1062b72c0b87226a3a8fd986d6927aa16904a883a566fcc1cd4d8f4684d276/policy-development-slide1.png.

Legislative policies encouraging investment in research scholarships, research grants for new technology, and public-private partnerships can catalyze this critical work. Merging findings from ongoing research can inform legislative action to enhance air quality and, by extension, overall environmental health (Michael et al., 2019). The involvement of citizens is also required for effective policy implementation on these scientific recommendations. Public engagement platforms need to be considered by policymakers when addressing air pollution problems, maintain transparency, and ensure trust in policies developed. By creating avenues for dialogue among scientists, the public, and government officials, the development of workable policies can be made more inclusive and responsive to local needs. Moreover, incorporating public sentiments can enhance funding strategies through the targeting of specific community-level problems and priorities. Shaping policy with such information not only facilitates the mobilization of the needed financial resources but also aids in establishing the measures implemented as usable and efficient. Thus, efficient communication and coordination can maximize the possibilities of innovative photochemical air pollution control methods (Crawford et al., 2021) (Figure 6.15).

SUMMARY

The research on photochemical methods in air pollution control exemplifies a critical point of overlap between technology innovation and sustainability in the environment. With rising industrialization, it is at the center of mounting air quality issues that these methods have ready solutions to mitigate the negative effect of pollutants. They harness sunlight energy to induce chemical reactions that can break down toxic materials, particularly nitrogen oxides and volatile organic chemicals, the precursors to tropospheric ozone development. Efficient use of these methods relies not just on the creation of photochemical technologies but on understanding the atmospheric implications behind their application. As evidenced by historical data in the past, little residual regulations have been utilized previously to allow for excessive levels of ozone concentrations, leading to extreme public health issues, thus validating the importance of efficient photochemical intervention strategies (Yue et al., 2021). In assessing the effectiveness of photochemical techniques, the need for continuous innovation is therefore considered essential. The relationship between pollutants, NO_x and VOCs, in this instance, and solar irradiance means that it is essential to have robust techniques that can adapt to changing environmental conditions. As dictated in recent research, such techniques are essential in the prevention of ozone formation, especially as climate change continues to generate atmospheric conditions conducive to its formation. There exists a pressing need for ongoing studies to develop these technologies further, which means overcoming current inefficiencies, expense, and scalability of use (Yue et al., 2021). By working in concert between regulatory bodies, researchers, and industry stakeholders, one can enhance the integration of these methods into total air quality management systems. Ultimately, photochemical strategies can go so much further than mere compliance with air quality standards; they are directed toward a complete paradigm shift toward sustainable resource management in the face of the climate crisis. Emphasized through the disturbing statistics concerning ozone-related health impacts, it is essential to highlight efforts towards reducing the release of pollutants in tandem with protecting public health (Zhang et al., 2019). Future efforts will need to be focused on fostering a synergy between technological progress and environmental policy. The double-barreled thrust will accelerate progress toward the development of cleaner air, and consequently, environmental and human health gain. The way ahead for effective air pollution control by photochemical means goes on, but the foundation laid so far is an optimistic foundation for a healthier, sustainable future. In addressing air pollution's pressing issues, photochemical processes come forth as feasible methods of controlling atmospheric pollutants. Photochemical processes utilize the interaction of light with contaminants to initiate chemical reactions that transform harmful chemicals into less toxic materials. The persuasive argument for the widespread application of photochemical processes stems from their ability to lower environmental impacts and improve public health. For instance, research has shown that children and teenagers, as well as patients suffering from asthma, are adversely impacted by exposure to air pollutants, which highlights the importance of efficient air quality management strategies (Gudziunaite et al., 2025). Thus, through adoption of photochemical methods into conventional air pollution control systems, stakeholders can efficiently protect vulnerable groups against the adverse impacts of polluted air. Secondly, photochemical means can also become more efficient if standardized application means are created that control their application. Comprehensive details on emission sources and pollutant levels must be known for determining the optimal points of control so that photochemical measures can target the most critical areas of issues. An effective approach presented in recent studies creates a central approach to finding control points in the course of urban territories and applying its use in other control methodologies for air pollution (Ovodkov et al., 2024). With advanced statistical calculations, scientists are able to calculate the representativeness of established control points, and they can ultimately yield more effective and targeted solutions to combating air pollution with the use of photochemical processes. Finally, as air quality management becomes more relevant by the day, advanced photochemical methods will play an even greater role in building sustainable urban landscapes. With the emphasis placed on reduction of emissions, particularly in densely populated regions, there lies a promise for these methods to play a major role in reducing urban pollution. As demonstrated by the considerable effects of air pollution on lung function and overall health, especially in children (Gudziunaite et al., 2025), effective and standardized air quality measures still remain necessary. Through photochemical air pollution management using advanced technologies and scientific methodologies, local governments can enhance public health outcomes, promote environmental sustainability, and subsequently create the avenues for healthier future generations.

REFERENCES

1. Adikaram, W. D. C. N., & Arambepola, C. (2025). Occupational hazard to on-road air pollution within passenger transport micro-environments: Evidence from traffic congested areas in Colombo, Sri Lanka. *BMC Public Health, 25.* https://www.semanticscholar.org/paper/d970f2eea8172fa68086980acc8c9bbbcbed8623 (accessed on 28th September 2025).
2. Ahn, J., Blake, D. R., Choi, W. C., Jung, J., et al. (2018). Meteorological and chemical factors controlling ozone formation in Seoul during MAPS-Seoul 2015. https://core.ac.uk/download/287622613.pdf (accessed on 28th September 2025).
3. Aniceto, C., Braz, P., Machado, A., & Matias-Dias, C. (2025). Mapping susceptibility to air pollution and its association with birth defects: A tool for public health intervention. *European Journal of Public Health.* https://www.semanticscholar.org/paper/353bffb0bc63a8de5bef961f1f86016236f62555 (accessed on 28th September 2025).
4. Armenta-Déu, C. (2025). Air pollution control engineering in urban and congested areas. *Advances in Environmental and Engineering Research.* https://www.semanticscholar.org/paper/fd38d4ba14e5c025925be0949a1bcef22045c1d4 (accessed on 28th September 2025).
5. Astini, B., Yasir, Y., & Hero, E. (2025). Analysis of local media content on the implementation of air pollution control policies. *Metacommunication: Journal of Communication Studies.* https://www.semanticscholar.org/paper/03ce1dfc8345684be0859f23ff609c686e4f166d (accessed on 28th September 2025).
6. Atavullayeva, M. (2024). Environmental pollution monitoring: Essential strategies for mitigating ecological degradation. *E3S Web of Conferences.* https://www.semanticscholar.org/paper/ffe2b60973f962f68c4c3e98f2458ce48009d7af (accessed on 28th September 2025).
7. Bhanderi, D., Lakhani, P., & Modi, C. K. (2023). Graphitic carbon nitride (g-C_3N_4) as an emerging photocatalyst for sustainable environmental applications: A comprehensive review. *RSC Sustainability, 2,* 265–287. https://doi.org/10.1039/d3su00382e (accessed on 28th September 2025).
8. Blake, D. R., Cheng, H. R., Guo, H., Jiang, J., et al. (2010). Assessing photochemical ozone formation in the Pearl River Delta with a photochemical trajectory model. https://core.ac.uk/download/286361448.pdf (accessed on 28th September 2025).

9. Boente López, C., Gómez Galán, J. A., Medina García, J., Rosa Díaz, I., et al. (2024). Designing a low-cost wireless sensor network for particulate matter monitoring: Implementation, calibration, and field-test. https://core.ac.uk/download/618762690.pdf (accessed on 28th September 2025).

10. Broeske, R., & Schwarzbözl, P. (2024). Numerical optimization and analysis of 3D-printed porous structures for open volumetric receivers. *ASME 2024 18th International Conference on Energy Sustainability*. https://www.semanticscholar.org/paper/744a403989799207600228bd99e000e563d679fb (accessed on 28th September 2025).

11. Chandran, A., Narasimman, L., Leena, T., Raj, J. J. J., & Veerapandian, V. R. (2024). Artificial intelligence driven visualization for enhanced waste management and air pollution control in smart cities. *2024 3rd International Conference on Automation, Computing and Renewable Systems (ICACRS)*. https://www.semanticscholar.org/paper/ec7a15663d901232d879cb29db5ff2f59a2eae87 (accessed on 28th September 2025).

12. Chen, S., Liu, Y., Lin, J., Shi, X. D., Jiang, K., & Zhao, G. L. (2021). Coordinated reduction of CO_2 emissions and environmental impacts with integrated city-level LEAP and LCA method: A case study of Jinan, China. *Advances in Climate Change Research*, *12*, 848–857. https://doi.org/10.1016/j.accre.2021.08.008 (accessed on 28th September 2025).

13. Crawford, J. H., Ahn, J., Al-Saadi, J. A., Chang, L. S., Emmons, L. K., Kim, J., Lee, G., et al. (2021). The Korea–United States Air Quality (KORUS-AQ) field study. *Elementa: Science of the Anthropocene*, *9*. https://doi.org/10.1525/elementa.2020.00163 (accessed on 28th September 2025).

14. Dhanuk, T. M., Khanal, R., Sah, S., & Ghimire, R. (2025). Advancing air quality management: A comprehensive review of UV technology, pyrolysis, and their integration. *OODBODHAN*. https://www.semanticscholar.org/paper/5395c55526b546ebf78526c459a15244f66170de (accessed on 28th September 2025).

15. Escobar, J. M., Montenegro, R., Montero, G., & Oliver, et al. (2012). Three-dimensional finite element modelling of stack pollutant emissions. https://core.ac.uk/download/41777955.pdf (accessed on 28th September 2025).

16. Feng, T., Wang, Y., Hu, W., Zhu, M., Song, W., Chen, W., Sang, Y., et al. (2023). Impact of aging on the sources, volatility, and viscosity of organic aerosols in Chinese outflows. *Atmospheric Chemistry and Physics*, *23*, 611–636. https://doi.org/10.5194/acp-23-611-2023 (accessed on 28th September 2025).

17. Gao, Y., Li, Z., & Sun, F. (2025). Comparative analysis of grid-connected inverter for photovoltaic generation. *Applied and Computational Engineering*. https://www.semanticscholar.org/paper/c7a769454866872648a95dd30f4b73addb9353f8 (accessed on 28th September 2025).

18. Gautam, S., Paudel, M., & Devkota, A. (2025). Heavy metal pollution and phytoremediation – A review. *Nepal Journal of Botany*. https://www.semanticscholar.org/paper/c14aed13b1e2dbdcb5bf40d654a59deefe8137da (accessed on 28th September 2025).

19. Giovia, G. (2024). Leveraging deep learning and feature importance methods to study the influence of traffic and weather on air quality in Milan. https://core.ac.uk/download/650037945.pdf (accessed on 28th September 2025).

20. Gudziunaite, S., Mackintosh, K., Davies, G. A., Jordan, K. A., Lewis, P. D., Griffiths, C. J., Swain, T. A., et al. (2025). Global trends in the relationship between chronic air pollution exposure, physical activity and lung function in youth aged 5–18 years with and without asthma: A systematic review. *Sports Medicine - Open, 11*. https://www.semanticscholar.org/paper/d61d8c16392f94641de2b882e64098bb26899160 (accessed on 28th September 2025).

21. Hassaan, M. A., El-Nemr, M. A., Elkatory, M. R., Ragab, S., Niculescu, V. C., & El Nemr, A. (2023). Principles of photocatalysts and their different applications: A review. *Topics in Current Chemistry*, *381*. https://doi.org/10.1007/s41061-023-00444-7 (accessed on 28th September 2025).

22. Hilser, J. R., Han, Y., Yang, Y., Zhang, H., Finch, C., Mack, W., Zhang, G., et al. (2025). Abstract Tu0086: Gene-environment interaction between the TRIP4 locus and air pollution exposure increases risk of coronary artery disease. *Arteriosclerosis, Thrombosis, and Vascular Biology*. https://www.semanticscholar.org/paper/9287c14537552a36dc35052be18b486ad8332f86 (accessed on 28th September 2025).

23. Hochella, M. F., Mogk, D. W., Ranville, J. F., Allen, I. C., Luther, G. W., Marr, L. C., McGrail, B. P., et al. (2019). Natural, incidental, and engineered nanomaterials and their impacts on the Earth system. *Science*, *363*. https://doi.org/10.1126/science.aau8299 (accessed on 28th September 2025).

24. Hotamova, M., & Dilova, N. (2024). Addressing environmental pollution: Solutions for the Bukhara region at local and regional levels. *E3S Web of Conferences*. https://www.semanticscholar.org/paper/2562c610984f964137561c4af9b4b80497dd4996 (accessed on 28th September 2025).

25. In Memoriam, P. M. (2023). In memoriam: Peter K. Mueller (1926–2023). *Journal of the Air & Waste Management Association*, *73*, 333–334. https://www.semanticscholar.org/paper/02931 8dbd589424d1f8e8acd3cfeb3ed7fbf64ab (accessed on 28th September 2025).
26. Jeevanandam, J., Barhoum, A., Chan, Y. S., Dufresne, A., & Danquah, M. K. (2018). Review on nanoparticles and nanostructured materials: History, sources, toxicity and regulations. *Beilstein Journal of Nanotechnology*, *9*, 1050–1074. https://doi.org/10.3762/bjnano.9.98 (accessed on 28th September 2025).
27. Johnson, A. C. (2025). Air pollution dynamics in Shanghai: Trends from 2014 to 2024. *International Journal of Scientific Research in Science and Technology.* https://www.semanticscholar.org/paper/a6ab192f69dfd7ddfac2fcb9dab06d5e5b7a11c9 (accessed on 28th September 2025).
28. Khanali, M., Elhami, B., Eslami, H., & Hosseinpour, S. (2016). Evaluation and comparison of environmental indices of hybrid corn (*Zea mays L.*) production using three different harvesting methods in Alborz Province based on life cycle assessment. *9*, 892–909. https://www.semanticscholar.org/paper/e86230064b8aedcd750a598db64f97bd8ea5ce51 (accessed on 28th September 2025).
29. Lama, S., Saundersa, S. M., Chengc, H., & Guod, H. (2013). Examination of a tropospheric ozone control methodology from the explicit representation of POCPs across varying temporal and continent spatial domains. https://www.semanticscholar.org/paper/d5b318a3bdf4e35f20226faf736c2d4185b04b4a (accessed on 28th September 2025).
30. Li, T. (2025). Problems existing in urban air pollution control and their optimization measures. *MATEC Web of Conferences*. https://www.semanticscholar.org/paper/e144c09cb585ce2b544ab255cd2e805fae6b0ee7 (accessed on 28th September 2025).
31. Li, Y., Wang, J., Li, L., Bai, Y., Gao, J. Y., He, L., Tang, M., et al. (2022). Research on the influence of weather patterns on ozone concentration: A case study in Tianjin. *Atmosphere*. https://www.semanticscholar.org/paper/6e02bd2016468072d7013ea1e7714d429c228a4e (accessed on 28th September 2025).
32. Luo, R., Lun, X., Gao, R., Wang, L., Yang, Y., Su, X., Habibullah-Al-Mamun, M., et al. (2025). A review of biogenic volatile organic compounds from plants: Research progress and future

prospects. *Toxics*, *13*. https://www.semanticscholar.org/paper/fa5a5336601bb8126ab23f070d238aac3710c848 (accessed on 28th September 2025).

33. N/A. (2022). *Green City: Towards Low-Carbon Urban Planning and Governance. Guidebook to Carbon Neutrality in China.* https://www.semanticscholar.org/paper/3d21cd1283a326a05461d1511282a798af65db47 (accessed on 28th September 2025).
34. N/A. (2025). Environmental quality. https://core.ac.uk/download/pdf/42884433.pdf (accessed on 28th September 2025).
35. Ovodkov, M., Petrov, V., Tkachev, M., & Azarov, V. (2024). A methodological approach to determining the number and location of control points for conducting summary calculations and emission quotas calculation. *Problems of Risk Management in the Technosphere*. https://www.semanticscholar.org/paper/2216c4df38796ddcb5c07fcc27c3096125b9464e (accessed on 28th September 2025).
36. Pagola, S. (2023). Outstanding advantages, current drawbacks, and significant recent developments in mechanochemistry: A perspective view. *Crystals*, *13*, 124–124. https://doi.org/10.3390/cryst13010124 (accessed on 28th September 2025).
37. Peng, Z., Liu, H., Zhang, C., Zhai, Y. H., Hu, W., Tan, Y., Li, X., et al. (2024). Potential strategy to control the organic components of condensable particulate matter: A critical review. *Environmental Science & Technology*. https://www.semanticscholar.org/paper/d792c4f4e261b8da6864bf3c9532a3cbf6968e62 (accessed on 28th September 2025).
38. Priyadarshini, M. (2023). Evaluation of the development of landscape architecture using WSM method. *4*. https://www.semanticscholar.org/paper/5888c49e933d5ca45fafb90343849c2451b8a4fe (accessed on 28th September 2025).
39. Ranganathan, C., Mohamed Abuthahir Riazulhameed, A. A., Brindha, G., Mohankumar, N., Yuvaraj, S., & Murugan, S. (2024). Cloud-integrated wireless sensor networks for dynamic household air pollution management. *2024 Asia Pacific Conference on Innovation in Technology (APCIT)*. https://www.semanticscholar.org/paper/59d83c26ae1ce3b03e36385f7a872a5f96fbabd0 (accessed on 28th September 2025).
40. Ren, X., Qu, R., Liu, S., Zhao, H., Wu, W., Hao, S., Zheng, C., et al. (2020). Synthesis of zeolites from coal fly ash for removal of harmful gaseous pollutants: A review. *Aerosol and Air Quality Research*, *20*, 1127–1144. https://www.semanticscholar.org/paper/

f85be21bd9ff202793ddf92602bd7fcff4c078ed (accessed on 28th September 2025).

41. Singh, J., Dutta, T., Kim, K. H., Rawat, M., Samddar, P., & Kumar, P. (2018). 'Green' synthesis of metals and their oxide nanoparticles: Applications for environmental remediation. *Journal of Nanobiotechnology, 16*. https://doi.org/10.1186/s12951-018-0408-4 (accessed on 28th September 2025).
42. Thomas, B., Raj, M. C., Athira, K. B., Rubiyah, M. H., Joy, J., Moores, A., Drisko, G. L., et al. (2018). Nanocellulose, a versatile green platform: From bio sources to materials and their applications. *Chemical Reviews, 118*, 11575–11625. https://doi.org/10.1021/acs.chemrev.7b00627 (accessed on 28th September 2025).
43. Venkataraman, C., Bräuer, M., Tibrewal, K., Sadavarte, P., Ma, Q., Cohen, A., Chaliyakunnel, S., et al. (2018). Source influence on emission pathways and ambient $PM_{2.5}$ pollution over India (2015–2050). *Atmospheric Chemistry and Physics, 18*, 8017–8039. https://doi.org/10.5194/acp-18-8017-2018 (accessed on 28th September 2025).
44. Wang, J. (2024). Overview and prospect of the history of air pollution control in the United States in the 20th century. *International Journal of Social Sciences and Public Administration*. https://www.semanticscholar.org/paper/69633a32734a975c120ad5cf7145877d5c6a19a9 (accessed on 28th September 2025).
45. Wang, L., Yang, H., Wang, Q., & Niranjan, S. (2013). Energy and environment: Challenges and achievements in rapid urbanization. *The Scientific World Journal, 2013*. https://www.semanticscholar.org/paper/55ce4372f67c40e4dbf1a02037703a05d280966b (accessed on 28th September 2025).
46. Wang, S., Jeong, S., & Ha, E. (2025). Spatiotemporal associations between air pollution and emergency room visits for cardiovascular and cerebrovascular diseases in Korea using a multivariate graph autoencoder modeling approach: An ecological study. *Ewha Medical Journal, 48*(3), e43. https://www.semanticscholar.org/paper/dad95c4a82302d5462ea0e34b30626f91b168a6b (accessed on 28th September 2025).
47. Wardana, L. M. F., Nurjazuli, N., Joko, T., & Rizaldi, M. A. (2025). Air pollution microplastics with the potential risk of lung disease: A systematic review. *Journal of Environmental Health and Sustainable Development*. https://www.semanticscholar.org/paper/34dc68059ccf81a2465dd273cadc19619fed74a3 (accessed on 28th September 2025).

48. Yan, X., Wang, J., Wu, F., Bai, J., Zhang, X., Li, G., & Fei, H. (2025). Spatial and temporal variation characteristics of air pollutants in coastal areas of China: From satellite perspective. *Remote Sensing*. https://www.semanticscholar.org/paper/f0da5981c5b1d6ed40435130cc8a957cd21eff28 (accessed on 28th September 2025).

49. Yue, M., Lambert, H., Pahon, E., Roche, R., Jemeï, S., & Hissel, D. (2021). Hydrogen energy systems: A critical review of technologies, applications, trends and challenges. *Renewable and Sustainable Energy Reviews*, *146*, 111180–111180. https://doi.org/10.1016/j.rser.2021.111180 (accessed on 28th September 2025).

50. Zhang, J., Wei, Y., & Fang, Z. (2019). Ozone pollution: A major health hazard worldwide. *Frontiers in Immunology*, *10*. https://doi.org/10.3389/fimmu.2019.02518 (accessed on 28th September 2025).

51. Živković, V. B., Kljajić, M. V., Milenković, A. D., & Jeremić, L. D. (2024). Reducing nitrogen oxides on units A3–A6 of the thermal power plant "Nikola Tesla A" by implementing secondary methods. *IEEP2024: Zbornik Radova*. https://www.semanticscholar.org/paper/05389a32efb8747015e3d803f80d2b63d173faa7 (accessed on 28th September 2025).

CHAPTER 7

POLLUTANT DEGRADATION PATHWAYS AND KINETICS

Contents

Introduction 267
7.1. Types of Pollutants 272
7.2. Mechanisms of Degradation 278
7.3. Factors Influencing Degradation Rates 283
7.4. Kinetic Models in Pollutant Degradation 288
7.5. Analytical Techniques for Studying Degradation 293
7.6. Case Studies of Pollutant Degradation 298
Summary 303
References 306

INTRODUCTION

The omnipresent availability of pollutants in various ecosystems places the emphasis on effective degradation processes and kinetics to mitigate their impact on the environment. As industrial activities increase and urbanization ensues, the release of harmful substances into the environment continues to be a significant threat to human health as well as ecological harmony. Various organic contaminants such as 4-nitrophenol and bromoxynil have been in the spotlight due to their toxicity and recalcitrance, and hence, researchers have shown particular interest in studying new improved degradation processes. Knowing the mechanisms and kinetics of pollutant degradation is a necessity for designing efficient treatment processes and regulatory schemes for handling pollutants. One of these ways of degrading pollutants is through the application of thermal plasma technology, which has been shown to be very effective in degrading various toxic compounds.

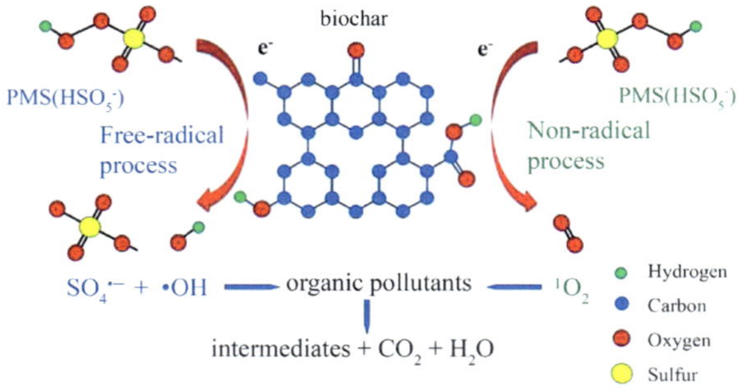

Figure 7.1. *Pollutant degradation pathways and kinetics.*

Source: https://www.researchgate.net/publication/350825090/figure/fig1/AS:1011928587567104@1618273984469/The-mechanism-of-organic-pollutants-degradation-using-biochar-PMS.png.

For instance, submerged thermal plasma degradation of 4-nitrophenol resulted in a degradation efficiency of 94% after 30 minutes, which is indicative of the process's potential for the elimination of waterborne toxic organic pollutants. Degradation pathways in such situations generally involve reaction rates according to pseudo-first-order kinetics, so modeling and reaction mechanism evaluation become essential for design and optimization. Identification of intermediates in degradation can also provide indications about the safety and efficiency of the treatment processes and aid in a comprehensive understanding of the pollutant behavior under actual conditions (Nandy et al., 2025). Further, ozonation is also a possible means for the degradation of the contaminants such as bromoxynil, an agricultural herbicide that has severe implications on the environment. Research indicates that the ozonized reaction of bromoxynil follows the Criegee mechanism, leading to primary ozonides that further decompose into less toxic compounds. The kinetics of such an ozonation reaction is very important for understanding the degradation mechanisms and the formed byproducts which are required to understand the environmental fate and potential toxicity of these intermediates. High-level quantum chemical calculations and reaction pathway simulations allow researchers to predict the effectiveness of ozonation more accurately and explore the implications of the resulting degradation products. It is essential to comprehend the mechanism for not only minimizing pollution but also for formulating environmental sustainability procedures on the basis of degradation reaction kinetics (Chu et al., 2025) (Figure 7.1).

Definition of Pollutants

Knowing what pollutants are is fundamental to knowing their degradation processes and kinetics. Pollutants generally refer to any substance that, when emitted into the environment, can adversely impact ecosystems, human health, or natural resource quality. Such contaminants can arise from various sources including industrial processes, agricultural processes, and urban runoffs. Their impact is multifaceted; pollutants can disrupt biochemical processes, compromise wildlife health, and contaminate water and terrestrial sources.

Their multifaceted nature stresses the importance of knowing not just what qualifies as a pollutant but also how it will act in ecosystems, since environmental parameters can influence the toxicological effects of such substances. Pollutants can be categorized into various classes based on source, chemical character, and state. Major groups include organic and inorganic pollutants, where the organic contaminants often come from carbon-based compounds such as pesticides, plastics, and drugs. On the other hand, inorganic pollutants entail heavy metals and minerals that are difficult to manage due to the persistence of these contaminants in the environment. Furthermore, pollutants can be both point source or non-point source, depending on whether they originate from a single point or from widespread, diffuse sources. These distinctions are crucial to appreciating if one is to solve the complexities that include pollutant pathways and degradation kinetics, as different categories require special mitigation and remediation solutions.

Moreover, the interaction between pollutants and the environment also has a great influence in determining their degradation processes and rates. For example, the inert nature of some nanomaterials and polymers can limit their biodegradation, with additional challenges in pollutant control, especially in the fields of medicine and waste treatment. Development of methods to functionalize these materials can enhance their reactivity and facilitate degradation under specific environmental conditions, reducing their long-term environmental footprint. Aside from this, further evidence for the importance of understanding surface alterations in pollutant behavior, materials science research on materials such as titanium dioxide nanoparticles holds promise for generating reactive oxygen species to be tapped for pollutant degradation or in therapeutic applications, demonstrating the cross-fertilization between pollutant definitions and their degradation mechanisms (Makvandi et al., 2020; Ziental et al., 2020).

Importance of Studying Degradation Pathways

Understanding degradation processes is central to effective management of pollutants in various environmental contexts. Given that industries continue to produce and release countless contaminants into the environment, uncovering the mechanisms through which these pollutants degrade gains more importance. Not only does it facilitate assessment of the environmental impact but also enhances the development of remediation strategies. The degradative processes of the pollutants can be very different based on the chemical structure of the pollutants, environmental conditions, and availability of other chemicals. Knowledge of these pathways enables researchers to identify the best and environmentally friendly methods of eliminating pollutants and hence minimizing injurious effects on ecosystems and human health. Further, an understanding of degradation mechanisms allows treatment technologies to be tailored to maximize their use in the specific pollutants present in various environments. More specifically, a study of pollutant degradation kinetics provides further insight into the effectiveness and efficiency of such processes. Kinetic studies determine the rate of pollutant degradation and supply essential information required for modeling and simulation of the environmental fate of these chemicals. For instance, photocatalytic dye decolorization is a promising wastewater treatment technology, demonstrating the influence of operational factors such as light intensity, dye concentration, and pH on overall decolorization efficiency. As witnessed in the photocatalysis discussion, understanding charge-carrier characteristics and formation of reactive species is crucial to optimize such systems (Chiu et al., 2019). Kinetic information also enables researchers to design more effective environmental policies and practical solutions for pollution reduction, as they provide data on the rate and extent of various pollutants' removal. Furthermore, the implications of studying routes of degradation can also be transferred to energy applications, particularly in alternative energy carriers. The study of such compounds as ammonia as a carrier for hydrogen illustrates the application of pollutant degradation in the broader strategy of sustainable energy. Ammonia, which has well-established transport infrastructure and high energy density, poses issues that could be softened by exploring its degradation tendencies across various combustion technologies (Valera et al., 2018). Study of such routes allows one to differentiate byproducts and their effects on the environment, contributing to the feasibility of using ammonia as a carbon-free energy carrier. For this purpose, pollutant degradation is not just vital research in the realm of environmental remediation but also proves to be critical in enhancing green energy alternatives, thereby pointing to its significance in general applications across domains.

Overview of Kinetics in Pollutant Degradation

The kinetics principle is critical in the description of pollutant degradation processes since it assists in explaining the rates and mechanisms of reactions. Through kinetic studies, one can have information on the time-dependent behavior of pollutants within various environmental media, including soil, water, and air. By quantifying the degradation reaction rates, researchers are able to determine the efficiency of different remediation processes and establish the function of environmental conditions such as temperature, pH, and concentration of the reactant. Kinetic modeling often employs various orders of reaction, with first-order kinetics being a universal model that simplifies analysis of pollutant degradation with time. Understanding the kinetics of these processes is not only important for determining the persistence of the pollutants but also for the design of effective treatment processes to counteract their environmental effects. Various pollutant degradation processes can have significantly different kinetic behaviors, reflecting the diversity of chemical and biological processes involved. For instance, evidence has been reported where catalytic ozonation process (COP) can enhance the degradation of pollutants, say 2,4-dinitrophenol (2,4-DNP), compared to single ozonation processes (SOP) on their own. Reports indicate that COP's effectiveness in degradation was as high as 98.2% under optimal conditions, way ahead of 75% efficacy in SOP. These improvements are attributable to the improved reaction kinetics with catalytic interactions, which facilitate pollutant degradation with lower doses and alter the mechanisms of reaction in degradation. The potential for reaction rate optimization by catalytic systems renders kinetic investigations crucial to progressing technologies for pollutant removal in industrial wastewater contexts (Asl et al., 2016). Furthermore, time-dependent behavior of degradation products can also give important insights into secondary pollution and complex chemical substance formation. For example, organosulfate compounds derived from biogenic precursors such as isoprene have been identified to be significant contributors to secondary organic aerosol formation. Their kinetic control suggests a multi-aspect influence of various mechanisms, including gas-to-particle partitioning and initiation radical processes that occur under specific atmospheric conditions. Specifically, nighttime concentration measurements highlight stable boundary layer importance in facilitating these interactions. By observing in real time, scientists are able to follow such conversions and measure how air quality and climatic impacts over urban regions depend on formation and degradation kinetics (Ault et al., 2011). Information on these dynamics is essential for effective environmental management and pollution control measures in reducing primary as well as secondary effects of pollutants (Figure 7.2).

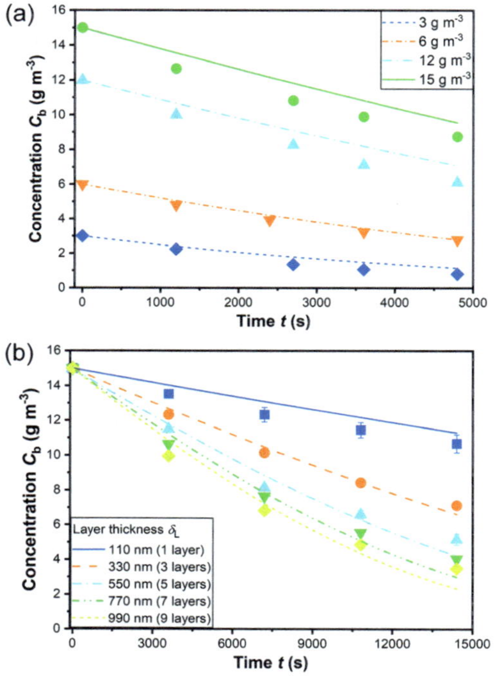

Figure 7.2. *Overview of kinetics in pollutant degradation.*

Source: https://pubs.rsc.org/image/article/2020/re/d0re00238k/d0re00238k-f8_hi-res.gif.

7.1. TYPES OF POLLUTANTS

The classification of the pollutants is of utmost significance in relation to the understanding of their pathways and kinetics of degradation, particularly in the realm of environmental science. Pollutants can be broadly classified into organic and inorganic chemicals, and each of them shows varying behavior within ecosystems. Organic pollutants such as pesticides, polycyclic aromatic hydrocarbons, and pharmaceuticals will be relatively less stable chemically, and hence microbial and photolytic degradation routes are possible. On the contrary, inorganic pollutants like salts and heavy metals are more likely to resist biodegradation and remain stable in the environment, causing continuous environmental hazards.

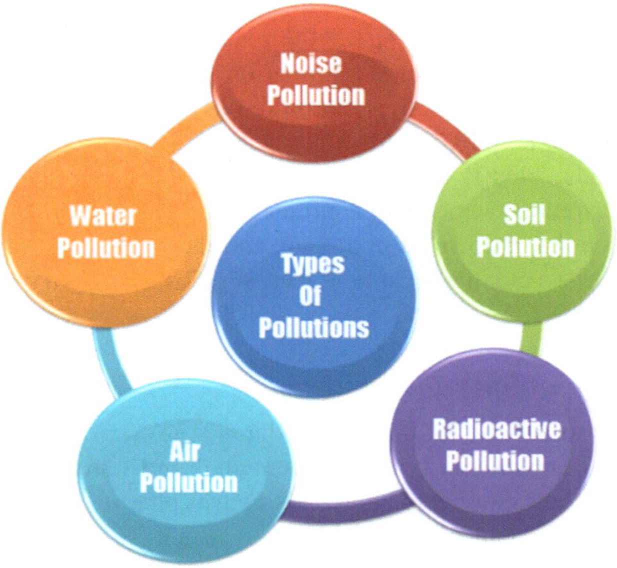

Figure 7.3. *Types of pollutants.*

Source: https://stylesatlife.com/wp-content/uploads/2016/08/Types-of-Pollution.png.

Determination of the types enables scientists to design remediation processes appropriately, adapting them to the specific nature and properties of contaminants involved. Advanced oxidation processes (AOPs) are now regarded as a central tactic to degrade organic pollutants in water supplies, giving highest priority to the efficiency of processes. The application of ultrasound, for instance, generates hydroxyl radicals that are implicated in the degradation of organic contaminants through reaction with them. Research has touched on the application of ultrasonic technologies to recalcitrant chemicals such as chlorophenols and polycyclic aromatic hydrocarbons. The effectiveness of degradation is based on various parameters such as initial substrate concentration, the frequency of ultrasound, and the presence of oxidizing agents, which can enhance the overall rate of degradation. This operational flexibility makes AOPs capable of degrading a wide range of organic pollutants while keeping their environmental impacts minimal, according to Bremner et al. (2011). When degrading new contaminants like clofibric acid (CFA), the degradation kinetics should be thoroughly understood in order to come up with effective treatment processes. Investigation using ionizing radiation and non-thermal plasma has confirmed quasi-exponential decay of CFA degradation, underscoring the relevance of kinetic modeling in the treatment of pollutants. Both methods yield significant reduction of the content of CFA, yet the mechanisms by which degradation occurs exhibit significantly varied formation of by-products. Although the ultimate product of degradation is carbon dioxide, the intermediates can have complex structures, and this captures the incomplete mineralization of certain organic pollutants. This is a nuanced understanding of pollutant degradation processes that captures the need for continued research to optimize technologies to effect efficient breakdown of pollutants and the full realization of limits of traditional as well as creative treatment technologies, as shown by Ceriani et al. (2017) (Figure 7.3).

7.1.1. Organic Pollutants

The ubiquitous presence of organic pollutants in various ecosystems has become a major concern to both environmental and human health. Organic pollutants, with a vast number of chemicals such as pharmaceuticals, pesticides, and industrial solvents, are particularly resistant to regular degradation mechanisms due to their complex molecular structures. Knowledge of organic pollutant degradation pathways and kinetics is essential for formulating effective remediation strategies. New methods, such as the combination of advanced oxidation processes with adsorption systems, have been discovered to be effective in enhancing removal efficiency of such pollutants. Through the optimization of reaction parameters and the employment of catalysts, researchers are able to maximize the degradation of organic pollutants into less toxic by-products, ultimately reducing their environmental impact and safeguarding water quality. The Fenton process, a well-known advanced oxidation method, has gained popularity because it can generate hydroxyl radicals, which are good oxidizing agents with the potential to mineralize the majority of organic pollutants. Traditional Fenton systems are, however, plagued by limitations such as slow iron ion cycling and saturation of adsorbents, complicating their use in water treatment on a long-term scale. Current studies reveal the potential of dynamic single-atom sites existing on heteroatom-doped carbons to efficiently reduce Fe^{3+} to Fe^{2+}, thereby sustaining the process of oxidation for subsequent cycles. The novel process not only speeds up the degradation kinetics of organic contaminants but also boosts the overall performance of the treatment system. By bridging Fenton chemistry with novel carbon materials, researchers have made it possible to develop new water treatment technologies that address the problem of persistence of organic pollutants in the environment in a holistic way

(Li et al., 2025). In addition to advanced oxidation processes, other treatments like UV irradiation also hold great promise for organic pollutant destruction. Interestingly, studies have demonstrated the concurrent removal of organic pollutants and inorganic substances with impressive removal efficiencies under the best conditions. For instance, UV treatment of dexamethasone phosphate (DexP) with chromium led to rapid degradation of the organic compound and significant precipitation of chromium by-products. This simultaneous removal method not only accelerated the breakdown of complex organic structures but also minimized energy consumption compared to individual treatment processes. The simultaneous focus on organic and inorganic pollutants in remediation processes highlights the need for holistic techniques that are not only efficient but also ecologically sound (Sarkhosh et al., 2024). These developments signify important advancements toward addressing the multifaceted problems of organic pollutants and their position on ecological health.

7.1.2. Inorganic Pollutants

Inorganic pollutants, including heavy metals and metalloids, have tremendous environmental and public health importance due to their durability and bioaccumulation potential. The pollutants enter the environment through industrial effluent disposal, agricultural practices, and urban runoff, into water and soil contamination. Some of the most notable inorganic pollutants include copper, chromium, and lead, each with characteristic toxicological attributes that can affect target harmful aquatic life and human health. The degradation pathways of inorganic pollutants are significantly different from their organic counterparts, which are controlled to a great extent by environmental factors like pH, temperature, and the presence of reactive species. It is possible, for instance, that the reduction of chromium in solution can proceed via redox reactions that allow its removal, the kinetics of which depend on several controlling factors. Advanced oxidation processes (AOPs) are reported efficient technologies for inorganic pollutant degradation, leveraging reactive oxygen species (ROS) to facilitate transformation and elimination. A study probing the efficacy of ozonation processes reported excellent degradation of tetracycline in water through an optimized O_3/peroxymonosulfate/FeMoBC system, focusing on the employment of multifunctional catalysts to boost reaction kinetics. Specifically, FeMoBC catalysis initiated the development of hydroxyl and sulfate radicals that resulted in full degradation in minimal reaction time. This is a demonstration of how the addition of catalysts can enhance the mechanisms of degrading inorganic pollutants, with the ultimate result being more efficient water treatment processes.

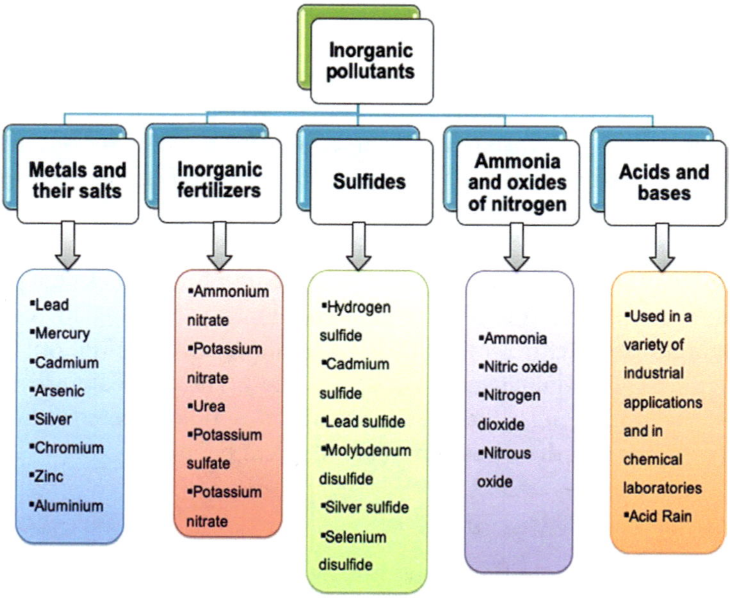

Figure 7.4. *Inorganic pollutants.*

Source: https://www.researchgate.net/profile/Bhushan-Vishal-3/publication/334148294/figure/fig2/AS:963448582180876@1606715450556/Categories-of-inorganic-pollutants.png.

The mechanism of this process demonstrates the complexity involved in degrading inorganic compounds, with significant impact on clean water technology plans. Moreover, concurrent treatment of over one pollutant, comprising both organic and inorganic substances, illustrates the potential of optimized remediation. An experimental study involving UV irradiation for the degradation of dexamethasone phosphate and chromium concomitantly documented that while degradation of the organic was rapid, chromium decomposition only ensued following destruction of organic counterparts. This discussion displays the complicated nature of pollutant degradation, showing that pollutants such as chromium can have their removal efficiency significantly impacted by the occurrence and degradation of organic matter. The results in the study demonstrate the intricacy of pollutant interaction and how it influences energy efficiency in remedial processes. Hence, it is imperative to understand these relationships in developing holistic strategies for pollutant regulation in contaminated ecosystems, laying the groundwork for continued improvement in environmental remediation techniques (Figure 7.4).

7.1.3. Emerging Contaminants

Emerging pollutants, as pharmaceuticals and personal care products, are increasingly threatening both human and environmental health since they are resistant and pervasive in aquatic environments. These chemicals are likely to avoid conventional wastewater treatment technologies, leading to accumulation in water bodies and thus garnering ecological attention.

Figure 7.5. *Emerging contaminants.*

Source: https://mytapscore.com/cdn/shop/articles/Emerging_Contaminants_DEC17.png?v=1678485773.

The degradation modes of such pollutants are exceedingly highly complex in nature, being likely to contain more than one intermediate byproducts with varied toxicity profiles. Advanced oxidation processes (AOPs) also have emerged as promising solutions for enhancing degradation of such pollutants. Employing oxidants like peroxymonosulfate (PMS) in combination with catalysts, scientists have observed increased degradation efficiency, which has allowed for better understanding of the kinetic behavior of such processes under real applications. Recent studies on degradation of some new emerging contaminants, such as Losartan, have improved our understanding of how these compounds can be effectively removed from water systems. In a recent assessment of Losartan degradation using PMS and ferrous iron catalysis, nine degradation byproducts were found, of which seven were new. Such precise examination unmasks the possible complexity of pollutant transformation in oxidative treatment, justifying the explanation of both kinetic rates and nature of product reaction. Utilizing mass spectrometry and nuclear magnetic resonance (NMR), this research, aside from illuminating the mechanisms of the reaction, provides a basis for optimizing AOPs. Such developments are essential to the development of targeted water system remediation approaches to addressing new contaminant burdens. Further research into the degradation pathways of contaminants such as tetracycline likewise underscores the potential of cobalt-based heterogeneous catalysts to enhance AOP effectiveness. Utilization of a g-$CoMn_2O_4$ spinelloid phase enabled high degradation rates, achieving 98.90% efficiency within a brief 10-minute reaction time. This study puts high emphasis on mechanism for pathway activation in which electron transfer is crucial to maximize oxidation. By employing advanced characterization techniques, the research identified intricate degradation pathways like demethylation and ring opening reactions. Such detailed investigation contributes towards overall knowledge regarding pollutant degradation kinetics and towards the rational design of catalysts and thus highlights innovative materials as a prospect to address the challenges posed by emerging pollutants in environmental scenarios (Figure 7.5).

7.2. MECHANISMS OF DEGRADATION

It is necessary to know the mechanisms of degradation in order to determine the environmental fate of pollutants. Degradation processes comprise complex reactions among pollutants and environmental constituents, which lead to their decomposition or transformation. The chemical degradation processes of pesticides in water and soil, for example, are representative of the processes facilitating the removal of such pollutants. Degradation of fungicide isotianil has been extensively studied, and pathways like hydrolysis and photolysis have been seen to produce several degradation products. These not only outline the rate at which isotianil is degraded in the environment but also determine the ecotoxicity of its byproducts. The identification and elucidation of such degradation routes through advanced techniques like ultrahigh-performance liquid chromatography–Q Exactive hybrid quadrupole–Orbitrap mass spectrometry highlight the intricacy of the fate of pollutants in environmental systems, and thus pesticide management and environmental conservation (Ai et al., 2024). Aside from chemical reactions, physical processes and biological processes also play pivotal roles in pollutant degradation in atmospheric environments. A paradigm example is tris(2-chloroisopropyl) phosphate (TCPP), which is a flame retardant with high atmospheric transport capability. Degradation of TCPP occurs primarily through hydroxyl radical (·OH) reactions, and prereactive complexes formation is significant in controlling the degradation kinetics. Addition of atmospheric water alters the reaction dynamics so that it causes TCPP to have a longer lifetime than it would without water. Specifically, the presence of atmospheric water has been found to favor secondary reaction routes of the TCPP radicals, suggesting the intercorrelation among environmental conditions in degradation reactions. Such findings underscore the need for comprehensive assessments to improve predictions of the environmental persistence of emerging pollutants (Li et al., 2017). The study of the degradation mechanisms extends beyond the mere establishment of the degradation products; it includes the investigation of each process's kinetics. Rates of degradation are established based on a number of factors, including temperature, pH, and the presence of inhibitors or catalysts. For instance, freshly produced chemicals may exhibit different degradation patterns for varying environmental conditions, further complicating risk assessments. Understandings about degradation pathways' kinetics assist scientists to predict how quickly the pollutants will degrade and what type of products of degradation they will produce in different ecological pools. Synergistic integration of understandings on environmental factors and chemical reactivity is central to developing effective strategies to counteract pollutant action. With continued research uncovering the intricacy of degradation processes, it is increasingly clear that an interdisciplinary effort involving both chemical and environmental interactions needs to be adopted to facilitate effective environmental management and remediation efforts.

7.2.1. Biodegradation Processes

Biodegradation processes are accountable for the natural degradation of contaminants by providing a mechanism whereby toxicants can be transformed into less or non-toxic substances. Biodegradation is facilitated by a diverse array of microorganisms that use contaminants as an energy source or nutrient to cleave complex organic material. Biodegradation is further divided into primary and secondary degradation. Primary degradation involves enzymatic breakdown of the pollutant into more straightforward chemical structures, while secondary degradation involves conversion of the resultant products into nontoxic compounds. The efficiency of these processes depends on various environmental parameters like temperature, pH, availability of oxygen, and nutrients. These conditions dictate the metabolic pathway employed by microbes, highlighting the need for optimizing environmental factors to optimize the effectiveness of biodegradation. One of the more recent fields of biodegradation involves the utilization of bioelectrochemical systems (BES), specifically oil spill remediation. BES functions through electrical currents in order to stimulate microbial processes to facilitate oxidation of pollutants as electron acceptors or donors. Importantly, it is possible to control electron flow in BES to facilitate particular stimulation of microbial communities and thus facilitate increased degradation rates of petroleum hydrocarbons. As the complexities of the sulfur cycle in sulfate-rich systems are uncovered, it is evident that it is pivotal in hydrocarbon oxidation. While the integration of BES in biodegradation systems has tremendous potential, hurdles are imminent, such as system design, electrode material interactions, and operating parameters (Arends et al., 2017). Continued research is called for to improve the understanding of the mechanisms involved and to make such systems useful in field situations. Furthermore, biodegradation of trace organic contaminants (TOCs) has been of concern due to their bioaccumulation and potential toxicity in the environment. Microbial breakdown of TOCs such as polycyclic aromatic hydrocarbons (PAHs) and polychlorobiphenyls (PCBs) has been found to include long adaptation times, primarily due to limitation in bioavailability as well as the complex relationships between microbial populations and organic matter in the environment (Barret et al., 2012). These relationships are important as they define the sorption behavior of TOCs and hence their bioavailability to microorganisms (Barret et al., 2012). The present study emphasizes the importance of microbial cometabolism, whereby TOC biodegradation depends on co-metabolic activities of other carbon substrates. The information also sheds more light on biodegradation kinetics and highlights the need for new technology strategies towards optimizing TOC removal efficiency in dynamic environments, thus creating avenues towards the improvement of pollutant management regimes (Figure 7.6).

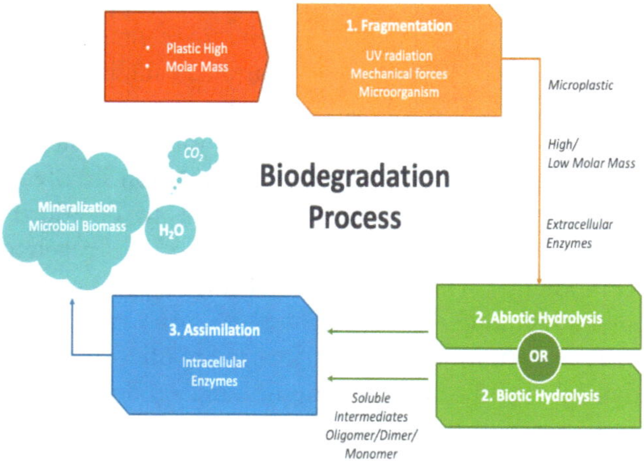

Figure 7.6. *Biodegradation processes.*

Source: https://www.collidu.com/media/catalog/product/img/2/a/2a6c5dafec4af32eb08875ae46024c9eb00a12cd21a6da785e62543e636fd024/biodegradation-slide1.png.

7.2.2. Chemical Degradation Processes

Chemical degradation reactions are at the heart of environmental control against pollution, significantly enabling toxic material degradation to less dangerous or harmless products. Among the many technologies employed, advanced oxidation techniques (AOPs) have proved particularly valuable. AOPs utilize extremely potent oxidizers such as hydroxyl radicals to catalyze the degradation of recalcitrant organic pollutants. Such methods are of especially relevant use in treating recalcitrant compounds—substances that are resistant to degradation by routine biological treatment processes. Recent advances have demonstrated that the use of ultrasound as a source of hydroxyl radical production is used to significantly enhance the effectiveness of pollutant abatement in water, particularly in the removal of recalcitrant pharmaceutical residues and chlorinated organics from wastewater streams, determining the importance of chemical degradation reactions in the environmental remediation scheme. The degradation kinetics of contaminants under various conditions determines the selection and optimization of treatment technologies.

It has been proven by research that ultrasound-assisted AOPs efficiency relies on a myriad of parameters such as ultrasonic frequency, pollutant concentration, and the use of additives such as hydrogen peroxide. Studies indicate that pharmaceuticals degradation follows pseudo-first-order reaction kinetics under the best conditions, with total removal of toxic substances. However, it is important to note that there is always residual dissolved organic carbon, which lingers, a sign that there is the formation of long-lived intermediates that are resistant to other treatment breakdown. This acknowledgment emphasizes the need for a broad perspective of the mechanisms of degradation in order to better design complex treatment systems. In chemical degradation, the role of radical reactions cannot be overemphasized since they are the agents responsible for degrading organic pollutants in aqueous phases. The application of ultrasound not only generates hydroxyl radicals but also stimulates the formation of secondary reactive species, enhancing the oxidation capability of the system. Substrate concentration as well as the character of ultrasonic frequency directly control the rates of degradation observed. For instance, it has been confirmed that initial concentrations tend to inhibit degradation efficiency, while certain frequencies optimize radical generation. Hence, the crosstalk between the chemical nature of pollutants and operation conditions depicts the complexity involved in degradation mechanisms, necessitating research towards continually optimizing processes in handling real-life water treatment problems, as by Delmas et al. (2009) and Bremner et al. (2011) (Figure 7.7).

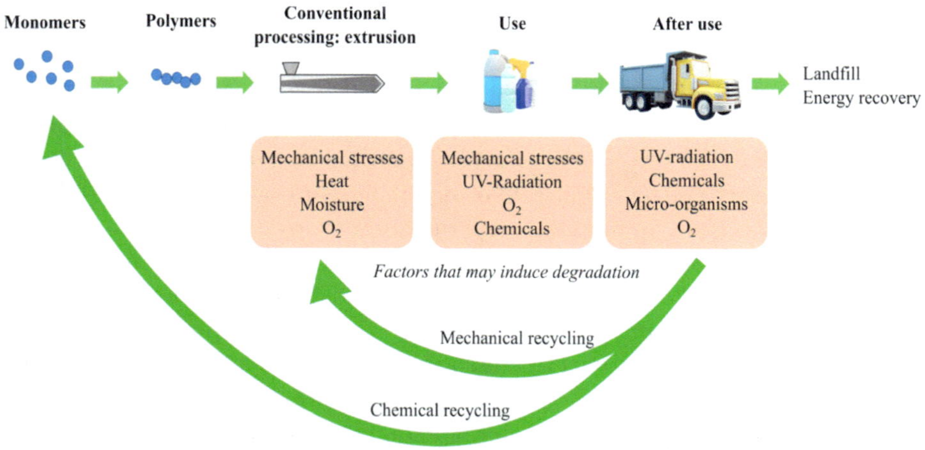

Figure 7.7. *Chemical degradation processes.*

Source: https://www.mdpi.com/molecules/molecules-28-02344/article_deploy/html/images/molecules-28-02344-g001.png.

7.2.3. Photodegradation Processes

Photodegradation reactions are a significant process for the removal of organic pollutants from water environments via chemically degrading contaminants by the energy of light. It is utilized extensively to counteract toxicities of various compounds, including organophosphate esters and antibiotics, which have become well-known environmental concerns. Triphenyl phosphate (TPHP), an organophosphate ester, demonstrated extreme persistence in aquatic environments. Recent studies revealed that effective TPHP degradation is achievable under different advanced oxidation processes, such as ultraviolet-hydrogen peroxide (UV-H_2O_2), whereby photodegradation kinetics are consistent with first-order reaction and half-lives of about 74 minutes under optimal conditions. Such studies demonstrate the importance of understanding photodegradation mechanisms in reducing pollutant concentrations and preventing their environmental impact (Xu et al., 2022). Similarly, metronidazole (MNZ), an antibiotic that has broad usage in human medicine and animal husbandry, is very toxic to the environment. Photodegradation methods involving catalysts like magnesium oxide nanoparticles (MgO NP) have been very efficient in decomposing MNZ in aqueous solutions. In comparative investigations, MNZ photodegradation was marked by significant reductions in toxicity after treatment, with a near complete degradation under specific operating conditions. The process that causes such degradation shows a mechanism through an adsorption phase and photolytic disintegration, indicating that studies on these processes are essential for improved control of pollutants. The findings validate the use of photocatalytic processes as a potential treatment approach for emerging organic pollutants in various environmental contexts (Mohamed et al., 2020). The complexities of photodegradation processes highlight the need for ongoing investigation of reaction mechanisms, kinetics, and efficient catalyst design. Knowledge of how different materials and wavelengths of light influence degradation kinetics has the potential to lead to more effective targeted remediation against diverse pollutants. Application of advanced characterization techniques is required not just for tracing degradation pathways but also for optimizing reaction conditions to ensure maximal efficiency rates. As evident in TPHP and MNZ studies, photodegradation upon UV light exposure is bound to reduce the levels of pollutants and reduce potential ecological hazards. Hence, further research developing the mechanisms and effectiveness of photodegradation is essential to improve innovative approaches to pollution control and environmental health protection (Xu et al., 2022; Mohamed et al., 2020).

7.3. FACTORS INFLUENCING DEGRADATION RATES

Pollutant degradation rates are influenced by an extensive array of variables, from the pollutant itself to environmental parameters and the availability of catalytic agents. Physical characteristics of the pollutant, including molecular structure, polarity, and solubility, influence susceptibility to degradation processes directly. For instance, larger and more complicated organic molecules are more stable to degradation than compact molecules due to steric hindrance and stability to reactive species. Additionally, environmental parameters such as temperature, pH, and light exposure have a profound influence on degradation kinetics. Higher temperatures will accelerate reaction rates, while pH can affect ionization status of the pollutants and therefore sensitivity to reactivity with oxidants. A comprehension of these diverse influences is essential to successfully devise optimal strategies for maximizing degradation processes in real-world applications. Catalysts are central to promoting the degradation of pollutants, enabling more efficient channels for their destruction. Recent research demonstrates that an optimized catalyst could multiply degradation rates exponentially by producing active species that attack pollutants at higher speeds than would take place if they were not present. As an example, the addition of a low-cost iron oxide catalyst during ozonation of sulfamethoxazole (SMX) efficiently enhances its degradation rate from 35.1% to 88.1% within just 30 minutes, which indicates the catalytic effect on reaction kinetics (Luo et al., 2023). Catalyst surface properties such as hydroxyl group availability control the initiation of hydroxyl radical ($^{\bullet}OH$) formation and subsequent pollutant degradation. This highlights the importance of catalyst selection and optimization in formulating efficient pollutant treatment processes. In addition, the interaction among numerous oxidants with pollutants makes the degradation rates complex. Studies indicate that different oxidizing agents, such as ozone, chlorine, and hydrogen peroxide, can have selective efficiencies in breaking down specific contaminants based on numerous factors, such as solution pH. For instance, β-N-methylamino-l-alanine (BMAA) degradation has been shown to be close to total reduction rates in ultra-pure water under some conditions, but the efficiency is considerably reduced when in actual water conditions because of interfering reactions and the formation of by-products (Yan et al., 2023). Ozone, particularly under near-neutral pH conditions, has proved to degrade BMAA effectively, highlighting the need for process optimization of chemical treatment to factor in environmental variability. This is crucial when coming up with effective interventions to reduce pollutant concentrations in environmental systems.

7.3.1. Environmental Conditions

Environmental parameters should be taken into account while evaluating pollutant degradation route kinetics. Influencing the degradation process efficiency can be pH, temperature, and the presence of co-solutes. Fluctuations in pH levels, for instance, can modify the reactivity of pollutants and the activity of degradation agents. In a recent study, it was found that the pH of the environment profoundly influenced the removal efficiency of antiretroviral drug residues by employing a UV-Vis/TiO_2/H_2O_2 hybrid system and that under optimal conditions gave excellent degradation values of 92.56% for emtricitabine, 91.67% for lamivudine, and 94.83% for tenofovir. These variations point to the importance of optimizing treatment conditions depending on specific environmental factors in order to achieve maximum pollutant removal and minimize secondary pollution impact in water treatment systems (Ncube et al., 2025). Temperature is also a crucial parameter in pollutant degradation kinetics. In most chemical processes, increased temperature tends to accelerate the rates of reactions, which can lead to enhanced degradation of contaminants. This temperature-degradation efficiency relationship highlights the importance of monitoring thermal conditions in treatment processes. The O_3/PAA system for iopamidol degradation illustrated how maintaining an alkaline pH not only enhanced the degradation rate constant but also provided stability in the presence of interfering coexisting anions. The technology exhibited the highest rate constant of 0.23 min−1 under alkaline conditions, showing that environmental conditions play an immediate role in determining both degradation efficiency and treatment viability (Bai et al., 2025). Therefore, temperature and pH identification and regulation are necessary while optimizing and developing the process of pollutant degradation. Organic and inorganic co-contaminants can complicate the degradation processes of target pollutants. These compounds can inhibit or enhance the effectiveness of degradation pathways, and it is therefore of prime importance to identify and control them in environmental remediation. In researching the O_3/PAA treatment system, it was discovered that the process was very resistant to coexisting anion interference, with high removal efficiencies for iopamidol. Furthermore, from the quenching experiments, the dominance of hydroxyl radicals in the degradation process was established, showing the importance of interaction studies between different contaminants. Such studies confirm the need for thorough assessments of environmental conditions, as they necessarily establish the performance and reliability of pollution abatement technologies in water ecosystems (Bai et al., 2025).

7.3.2. Chemical Structure of Pollutants

The chemical structure of the pollutants is a critical factor in determining their degradation rates and mechanisms in environmental cleanups. Various classes of pollutants, such as pharmaceuticals, pesticides, and industrial chemicals, possess varying molecular structures that determine their reactivity and biodegradability.

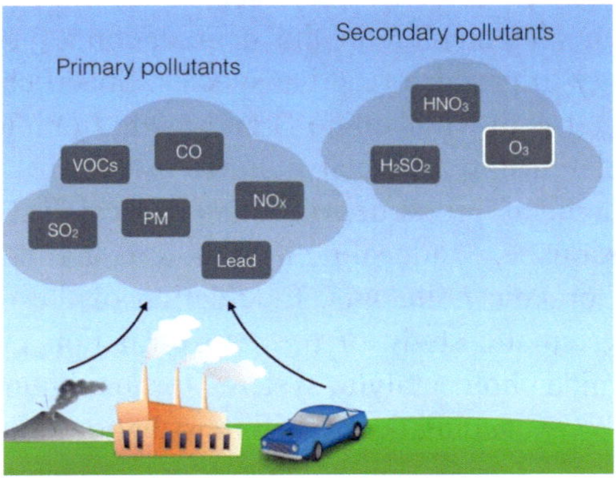

Figure 7.8. *Chemical structure of pollutants.*

Source: https://wou.edu/chemistry/files/2018/08/air_pollution_module-768x616.jpg.

For instance, organic substances with several functional groups are resistant to degradation due to steric hindrance or chemical bonds and therefore will not react to conventional treatment. Structural features of the pollutants can be identified only in this manner, and therefore, it is possible to formulate special measures that enhance degradation efficiency. As witnessed in the case of iodinated disinfection by-products, the insensitivity of contaminants like Iopamidol to traditional treatment processes highlights the need for innovative oxidation methods to achieve effective removal, thereby broadening our approach to addressing these complex entities (Bai et al., 2025). In the past decade or so, advanced oxidation processes (AOPs) have been recognized as a promising approach to the degradation of persistent pollutants, leveraging new chemical structures for improved efficacy. AOPs utilize strong oxidizers such as ozone and peracetic acid to generate hydroxyl radicals capable of breaking down most organic pollutants into less hazardous products. Chemical composition of pollutants is an important determinant in influencing reaction pathways and rate kinetics of treatment. For example, Iopamidol ozonation kinetics using a mixture of ozone and peracetic acid exhibited differential efficacy based on the pH of the solution, thus illustrating the way molecular structure interacts with environmental conditions to determine the effectiveness of the treatment. These studies show the value of structural examination in the optimization of degradation routes and accelerating the kinetics of pollutant conversion (Bai et al., 2025). Other than oxidation strategies, photocatalysis is a helpful process for organic pollution degradation, relying on the knowledge of their chemistry of structure. The use of photocatalysts, such as bismuth tungstate, has been found to possess much potential in the degradation of complex pollutants with photonic energy to produce reactive species. Specifically, efficiencies of degradation produced with Rhodamine B and methyl orange demonstrate the very significant relationship between pollutant structure and photocatalytic activity. Physicochemical properties of the photocatalyst are regulated by structural modifications, e.g., annealing temperature, and hence its activity for the degradation of organic pollutants. Elucidation of the nature of chemical interactions and systematic study of the transformation products will allow researchers to optimize photocatalytic systems for maximum possible removal of pollutants, thereby enhancing the overall remediation efficacy (Figure 7.8).

7.3.3. Presence of Microbial Communities

The presence of microbial communities is an important determinant of contaminant degradation processes and the rates of degradation, particularly in polluted environments involving chemicals such as tetracycline and hydrocarbons. Microbial communities, consisting of various species of bacteria, function to break down complex contaminants into less toxic materials, hence influencing the overall efficiency and rate of degradation.

Diversity among these populations allows for a range of metabolic capability, which may be applied to a range of contaminants. For example, certain bacteria are able to metabolize tetracycline, permitting its ecological footprint to be diminished gradually, while others are able to thrive in hydrocarbon-supported environments, utilizing such chemicals as a source of energy. Understanding microbial community function and structure is essential to the optimization of bioremediation, since it is able to inform policymakers on effective solutions to polluted ecosystems, as by Ahmad et al. (2021). In determining pollutant degradation, interactions within microbial communities convey information regarding their stability and functionality. Factors such as the availability of nutrients, environment, and presence of some pollutants can affect microbial community behavior significantly, finally controlling degradation rates. It has been experimentally proved that the addition of nutrients increases growth of biomass, leading to higher degradation performance, particularly in cases where there is a requirement to increase microbial activity, as put forth by Łukasz et al. (2020).

However, shortcomings in key nutrients will suppress microbial activity, pointing to the requirement of a balanced nutrient loading for supporting active community processes. A sympathetic strategy for the assessment and handling of such communities can then improve more productive bioremediation schemes, facilitating goal-specific applications with the exploitation of microbial diversity and versatility towards specific pollutant substrates. In addition, microbial community function is more than degradation; microbial communities can also affect pollutant bioavailability, and this, consequently, influences degradation efficiency.

For instance, while some communities will rapidly degrade hydrocarbons, development of biofilms and the production of biosurfactants can enhance pollutant solubility, hence enhancing accessibility by microorganisms to degrade such pollutants. This is particularly pertinent for hydrocarbon-polluted soils, where microbial populations must acclimatize to fluctuating environmental factors if they are to maintain their activity. The interaction between microbial communities and environmental characteristics therefore highlights the requirement to completely understand their dynamics and interactions. Through the evaluation of aspects like community composition, metabolic pathways, and environmental interplay, scientists are able to have a better prognosis of pollutant degradation kinetics and create more efficient remediation methods, as reflected by recent studies looking into microbial interactions within pollutant-contaminated environments like the ones discussed by Ahmad et al. (2021) and Łukasz et al. (2020).

7.4. KINETIC MODELS IN POLLUTANT DEGRADATION

Kinetic modeling of contaminant degradation is a significant parameter in the study of the processes by which various contaminants are processed and eventually eliminated from matrices within the environment. Kinetic modeling involves the application of mathematical equations in describing the degree to which contaminants are degraded and defining the parameters governing their processes of degradation. By rigorous analysis of the kinetics of breakdown, researchers have identified key parameters such as the activation energy, a central factor in defining the efficacy of different reactions under specific conditions. Not only does this numerical system predict the long-term fate of pollutants but also enable building tailored remediation strategies that can be effectively utilized in diverse ecosystems. Thus, kinetic models are great tools bridging research theory and practical applications into the environment. Recent years have been driven by the need to explain complicated mechanisms of pollutant degradation, and kinetic model development has been stimulated by this. Various methodologies have emerged that incorporate both model-independent and model-dependent approaches and understanding of degradation pathways is now very complete. For instance, the linear Coats–Redfern and nonlinear Masterplots models have assisted in elucidating the mechanisms of reactions governing thermal and chemical degradation processes. The models allow for further investigation of thermal activation, focusing on certain degradation steps which can be critical for the prediction of pollutant performance under diverse environmental circumstances (Shang et al., 2024). By depicting the steps and their corresponding activation energies, such models present a system that allows for delimitation of the optimal conditions under which pollutant degradation can occur, thereby informing remediation and policy development by regulators. Beyond pollutant degradation, the implications of kinetic modeling extend into more universal environmental issues. Analysis of degradation pathways informs policymakers on how various environmental parameters interact to influence land and water quality. By considering multi-simultaneous routes of advanced socio-environmental hazards such as land degradation, specialists are able to employ kinetic modeling to build comprehensive metrics and indices that examine the overall ecosystem health (Prăvălie et al., 2024). This not only serves for high-priority areas detection that need intervention but also for the establishment of strategic sustainability targets. Lastly, the use of kinetic models in cases of pollutant degradation offers a general overview of environmental health, which facilitates informed decision-making towards efficient control against pollution (Figure 7.9).

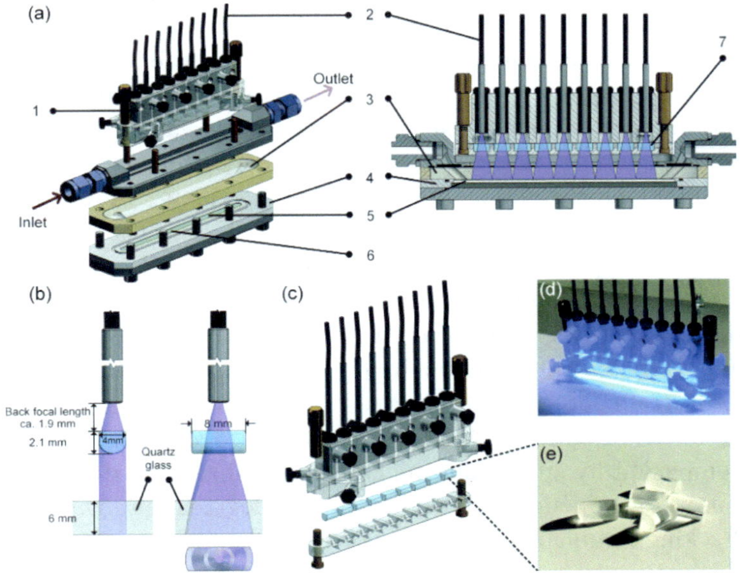

Figure 7.9. *Kinetic models in pollutant degradation.*

Source: https://pubs.rsc.org/image/article/2020/re/d0re00238k/d0re00238k-f2_hi-res.gif.

7.4.1. Zero-Order Kinetics

The zero-order principle is significant in the kinetics of degradation reactions for some pollutants because it determines a reaction rate that is not dependent upon the concentration of the reactant. This characteristic is particularly valid where the toxic contaminant degradation occurs in a saturated system where active site or reactive species availability is the limiting factor. For instance, in thermal plasma treatment of hazardous materials like 4-nitrophenol, kinetics can exhibit zero-order characteristics initially before the sites become depleted. In such systems, the rate of degradation never decreases with lowering reactant concentration, typically leading to better treatment in shorter time spans. Such an insight helps the researchers design maximum treatment schedules so that degradation of the contaminants is at optimal pace throughout.

The relevance of zero-order kinetics is also illustrated in electron beam irradiation (EBI) processes for the removal of contaminants, particularly with compounds such as sulfamethoxazole. From experiments, it has been established that under some conditions, the pseudo-first-order kinetics closely describe the degradation of sulfamethoxazole, but when aqueous conditions and concentrations change, the system might become zero-order, particularly if the reactive species are saturated. Furthermore, EBIs performance reflects high dependence on operating parameters such as pH and competitive existence of other wastewater constituents. This is a critical observation in characterizing the effective operating parameters to obtain optimal pollutant degradation rates, hence highlighting the significance of investigating kinetics as well as pollutant degradation pathways in wastewater treatment processes. The linkage of zero-order kinetics and pollutant degradation pathways poses critical challenges and opportunities for future investigations. For example, even while certain degradation mechanisms can favor a zero-order regime due to saturation effects, the involvement of toxic by-product formation makes the overall efficiency of treatment processes questionable. For thermal plasma technology employed for 4-nitrophenol degradation, enhanced degradation efficiencies were accompanied by significant mineralization; this is an illustration of the bivalent impact of kinetics and mechanisms of degradation towards pollution control. Lacking complete investigation of kinetics and the subsequent products of degradation, treatment processes may not respond to potential environmental impacts from intermediate species. Thus, relating kinetic models, say zero-order dynamics, with an advanced understanding of degradation mechanisms dictates the domain of effective pollutant control strategies in environmental science (Figure 7.10).

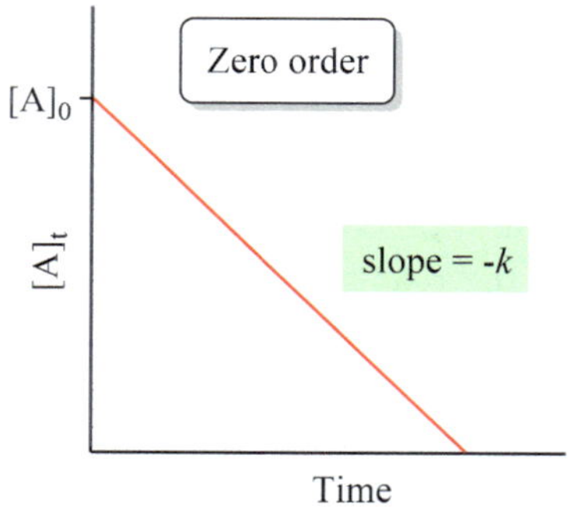

Figure 7.10. *Zero-order kinetics.*

Source: https://general.chemistrysteps.com/wp-content/uploads/2022/04/Zero-order-rate-law-liner-equation-graph.png.

7.4.2. First-Order Kinetics

First-order kinetics refers to a fundamental principle of pollutant degradation studies where the reaction rate is proportional to the concentration of one reactant. This feature implies that as the concentration of the pollutant reduces, the rate of degradation reduces in a uniform manner. A majority of environmental processes including photodegradation and oxidation reactions exhibit first-order kinetics, which enables scientists to forecast and simulate the fate of pollutants over time. For instance, studies on iron-terephthalate metal-organic frameworks (MOFs) illustrate how varying structures influence degradation rates, with studies indicating that one of the frameworks, MIL-101, exhibits a considerably higher pseudo-first-order rate constant than MOF-235 when exposed to contaminants like methylene blue. Such perceptions are critical in developing effective pollutant treatment processes in varied environmental conditions.

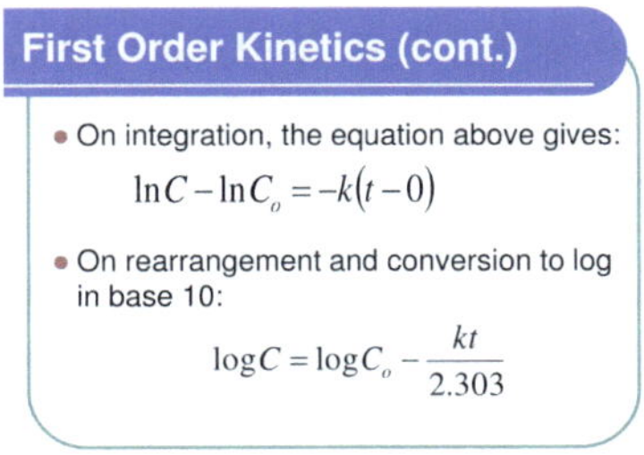

Figure 7.11. *First-order kinetics.*

Source: https://image.slideserve.com/999616/first-order-kinetics-cont-l.jpg.

Empirical modeling with first-order kinetics is widely used with the combined rate equation that models the natural logarithm of the ratio between the original concentration and the concentration remaining at time t. This interaction allows kinetic data analysis with researchers being able to acquire useful parameters, such as the decay constant (k), which shows efficiency in the process of degradation. An appropriate example is tetracycline degradation through interaction with graphene oxide. The kinetic study found a first-order reaction consistent with the Langmuir–Hinshelwood mechanism and demonstrated that the degradation rate is not only a function of the pollutant's concentration but also of ROS generation under the photocatalyst action. This indicates the relevance of both degradation pathways and reaction kinetics in designing effective treatment strategies. The understanding of first-order kinetics is fundamental in the broader context of pollutant degradation processes because it is the foundation of effectiveness and operating mechanism of various catalytic systems applied in remediation. The correlation between kinetic parameters and the efficacy of pollutant degradation makes it easier to discover new materials and means for environmental remediation. Notably, more recent advances in photocatalysis have reported more rapid degradation rates via ROS production, as with tetracycline degradation using graphene oxide. In addition to accelerating the breakdown of contaminants, these processes enhance the range of contaminant that is successfully treated. Through an understanding of the kinetic properties of such systems, scientists are able to determine the best conditions for pollutant breakdown, ultimately working towards creating environmentally friendly management practices (Figure 7.11).

7.4.3. Higher-Order Kinetics

The higher-order kinetics theory is core in the determination of degradation modes of pollutants in various environmental situations. Reactions are usually described with respect to zero, first, and second-order kinetics, but most complex processes have higher-order characteristics where rates of reactions are a function of concentrations of reactants to powers greater than two. These are higher-order reactions in more sophisticated photocatalytic systems, where there are multiple active sites or processes influencing the degradation rate. For example, when photocatalytic degradation of organic pollutants is being carried out, the interaction among multiple catalysts and pollutants may create a reaction order higher than second order, and the analysis of kinetic data becomes difficult. By exploring these higher-order kinetics, researchers are able to better explain the mechanisms of pollutant degradation and therefore create more efficient remediation processes. In photocatalytic degradation, for instance, a number of authors have indicated the predominance of higher-order kinetics over some reactions. For instance, degradations of organic pollutants like 4-nitrophenol have been reported to exhibit pseudo-first-order kinetics, but only in conditionally dependent on the catalyst composition and pollutant concentration (Yadav et al., 2023). Such observations suggest a sophisticated interaction where the effectiveness of photocatalyst-pollutant interactions can result in complex kinetic behaviors. Catalysts also tend to promote photon absorption and electron-hole pair generation, which further tends to alter reaction mechanisms. These changes can drive the process into higher-order regimes, demonstrating the way that both the chemistry of the pollutant and the catalytic material control rates and pathways of reaction. Knowledge of these phenomena is important for the optimization of degradation processes and for the prediction of the effectiveness of different systems of treatment. In addition, research using doped catalysts has also provided insights into higher-order kinetics, particularly in terms of photodegradation of dye pollutants. The catalytic performance of Mn-doped zinc oxide in methyl green degradation is a typical example of how the dopants control kinetic orders (Akram et al., 2023). The enhancement is attributed to improved charge separation and increased surface area, which in turn facilitate more active reaction sites for the reaction and adsorption of the lower-bound pollutant. Interestingly, kinetics in these cases may shift from pseudo-second order to higher-order kinetics, suggesting the influence of doped materials on reaction mechanisms. As such complex interactions are better understood, researchers are in a position to optimize photocatalytic systems to guarantee better degradation efficiencies, supporting the relevance of higher-order kinetics in designing future environmental remediation strategies.

7.5. ANALYTICAL TECHNIQUES FOR STUDYING DEGRADATION

In the degradation of pollutants, analytical techniques play a key role in the determination of routes and kinetics. Techniques such as high-performance liquid chromatography (HPLC) and gas chromatography-mass spectrometry (GC-MS) are central to the determination of amounts of pollutants and their breakdown byproducts. HPLC, indeed, enables scientists to separate and identify complex mixtures, thereby allowing them to have a good understanding of how specific compounds are broken down with time. This method has been rendered more specific with the introduction of advanced detection instruments, increasing the sensitivity and specificity of analysis. Additionally, this interfacing of chromatographic technologies with mass spectrometry increases the identification of unknown degradation intermediates, as needed in determining the effectiveness of many remediation options. Using such methods, researchers can uncover the intricate processes in pollutant degradation, resulting in evidence-based decision-making in environmental management. Spectroscopic techniques yield more information about the degradation of pollutants by providing insight into molecular structures and functional groups. Techniques such as nuclear magnetic resonance (NMR) spectroscopy and infrared (IR) spectroscopy are especially useful in analyzing changes in chemical bonds over the period of degradation. For instance, NMR spectroscopy can generate quantitative data about the concentrations of specific isotopes along degradation pathways, which serves to disclose kinetic parameters governing these processes. Similarly, IR spectroscopy can render detection of changes in functional groups feasible, thus making it possible to monitor in real-time the manner in which the pollutant evolves chemically during degradation. Synergistically, UV-Vis spectrophotometry allows monitoring of pollutant concentration variations with time, further aiding in the understanding of degradation kinetics. In combination, these analytical methods form a multidisciplinary method for studying pollutant degradation. Combination of advanced analytical methods with experimental methods helps in the increased robustness of degradation research. The application of thermal plasma technology, for example, has proved to be highly efficient in the degradation of numerous pollutant types, as in recent studies of 4-nitrophenol degradation, where removal efficiencies of compounds were achieved within a brief time frame. The optical emission spectroscopy of the characterizing plasma jets has provided insight into the dynamic process behavior, lending justification to thermal plasma as a viable remediation technology. Furthermore, investigations of chiral pesticides like profoxydim emphasize the significance of stereoselectivity in degradation kinetics, whereby certain isomers are found to have differential persistence in the environment, hence making risk assessment challenging. Such investigations point towards the utilization of an array of analytical techniques to effectively evaluate both the degradation pathways and kinetics of degradations of pollutants, thereby making enhanced environmental management practices and green agronomy feasible.

7.5.1. Chromatography Methods

In studies of pollutant degradation, chromatography methods have proven to be essential analytical tools for understanding the patterns of degradation and evaluation of the kinetics of pollutants. Through chromatography methods, separation and identification of complex mixtures are enabled such that analysis of intermediate compounds and final products of degradation becomes possible. For instance, liquid chromatography coupled with mass spectrometry (LC-MS) has been highly relevant for tracking degradation of toxic pollutants such as 4-nitrophenol (4-NP), a frequent offender for large-scale environmental problems. As depicted in various studies, the efficacy of chromatographic techniques allows comprehensive assessment of how pollutants degrade under varied conditions, as exemplified by ascertaining both concentration dynamics and transformation products throughout the degradation process (Nandy et al., 2025). The use of these techniques therefore lays doors ajar to optimizing treatment procedures for hazardous waste management. Chromatographic methods also play a vital role in describing pollutant kinetic behavior during processes of degradation. With various configurations such as high-performance liquid chromatography (HPLC) and gas chromatography (GC), researchers can efficiently quantify the rates at which pollutants degrade and assist in describing their pathways. Application of these methods enables derivation of kinetic models that describe pollutant concentration-degradation time relationships. The pseudo-first-order kinetic model, in turn, provides an effective model of the description of toxic compound degradation, for instance, 4-NP in aqueous media, highlighting the time-dependent nature of chemical removal processes (Nandy et al., 2025). On the other hand, the precise quantification with chromatographic analysis is indispensable for validating theoretical models and justification of more efficient technologies for degradation particular pollutants. Moreover, the information gained through chromatography not only informs the degradation kinetics but also assists in the process of optimizing the treatment protocols of the pollutant across different environmental conditions. The use of advanced chromatographic techniques ensures that researchers can identify the thousands of degradation products that are generated, revealing potential toxicity or environmental persistence of metabolites. For instance, in thermal plasma treatment of the contaminants, chromatographic methods allow for the systematic collection and explanation of reaction by-products and their corresponding kinetics, thus facilitating a qualitative understanding of their degradation processes. This reiterates the importance of chromatography in environmental chemistry and microbiology, which allows for innovative solutions that can reduce chemical contaminant ecological impact. With advancing science, the inclusion of chromatography techniques will be at the center of continuing development of our knowledge of pollutant degradation pathways and kinetics within various ecosystems (Figure 7.12).

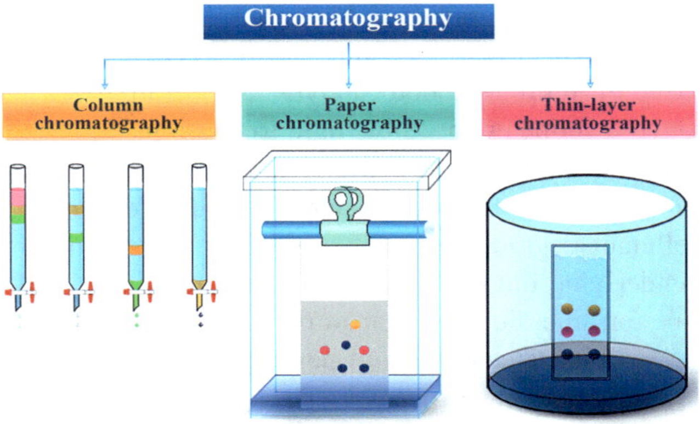

Figure 7.12. *Chromatography methods.*

Source: https://www.priyamstudycentre.com/wp-content/uploads/2021/11/chromatography-techniques-681x478.png.

7.5.2. Spectroscopy Methods

The use of spectroscopic methods is of utmost significance in interpreting the mechanism and kinetics of pollutant degradation, particularly when advanced oxidation processes are implicated. Methods such as Fourier Transform Infrared Spectroscopy (FTIR) and UV-Vis spectroscopy are significant in that they provide key information on chemical change during contaminant degradation. For instance, FTIR facilitates researchers to monitor changes in pollutant molecules' functional groups, hence enabling them to detect end degradation products and intermediate products. This information is essential in determining the efficiency of different processes of treatment and also in optimizing operating conditions to obtain maximum pollutant removal. By describing chemical change at the molecular scale, spectroscopy methods offer a comprehensive approach to identifying pollutant degradation mechanisms that is critical for designing adequate environmental remediation procedures. Monitoring the kinetics of pollutant degradation is another integrated field where spectroscopy methods are highly useful. UV-Vis spectroscopy, in particular, can be applied to monitor the concentration of impurities with time for reaction rate calculations and assessment of efficacy of degradation. This is particularly important when investigating processes involving reactive oxygen species generated through photocatalytic systems as it is capable of providing an account of the influence of variables such as intensity of light, catalyst, and concentration of contaminants on rates of degradation. For example, use of highly photoactive materials like the $Cu/Bi_2Ti_2O_7/rGO$ composite allows better dynamics of charge carriers by virtue of improved photocatalytic activity in drug degradation like ibuprofen and tetracycline upon visible light irradiation (Acevedo et al., 2019). This information is useful for optimizing treatment systems and adapting them to different contaminants and wastewater types. In addition to assessing the kinetics of pollutant degradation, spectroscopic methods enlighten us on the mechanisms underlying the processes. For example, *in situ* analysis of ozone oxidation using on-line Fourier Transform Infrared Spectroscopy provides details on the evolving decomposition pathways of species like 1,4-dioxane during wastewater treatment processes. The study indicates that at alkaline pH, breakage is faster through the intermediates such as ethylene glycol, while acidic environments lead to breakage at reduced rates and other end products (Barndok et al., 2014). Scientists can plan means to achieve maximum destruction efficiency of harmful agents by identifying the reaction channels. Lastly, application of spectroscopy methods increases our knowledge of the pollution degradation pathways, thus supporting the development of more effective technologies aimed at minimizing the environmental footprint.

7.5.3. Mass Spectrometry (MS)

Mass spectrometry (MS) is a crucial analytical tool in pollutant degradation pathways and kinetics studies, with the ability to deliver precise molecular information on a broad variety of contaminants. Mass spectrometry informs scientists of very precise identification of parent compounds as well as their breakdown products through detection of the mass-to-charge ratio of ions. This becomes critical in dealing with complicated environmental matrices where pollutants such as pharmaceuticals and their metabolites coexist. For instance, identification of byproducts caused by degradation processes accounts for pathways along which pollutants are degraded and eliminated to inform remediation strategies.

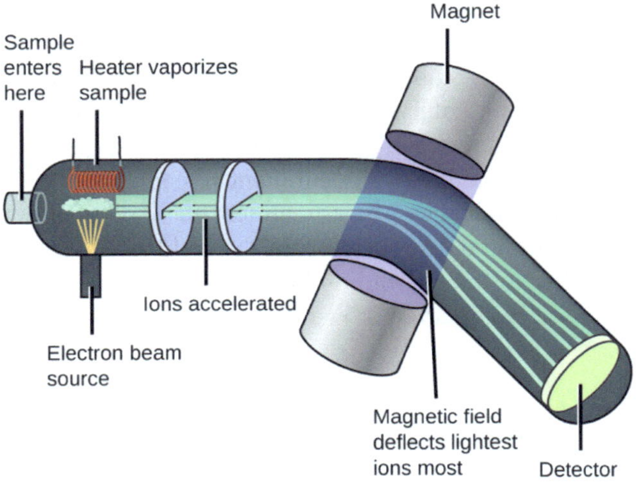

Figure 7.13. *Mass spectrometry (MS).*

Source: https://i.pinimg.com/originals/06/49/13/0649133927da8d6cb5fee9651957eb23.png.

The integration of mass spectrometry with other analytical techniques allows for thorough description of pollutant mobility in diverse environments, revealing not only initial degradation steps but also intermediates generated during such degradation processes, eventually aiding in effective pollution management. Mass spectrometry was also instrumental in recent studies to reveal the degradation mechanisms of single pollutants like diclofenac and losartan, whose environmental relevance is significant. As can be observed in a study on the degradation of diclofenac, mass spectrometry was employed to authenticate various byproducts that are formed in the course of catalytic degradation with terpyridine-based metal complexes. The results revealed the presence of intermediates that can further be used for developing more efficient photocatalysts to eliminate contaminants (Mughal et al., 2025). Similarly, research focused on losartan degradation by advanced oxidation processes used mass spectrometry to identify nine various byproducts, seven of which were discovered for the first time. This advancement not only enhances the understanding of losartan degradation pathways but also aids the maximization of the treatment protocols for water systems affected by pharmaceutical pollutants (Medici et al., 2024). Moreover, the evaluation of pollutant degradation kinetics is significantly enhanced by the integration of mass spectrometry data to create knowledge on reaction mechanisms and rates. Accuracy of mass spectrometry enables real-time monitoring of pollutant concentrations and their by-products, which are useful for simulating reaction kinetic models. With pseudo first-order kinetics, as various studies have observed, researchers can ascertain the performance of different catalytic reactions in decomposing target pollutants, such as reactive iron catalyst with peroxymonosulfate for losartan elimination. This method not only facilitates contaminant remediation strategy evaluation but also advances sustainable water treatment technologies development. In general, mass spectrometry application in the field of environmental science enhances our ability to trace pollution breakdown processes for contaminants, enabling more effective intervention strategies against environmental pollution (Figure 7.13).

7.6. CASE STUDIES OF POLLUTANT DEGRADATION

Degradation of pollutants is a subject of environmentally important research, and particularly for harmful organic pollutants in industrial wastewaters. One particular study focuses on the use of thermal plasma technology in degrading 4-nitrophenol (4-NP), which is a highly toxic chemical compound. With a 10 kW DC non-transferred arc thermal plasma torch purpose-built to meet the specifications, the scientists achieved a very good degradation efficiency with 94% removal and 58% mineralization of 4-NP in 30 minutes at an optimum input power of 6.2 kW. Reaction kinetics followed pseudo-first-order models, which showed an established degradation path and enabled compound transformation predictions. This case study illustrates the effectiveness of plasma technology not just in the abatement of specific pollutants but also in furthering our understanding of highly evolved degradation mechanisms that occur in advanced oxidation processes, setting a precedent for future research on similar technologies (Nandy et al., 2025). A second compelling case study employs cobalt-based catalysts to activate peroxymonosulfate (PMS) for use in the treatment of the degradation of pharmaceutical contaminants, namely tetracycline (TC). The research documents a novel morphology-designed spinelloid phase (g-$CoMn_2O_4$), having the 3D hierarchical nanofloral morphology to enhance the accessibility of active sites. Therefore, it attained a superior degradation efficiency of 98.90% within only 10 minutes, with a remarkable 3.6-fold enhancement in degradation capacity compared to traditional PMS systems. This study emphasizes the significance of material engineering and design for catalysis, demonstrating intricate biochemical interactions that drive pollutant degradation. Dramatic enhancement of reaction kinetics was illustrated, most importantly the identification of nonradical oxidation processes, which are an essential factor in the degradation process, thereby enlarging our portfolio of understanding how to optimize catalysis for environmental cleansing (Zhang et al., 2025). The examination of pollutant degradation pathways and kinetics by way of these case studies not only demonstrates the promise of novel technologies but also emphasizes the necessity for continued innovation in the remediation of the environment. Using the implementation of plasma treatment technology as well as advanced catalysts, researchers are able to design more effective ways of degrading harmful chemicals, leading to improved water quality and reduced ecological risk. These studies provide us valuable information regarding the reaction mechanisms and rates behind and pave the way for the optimization of current practices and new methods for pollutant degradation. The application of such new strategies becomes inevitable in the long run to address the growing effects of environmental pollutants and ensure sustainable management of our water resources in the wake of global industrialization and pollution problem.

7.6.1. Pesticide Degradation in Soil

Soil degradation of pesticides is a significant aspect of environmental management and agricultural sustainability. Persistence of pesticides in the soil has severe ecological consequences that disrupt biodiversity and food safety. Most commonly used pesticides such as pyrethroids act rapidly against insects but are resistant and thus remain as environmental contaminants. This accumulation can harm non-target organisms and further pollute water bodies through runoff, making it important to understand degradation pathways. Research has focused on the role of microbial communities in biodegradation due to the fact that some microorganisms possess the enzymatic ability to transform and degrade pesticide chemicals, thereby lowering their impact on the ecosystem and enhancing soil quality. Microbial degradation of pesticides involves a series of mechanisms like hydrolysis, oxidation, and ring cleavage that can lead to the formation of less toxic metabolites. Certain microbial strains have been shown to degrade a large number of pesticides, for example, the degradation of fenvalerate by *Citrobacter freundii* strain CD-9. Under optimized conditions, the strain exhibited an appreciable rate of degradation efficiency by degrading nearly 88% of fenvalerate in 72 hours. The characterization of metabolites generated in this degradation process, such as o-phthalaldehyde, helps further the understanding of the metabolic processes involved. More importantly, the understanding of these products of degradation can be used in guiding the formulation of remediation techniques that utilize these microorganisms in bioremediation applications, being a sustainable method of pesticide pollution control in agricultural soils (Tang et al., 2020). Pesticide degradation is also a subject that intersects with more extensive water treatment strategies, since runoff residues from agriculture create residues that must be removed effectively to help safeguard aquatic environments. Advanced oxidation processes, adsorption technologies, and membrane technology have all been tested for efficacy in removing pesticide residue from treated water sources. These removal methods can be coupled with soil-based degradation processes, and thus there may be hybrid removal systems that optimize efficiency. An understanding of how mechanisms of soil degradation interact with water treatment operations is imperative in developing comprehensive strategies that address pesticide contamination. By identifying the best removal techniques for specific types of pesticides, it is possible to enhance overall ecological quality and ensure food safety while minimizing the economic cost of pesticide contamination (Saleh et al., 2020) (Figure 7.14).

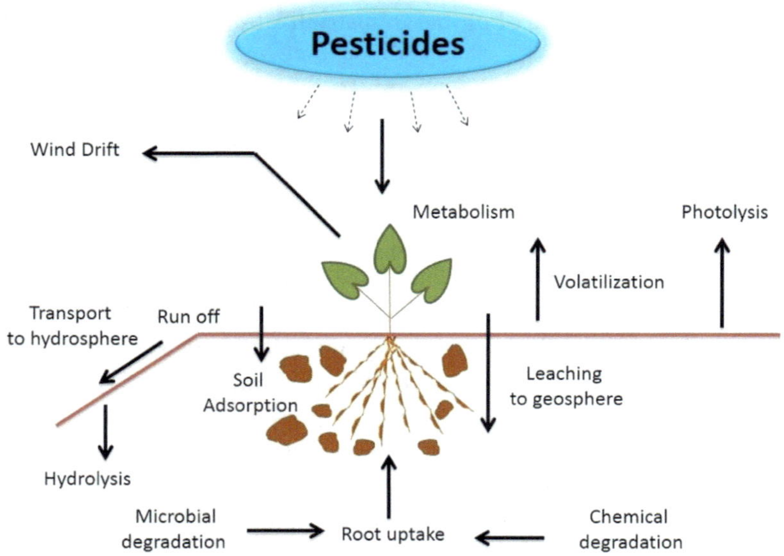

Figure 7.14. *Pesticide degradation in soil.*

Source: https://www.researchgate.net/profile/Munees-Ahemad/publication/259503288/figure/fig1/AS:340848328888324@1458275982055/The-environmental-fate-of-pesticides.png.

7.6.2. Heavy Metal Remediation in Water

Contamination of water sources by heavy metals is of significant concern to the environment and human health and needs to be effectively remediated. The chronic toxic potential combined with the bioaccumulation property in living organisms renders heavy metals, such as lead, mercury, and cadmium, of special interest. The need for remediation is emphasized by the growing lack of pure water, which increases the necessity of eliminating such contaminants from wastewater before it is released into broader ecosystems. Conventional remediation techniques, such as chemical precipitation and ion exchange, have been most commonly used; however, these methods have shortcomings like inordinate cost and the generation of secondary waste. Therefore, the development of sustainable, eco-friendly heavy metal remediation technologies has been among the top focuses in environmental science and policy, advancing various Sustainable Development Goals that strive for universal access to clean and safe water. Among the emerging technologies for heavy metal remediation, adsorption has been among the most favored processes due to its simplicity and efficiency. This method employs a variety of materials, including activated carbon, agricultural residues, and even biogenic sources like microorganisms, to adsorb and remove heavy metals from contaminated water. Versatility in adsorbents allows customization based on specific pollutants, which optimizes overall treatment effectiveness. Adsorption kinetics have been found to influence the process, which is a critical component in determining the rate and capacity uptake of metal. Understanding of the adsorbent-metal ion interaction makes it possible for scientists to design material choice and material design for improved performance in real application. These advancements not only reduce pollution but also promote sustainable technologies in water treatment technologies, supplementing the initiative towards environmental protection. Future advancement of photocatalytic processes has the potential of enhancing the efficiency of heavy metal removal by transition metal dichalcogenides (TMDCs). TMDCs possess physicochemical properties inherent to the induction of light-driven redox processes that can be responsible for the degradation of heavy metal impurities. Scaffolding TMDCs with other semiconductors is a method aimed at enhancing charge separation and broadening spectral response, thus improving photocatalytic activity. A deep understanding of the mechanistic aspects of charge carrier generation and interfacial electron transfer is critical in order to optimize these materials under real-world conditions. As highlighted in recent review studies, though TMDCs are highly promising for pollutant degradation, photo-corrosion and stability under working conditions remain limitations to date. These limitations are overcome by new heterojunction architecture and application of cocatalysts but remain central to the development of scalable solutions for the imperative task of heavy metal pollution in water systems (Takhar et al., 2025; Sheraz et al., 2024) (Figure 7.15).

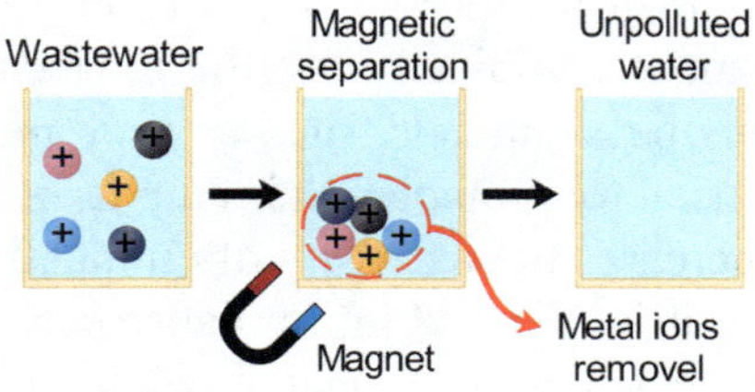

Figure 7.15. *Heavy metal remediation in water.*

Source: https://media.springernature.com/lw1200/springer-static/image/art%3A10.1038%2Fs41545-021-00127-0/MediaObjects/41545_2021_127_Fig2_HTML.png.

7.6.3. Pharmaceutical Degradation in Wastewater

The issue of pharmaceutical degradation in wastewater gains more significance given the environmental and public health effects resulting from leftover pharmaceutical pollutants. Discharge of different compounds such as nonsteroidal anti-inflammatory drugs and antibiotics into water bodies occurs predominantly via wastewater effluent discharge. Conventional treatment processes lack effectiveness in degrading such pollutants, indicating potential for advanced oxidation processes (AOPs) to provide better mechanisms for degradation. AOPs are capable of yielding very reactive species that facilitate the breakdown of intricate pharmaceutical compounds, thus highlighting the need for new degradation methods. Most important in understanding how the procedures operate is the determination of the reaction kinetics, which provides insight into how effectively and quickly individual pharmaceuticals may be targeted and broken down within wastewater treatment. Recent studies have addressed the degradation of some drugs using various new photocatalysts and proved their efficiency in solving pharmaceutical wastewater issues. For instance, terpyridine metal complexes have been very promising in degrading diclofenac potassium (DCF), which is a common drug pollutant. The complex process by which the complexes result in degradation involves the generation of reactive species under optimum conditions like pH and catalyst loading. Interestingly, the resulting degradation mechanisms have been shown to follow pseudo-first-order kinetics, demonstrating their efficacy in real-world applications. Besides the high efficiency of the photocatalysts in favoring the rapid degradation of pollutants, it also showcases their recyclability in several cycles, which is imperative in sustainable wastewater treatment processes as highlighted by studies into the synthesis and application of such compounds (Mughal et al., 2025). Further research on degradation mechanisms of other drug substances, such as clofibric acid (CA), in advanced oxidation processes has been conducted. Research of the Fe^{2+}/persulfate (PS) system indicated a high removal efficiency of CA but low mineralization, which indicates the complexity of drug degradation kinetics in seawater wastewater systems. Interestingly, chlorine was revealed to improve the efficiency of degradation, albeit at the expense of creating unnecessary chlorinated by-products with ecological danger potential. Realizing such processes makes it obvious that while AOPs are efficient in degrading drugs, the formed intermediates could vary in toxicity, thus complicating pollution cleanup. This points to a significant area for future studies—examining the complete environmental impact of the degradation products formed by such sophisticated processes (Wang et al., 2024) (Figure 7.16).

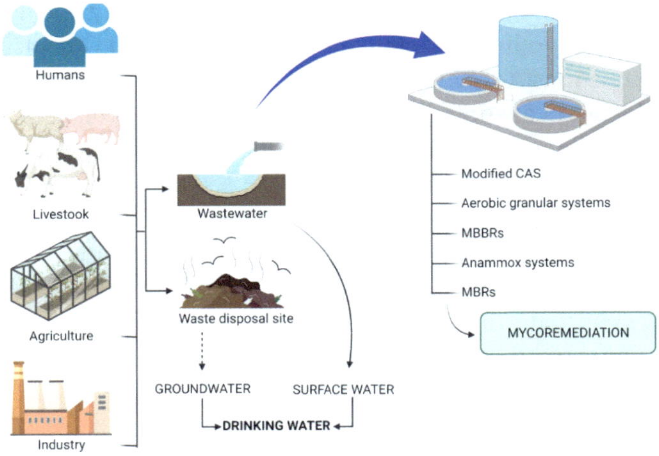

Figure 7.16. *Pharmaceutical degradation in wastewater.*

Source: https://www.frontiersin.org/files/Articles/869332/fmicb-13-869332-HTML/image_m/fmicb-13-869332-g001.jpg.

SUMMARY

In summary, research into pollutant degradation pathways and kinetics is essential in advancing our understanding of wastewater treatment processes. The research here reveals the sophisticated mechanism for contaminant degradation, particularly by presenting new methods such as the utilization of graphene oxide (GO) photogenerated electron–hole pairs. This approach not only improves the efficacy of pollutant removal like tetracycline but also points towards the importance of knowing reaction mechanisms involved. Such research is of immense value in the context of developing next-generation treatment technologies that can tackle advanced and emerging pollutants in wastewater treatment plants with a view to protecting the environment and public health. Moreover, identifying the products of degradation and reaction pathways is extremely important in the evaluation of the performance of diverse wastewater treatment processes. Through the employment of methods like compound similarity networks (CSN), researchers can track high volumes of data derived under nontargeted analysis and better understand the fate of pollutants. The results of such research can reveal more than one degradation pathway and potential biohazards of emerging pollutants. Thus, the use of advanced analytical techniques in pollutant degradation studies gives investigators the means to approach the complexities of real wastewater pollution effectively, and treatment can be conducted efficiently as well as safely. Lastly, advancing our knowledge of pollutant degradation kinetics and mechanisms has far-reaching implications for environmental sustainability. By applying novel materials and analytical strategies, researchers are able to make more efficient and long-lasting means of purifying wastewater, which is significant in relation to increasing rates of pollution and legislative standards. As demonstrated by the tetracycline removal research and application of CSN strategies, research in this field is still ongoing. More investment in research and technology needs to happen in order to further cultivate these methods, promote cleaner water sources, and contribute to a healthier society and environment overall in an increasingly polluted world.

In the context of the degradation pathways and kinetic processes of the pollutants, some significant observations have been made with respect to the impact of tire-derived pollutants (TDPs), namely *N*-(1,3-dimethylbutyl)-*N′*-phenyl-*p*-phenylenediamine (6PPD) and its oxidized derivative, 6PPD-quinone (6PPD-Q). Both compounds have been extensively found in urban areas, which subjects the public to serious health threats. Scientific research has shown that TDPs destabilize soil microbial communities, leading to reduced biodiversity and enhanced pathogenic species capable of deteriorating human health conditions, including respiratory diseases. The pathways of degradation of such pollutants must be identified within the framework of soil health in the urban setting, as soil quality degradation can hinder essential ecosystem functions such as nutrient cycling and water filtration, ultimately affecting urban resilience (Ihenetu et al., 2025). This necessitates an integrated treatment approach to pollutant management with environmental remediation solutions as well as public health consequences. Again, a crucial characteristic of pollutant degradation is the autooxidation of very oxygenated molecules involved in complex atmospheric processes. Studies have unraveled a common mechanism for autooxidation of peroxy radicals, which involve three successive steps of O_2 addition. The process is significant not only with regard to the knowledge of the mechanism of degradation of pollutants in the atmosphere but also for its implication towards the formation of secondary organic aerosols (SOAs). Observations show that these responses are controlled by highly oxygenated intermediates, which influence climate and air quality by their potential for generating climatically active SOAs (Wang et al., 2017). By the disclosure of the intrinsic kinetics and structural aspects of such autooxidation reactions, research acquires seminal insights relevant to technologies that aim to mitigate fuel combustion pollutants as well as to our comprehension of atmospheric chemistry. The combination of these findings underscores the immediate need for interdisciplinarity linking environmental health and atmospheric science. The chemical handling of TDPs and the complicated behavior of oxygenated intermediates in pollution degradation demonstrate the complicated dependencies between air quality and urban ecosystems. The evidence then shown suggests that autooxidation-based systematic degradation mechanisms may deeply influence the toxicity and persistence of most organic pollutants. Further, addressing these pollutants must be an effort in developing well-established policy frameworks that take into account the overall environmental impact, particularly in urban centers where human exposure risks are higher. Therefore, the findings above not only benefit science but also reflect the imperative for unified efforts at managing cities sustainably towards lessening threats to health from environmental contaminants (Ihenetu et al., 2025; Wang et al., 2017).

REFERENCES

1. Acevedo, R., Anbarasan, P. M., Gkanas, E., et al. (2019). High efficient catalytic degradation of tetracycline and ibuprofen using visible light driven novel Cu/ $Bi_2Ti_2O_7$/rGO nanocomposite: Kinetics, intermediates and mechanism. https://core.ac.uk/download/228156750.pdf (accessed on 28th September 2025).
2. Ahmad, F., Zhu, D., & Sun, J. (2021). Environmental fate of tetracycline antibiotics: Degradation pathway mechanisms, challenges, and perspectives. *Environmental Sciences Europe*, *33*. https://doi.org/10.1186/s12302-021-00505-y (accessed on 28th September 2025).
3. Ai, L., Guo, J., Chen, H., Hu, D. K., & Lu, P. (2024). Degradation of isotianil in water and soil: Kinetics, degradation pathways, mechanisms, and ecotoxicity assessments. *Journal of Agricultural and Food Chemistry*. https://www.semanticscholar.org/paper/00c44c757ba109ebd9ac4c661d57e27a2370c838 (accessed on 28th September 2025).
4. Akram, R., Fatima, A., Almohaimeed, Z. M., Farooq, Z., Qadir, K. W., & Zafar, Q. (2023). Photocatalytic degradation of methyl green dye mediated by pure and Mn-doped zinc oxide nanoparticles under solar light irradiation. *Adsorption Science & Technology*, *2023*. https://www.semanticscholar.org/paper/8226cd3bd7757cb759d1895ee46c3fbe9b3621dc (accessed on 28th September 2025).
5. An, L., Chen, B., Zhang, Y., Li, H., Huang, R., Li, F., & Tang, Y. (2024). Compound similarity network as a novel data mining strategy for high-throughput investigation of degradation pathways of organic pollutants in industrial wastewater treatment. *Analytical Chemistry*. https://www.semanticscholar.org/paper/453049fb8c5c04a7dd86a5da65b952f5cb897439 (accessed on 28th September 2025).
6. Arends, J. B. A., Aulenta, F., Bestetti, G., et al. (2017). Electro bioremediation of oil spills. https://core.ac.uk/download/287613479.pdf (accessed on 28th September 2025).
7. Ault, A. P., Chan, M. N., Creamean, J. M., Edgerton, et al. (2011). Measurements of isoprene-derived organ sulfates in ambient aerosols by aerosol time-of-flight mass spectrometry—Part 2: Temporal variability and formation mechanisms. https://core.ac.uk/download/4892628.pdf (accessed on 28th September 2025).
8. Bahrami Asl, F., Baneshi, M. M., Dadban Shahamat, Y., et al. (2016). Heterogeneous catalytic ozonation of 2,4-dinitrophenol in aqueous solution by magnetic carbonaceous nanocomposite: Catalytic activity and mechanism. https://core.ac.uk/download/52205519.pdf (accessed on 28th September 2025).

9. Bai, D., Liu, C., Zhang, S., Dong, H., Sun, L., & Yuan, X. (2025). Activation of peracetic acid by ozone for recalcitrant pollutant degradation: Accelerated kinetics, byproduct mitigation, and microbial inactivation. *Water*. https://www.semanticscholar.org/paper/ae7f8e2d21cf18fd166adc0e4e1d004c0865e821 (accessed on 28th September 2025).

10. Barndok, H., Cortijo, L., Hermosilla, D., et al. (2014). Removal of 1,4-dioxane from industrial wastewaters: Routes of decomposition under different operational conditions to determine the ozone oxidation capacity. https://core.ac.uk/download/33102935.pdf (accessed on 28th September 2025).

11. Barret, M., Braun, F., Cea-Barcia, G., et al. (2012). Anaerobic removal of trace organic contaminants in sewage sludge: 15 years of experience. https://core.ac.uk/download/33663602.pdf (accessed on 28th September 2025).

12. Bremner, D. H., Burgess, A. E., & Chand, R. (2011). The chemistry of ultrasonic degradation of organic compounds. https://core.ac.uk/download/228176790.pdf (accessed on 28th September 2025).

13. Ceriani, E., Madureira, J., Margaça, F. M. A., et al. (2017). Oxidation of clofibric acid in aqueous solution using a non-thermal plasma discharge or gamma radiation. http://arxiv.org/abs/1708.07483 (accessed on 28th September 2025).

14. Chiu, Y. H., Chang, T. F. M., Chen, C. Y., Sone, M., & Hsu, Y. J. (2019). Mechanistic insights into photodegradation of organic dyes using heterostructure photocatalysts. *Catalysts*, *9*, 430–430. https://doi.org/10.3390/catal9050430 (accessed on 28th September 2025).

15. Choi, H. I., Kim, T., Kim, J. W., Lee, G. J., Choi, J., Chae, Y., Kim, E., et al. (2024). Rat pharmacokinetics and *in vitro* metabolite identification of KM-819, a Parkinson's disease candidate, using LC-MS/MS and LC-HRMS. *Molecules*, *29*. https://www.semanticscholar.org/paper/88517891a9b1b252d474cc298ef40d9824dab052 (accessed on 28th September 2025).

16. Chu, R., Khan, Z. U. H., Zhu, Y., Zhao, R., Wu, W., & Sun, J. (2025). Theoretical simulation degradation of bromoxynil by ozonation in liquid phase: Mechanism pathways, kinetics, and ecotoxicity. *International Journal of Quantum Chemistry*. https://www.semanticscholar.org/paper/d7fd20f9bc4f35a941c1402c67194e4b9e5c1267 (accessed on 28th September 2025).

17. Delmas, H., Julcour-Lebigue, C., Jáuregui Haza, U. J., et al. (2009). Sonolysis of levodopa and paracetamol in aqueous solutions. https://core.ac.uk/download/12043344.pdf (accessed on 28th September 2025).
18. El Aouni, A., El Ouardi, M., Arab, M., Saadi, M., Haspel, H., Kónya, Z., Ben Ali, A., et al. (2024). Design of bismuth tungstate Bi_2WO_6 photocatalyst for enhanced and environmentally friendly organic pollutant degradation. *Materials*, *17*. https://www.semanticscholar.org/paper/3e99173d674b95839b38771dbefc6a44adfbabe4 (accessed on 28th September 2025).
19. El Bouraie, M. M., & Ibrahim, S. (2020). Differentiation between metronidazole residues disposal by using adsorption and photodegradation processes onto MgO nanoparticles. *International Journal of Nanomedicine*, *15*, 7117–7141. https://www.semanticscholar.org/paper/2baf33608cc14b61fca3d885e13c61afe3a942c0 (accessed on 28th September 2025).
20. Ihenetu, S. C., Xu, Q., Fang, L., Azeem, M., Li, G., & Enyoh, C. (2025). 6PPD and 6PPD-quinone in the urban environment: Assessing exposure pathways and human health risks. *Urban Science*. https://www.semanticscholar.org/paper/237eb02872bfe3d386cdcbf1b3a3ae6454223280 (accessed on 28th September 2025).
21. Kengne, B. T., Sun, Y., Wang, S., Wang, J., Bułka, S., Pyszynska, M., & Sudlitz, M. (2025). Kinetic analysis and transformation pathways of sulfamethoxazole degradation in water and wastewater under electron beam irradiation. *Water*. https://www.semanticscholar.org/paper/cbcd873370b4b075664d8db6ef327aecf467acf9 (accessed on 28th September 2025).
22. Ławniczak, Ł., Woźniak-Karczewska, M., Loibner, A. P., Heipieper, H. J., & Chrzanowski, Ł. (2020). Microbial degradation of hydrocarbons—Basic principles for bioremediation: A review. *Molecules*, *25*, 856–856. https://doi.org/10.3390/molecules25040856 (accessed on 28th September 2025).
23. Li, C., Chen, J., Xie, H., Zhao, Y., Xia, D., Xu, T., Li, X., et al. (2017). Effects of atmospheric water on ·OH-initiated oxidation of organophosphate flame retardants: A DFT investigation on TCPP. *Environmental Science & Technology*, *51*(9), 5043–5051. https://www.semanticscholar.org/paper/2e34a8685d8373b9a61274c71f2ba1435a91d19f (accessed on 28th September 2025).
24. Li, X., Li, Q., Chen, X., Yan, B., Li, S., Deng, H., & Lu, H. (2025). Efficacy, kinetics, and mechanism of tetracycline degradation in water by O_3/PMS/FeMoBC process. *Nanomaterials*, *15*. https://www.semanticscholar.org/paper/dbf17a7d5bd687489c95116e7ad9b5646672b933 (accessed on 28th September 2025).

25. Li, X., Liu, M., Wen, L., Li, R., Liu, Y., Yang, S., Ding, D., et al. (2025). Synergism between activated carbon and Fenton reaction for organic pollutant degradation: The hitherto overlooked role of dynamic single-atom sites. *Environmental Science & Technology*. https://www.semanticscholar.org/paper/e789de29ebf81ba2402049ef63ede496ef3bc27f (accessed on 28th September 2025).

26. Luo, L., Sun, Z., Chen, Y., Zhang, H., Sun, Y., Lu, D., & Ma, J. (2023). Catalytic ozonation of sulfamethoxazole using low-cost natural silicate ore supported Fe_2O_3: Influencing factors, reaction mechanisms and degradation pathways. *RSC Advances*, *13*, 1906–1913. https://www.semanticscholar.org/paper/6d0f914b9883ac3caa5cda59aae8f14522003c34 (accessed on 28th September 2025).

27. Makvandi, P., Iftekhar, S., Pizzetti, F., Zarepour, A., Zare, E. N., Ashrafizadeh, M., Agarwal, T., et al. (2020). Functionalization of polymers and nanomaterials for water treatment, food packaging, textile and biomedical applications: A review. *Environmental Chemistry Letters*, *19*, 583–611. https://doi.org/10.1007/s10311-020-01089-4 (accessed on 28th September 2025).

28. Medici, A., Luongo, G., Pedatella, S., Previtera, L., Di Fabio, G., & Zarrelli, A. (2024). Tackling losartan contamination: The promise of peroxymonosulfate/Fe(II) advanced oxidation processes. *Molecules*, *29*. https://www.semanticscholar.org/paper/86006dc8e0d12f0c94c3c120207a47fc36277e6e (accessed on 28th September 2025).

29. Moore, S., & Sarazen, M. L. (2023). Stability and kinetics of iron-terephthalate MOFs with diverse structures in aqueous pollutant degradation. *AIChE Journal*. https://www.semanticscholar.org/paper/b222e34f814eaf24449bc869f70b8be45c6c1714 (accessed on 28th September 2025).

30. Mughal, E., Naeem, N., Zainab, F., Javaid, A., Imran, M., Pandey, S. K., Sadiq, A., et al. (2025). Alkoxy-modified terpyridine complexes as efficient photocatalysts for diclofenac potassium degradation in pharmaceutical wastewater. *RSC Advances*, *15*, 10602–10620. https://www.semanticscholar.org/paper/3e302ed16527098255557c35335e8305990e19f63 (accessed on 28th September 2025).

31. Nandy, N., Pasupathi, A., & Subramaniam, Y. (2025). Degradation of 4-nitrophenol using Ar/air thermal plasma in submerged conditions: Plasma characteristics, degradation chemical kinetics and degradation pathways. *Journal of Physics D: Applied Physics*, *58*. https://www.semanticscholar.org/paper/68f3183fa2d0b2d8a5dd91f23f3f772104071c4d (accessed on 28th September 2025).

32. Ncube, P., Zvinowanda, C., & Ntuli, F. (2025). Photocatalytic degradation of emtricitabine, lamivudine, and tenofovir in water using the UV/TiO_2/H_2O_2 hybrid system. *International Journal of Environmental Research*. https://www.semanticscholar.org/paper/66f4908b029a0b1c4196e8bda66cf38b0724e18d (accessed on 28th September 2025).

33. Ni, K., Chen, Y., Xu, R., Zhao, Y., & Guo, M. (2024). Mapping photogenerated electron–hole behavior of graphene oxide: Insight into a new mechanism of photosensitive pollutant degradation. *Molecules*, *29*. https://www.semanticscholar.org/paper/41a8066a8f58adb3224fb97d93733ba90d744676 (accessed on 28th September 2025).

34. Prăvălie, R., Borrelli, P., Panagos, P., Ballabio, C., Lugato, E., Chappell, A., Miguez-Macho, G., et al. (2024). A unifying modelling of multiple land degradation pathways in Europe. *Nature Communications*, *15*. https://www.semanticscholar.org/paper/dc74a8c0abb50d716a7a8ebb3c3cd877341f0be0 (accessed on 28th September 2025).

35. Saleh, I., Zouari, N., & Al-Ghouti, M. A. (2020). Removal of pesticides from water and wastewater: Chemical, physical and biological treatment approaches. *Environmental Technology & Innovation*, *19*, 101026–101026. https://doi.org/10.1016/j.eti.2020.101026 (accessed on 28th September 2025).

36. Sarkhosh, M., Azizi, S., Mokrani, T., Seopela, M., & Maaza, M. (2024). Simultaneous Cr(VI) reduction and dexamethasone phosphate oxidation with organo-metallic sludge formation under UV irradiation: Kinetics, degradation pathways, and mechanism. *Arabian Journal for Science and Engineering*. https://www.semanticscholar.org/paper/bad3f492ca85b2bc0f3ff73f6b5223d8d1654048 (accessed on 28th September 2025).

37. Shang, F., Zhang, J., Zhu, T., Guo, Y., Fan, Y., Tao, S., Liu, R., et al. (2024). Combined kinetic analysis of thermal degradation characteristics and reaction mechanism of thin intumescent fire-retardant coating of steel structure. *International Journal of Polymer Analysis and Characterization*, *29*, 635–646. https://www.semanticscholar.org/paper/7965ddcdd96c624e90314a1328865209b0711a1c (accessed on 28th September 2025).

38. Sheraz, N., Shah, A., Haleem, A., & Iftikhar, F. (2024). Comprehensive assessment of carbon-, biomaterial- and inorganic-based adsorbents for the removal of the most hazardous heavy metal ions from wastewater. *RSC Advances*, *14*, 11284–11310. https://www.semanticscholar.org/paper/79fa82df3e49835501b9b188b5ec041267ec4c15 (accessed on 28th

September 2025).

39. Takhar, V., & Banerjee, R. (2025). Emerging photocatalytic applications of transition metal dichalcogenides and hybrid composites for energy applications and environmental remediation. *ChemCatChem*. https://www.semanticscholar.org/paper/362148a67765e0f9a9751ff7e975e54f89181ee2 (accessed on 28th September 2025).
40. Tang, J., Lei, D., Wu, M., Hu, Q., & Zhang, Q. (2020). Biodegradation and metabolic pathway of fenvalerate by *Citrobacter freundii* CD-9. *AMB Express, 10*. https://doi.org/10.1186/s13568-020-01128-x (accessed on 28th September 2025).
41. Valera-Medina, A., Xiao, H., Owen-Jones, M., David, W. I. F., & Bowen, P. (2018). Ammonia for power. *Progress in Energy and Combustion Science, 69*, 63–102. https://doi.org/10.1016/j.pecs.2018.07.001 (accessed on 28th September 2025).
42. Wang, H., Fan, S., Wen, H., Huang, Y., Gan, H., & Li, B. (2024). Degradation mechanism and toxicity assessment of Clo fibric acid by Fe^{2+}/PS process in saline pharmaceutical wastewater. *Environmental Technology, 46*, 2496–2508. https://www.semanticscholar.org/paper/3b407ab792c0ac4f096946bd5dfbed19ae1f6408 (accessed on 28th September 2025).
43. Wang, Z., Popolan-Vaida, D. M., Chen, B., Moshammer, K., et al. (2017). Unraveling the structure and chemical mechanisms of highly oxygenated intermediates in oxidation of organic compounds. *Proceedings of the National Academy of Sciences of the United States of America, 114*, 13102–13107. https://www.semanticscholar.org/paper/4d1d1c4cd81a9b9c1609a75ca934ff7ff1c9b411 (accessed on 28th September 2025).
44. Xu, Z., Yin, H., Xiong, Y., Song, J., & Qiao, Y. (2022). Degradation of triphenyl phosphate in water by UV-driven advanced oxidation processes. *Huan Jing Ke Xue, 43*(11), 4992–4999. https://www.semanticscholar.org/paper/57870c586e3191a3f83706c139b315b43a21eedc (accessed on 28th September 2025).
45. Yadav, V., Verma, P., Negi, H., Singh, R. K., & Saini, V. (2023). Efficient degradation of 4-nitrophenol using VO(TPP) impregnated TiO_2 photocatalyst: Insight into kinetics and mechanism. *Journal of Materials Research, 38*, 237–247. https://www.semanticscholar.org/paper/60ed467d4dd085eadc4d962b20744b8ffcab3d5f (accessed on 28th September 2025).
46. Yan, B., Han, C., Liu, Z., Wu, G., Wang, S., Li, J., Xia, W., et al. (2023). Degradation of cyanobacterial neurotoxin β-N-methylamino-L-alanine

(BMAA) using ozone process: Influencing factors and mechanism. *Environmental Science and Pollution Research, 30*, 47873–47881. https://www.semanticscholar.org/paper/52cfdfa60572a77a33f7644ccae0877717641252 (accessed on 28th September 2025).

47. Zhang, X., Gao, J., & Sun, Y. (2025). Tailored metal–oxygen interface engineering of self-assembled CoMn spinel nano catalysts for singlet oxygen-mediated peroxymonosulfate activation and pollutant degradation. *ACS Omega, 10*, 30728–30739. https://www.semanticscholar.org/paper/cfb190025871145b5c845129c78136c9fca7c950 (accessed on 28th September 2025).

48. Ziental, D., Czarczyńska-Goślińska, B., Młynarczyk, D. T., Glowacka-Sobotta, A., Stanisz, B., Gośliński, T., & Sobotta, Ł. (2020). Titanium dioxide nanoparticles: Prospects and applications in medicine. *Nanomaterials, 10*, 387–387. https://doi.org/10.3390/nano10020387 (accessed on 28th September 2025).

49. Zohora, F., Hore, S., Akter, H., Shapon, A. H., Ahmed, S., & Hore, R. (2025). Comparative analysis of pesticide degradation pathways and stereoselective behavior in agricultural soils. *International Journal of Professional Business Review*. https://www.semanticscholar.org/paper/4fac90a966d61bb9e020540b342309159edb1c2b (accessed on 28th September 2025).

CHAPTER 8

DESIGN AND OPERATION OF PHOTOCHEMICAL REACTORS

Contents

Introduction 313

8.1. Principles of Photochemistry 317

8.2. Types of Photochemical Reactors 322

8.3. Design Considerations for Photochemical Reactors 327

8.4. Operational Parameters 332

8.5. Applications of Photochemical Reactors 337

8.6. Challenges in Photochemical Reactor Design and Operation 342

Summary 347

References 350

INTRODUCTION

Structural and functional behavior of photochemical reactors is a pioneering field of development in chemical engineering and environmental sustainability. Photochemical reactions involve the application of light energy for catalyzing chemical transformations, thus playing an essential role in a variety of applications from clean energy production to waste treatment. With increasing global demand for sustainable solutions to problems such as energy depletion and pollution, it is essential to value the basic design aspects of such reactors. The application of photochemical reactors may lead to efficient processes giving rise to environmentally friendly end products, thereby complementing the overall sustainability agenda in industrial processes. Moreover, increased needs for cost-effective and efficient photochemical systems necessitate keen examination and innovation in this field, opening doors to significant advancements in reactor designs as well as operational efficiencies. One of the key aspects in the case of photochemical reactors is selection of materials and methods for optimizing their operating performance.

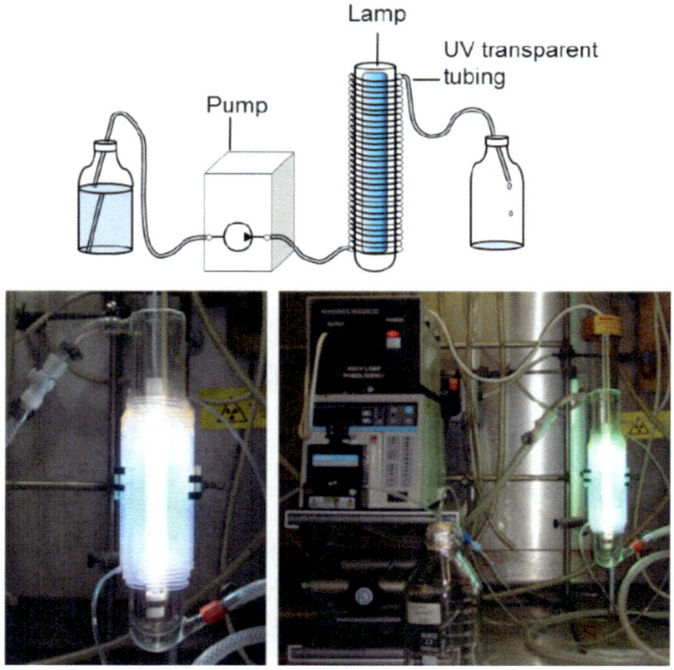

Figure 8.1. *Design and operation of photochemical reactors.*

Source: https://www.researchgate.net/publication/233840664/figure/fig4/AS:299992045637636@1448535085464/Schematic-of-a-typical-macroflow-photochemical-reactor-above-and-images-of-the-FEP.png.

The choice of photocatalysts, reactor design, and light sources can have profound impacts on the efficiency and effectiveness of photochemical reactions. For instance, semiconductor photocatalysis has been a leading method to water treatment by using the ability of materials like titanium dioxide to catalyze contaminant breakdown upon irradiation. These are significant contributors to

Langmuir–Hinshelwood-type kinetics to pollutant decomposition, enlarging the potential of photochemical reactors to ease environmental challenges (Lee et al., 2004). The inherent complexities associated with these design parameters demand sophisticated research into experimental and theoretical factors, thus making development rest on a solid scientific foundation. Economic soundness and efficiency are also equally important considerations in photochemical reactor design and operation strategies. Recent research has indicated that the overall plant efficiency is the sole most important factor of all the considerations that will determine the cost-effectiveness of photochemical systems (Biddle et al., 2025). This highlights the need for engineers to consider various design configurations, such as parabolic troughs and flat-plate collectors, which have a direct relationship with reactor performance and operational cost. Effective cooling reactor methods and energy consumption optimization also highlight the correspondence between economic performance and design in these systems. As research continues to advance, the integration of advanced technologies with the evolution of material science will further enhance the efficiency and affordability of photochemical reactors to keep them in the vanguard of sustainable technology solutions for energy and environmental applications (Figure 8.1).

Definition of Photochemical Reactors

A photochemical reactor is a dedicated device used to enhance chemical reactions induced by light energy. Its essence is the transformation of electromagnetic radiation, often ultraviolet or visible light, to chemical energy. Photochemical reactors usually consist of compositions that maximize light absorption and reaction kinetics with a view to obtaining high efficiencies in the chemical processes to be enhanced. Reactor design is critical in the sense that it must provide maximum exposure to light while efficiently managing heat and mass transfer. Solar applications, for example, hydrogen production from water, have specific designs—flat-plate and parabolic trough systems—that are especially employed for enhanced energy conversion efficiency and total chemical yield. This is significant since high efficiency is not just one of the key performance factors but also significantly contributes to the economic feasibility of the entire photochemical plant, as emphasized by recent studies on the efficiencies of plants and cost projections (Biddle et al., 2025). In the context of photochemical reactors, the nature of photochemical reactions is such that the effective conversion of light energy into chemically valuable products depends upon a sound understanding of the related photonic and chemical mechanisms. Photochemical reactors function through processes that make the production of reactive species possible upon interaction of light with reactants.

Optimal reactors would be at conditions where light utilization is optimized and losses by way of thermal effects and side reaction pathways are reduced. The greatest challenge lies in linking theoretical predictions of efficiency, which reflect potential thermodynamic limits, with actual reactor system performance in the real world. Current research identifies that efficiencies achieved for most conventional systems are significantly lower than calculated theoretically, necessitating new designs to reduce these discrepancies (Biddle et al., 2025). As photochemistry evolves, the operating strategies used in such reactors become more sophisticated. Most recent systems venture into hybrid designs by merging thermal processes with photochemical reactions to enhance overall efficiency. This is particularly relevant to uses like solar-driven hydrogen manufacture, where efficiency in photochemical systems can be dramatically increased through the use of sacrificial reagents or coupled quantum-thermal conversion processes. Such combined methods may have the potential to transcend some of the inherent limitations of standard photochemical processes. Not only do these strategies attempt to enhance the efficiencies observed in applications in reality, but they also testify to the ongoing progress in the field, showing a concerted effort to position photochemical reactors at the leading edge of sustainable solutions to energy (Biddle et al., 2025a, b).

Importance of Photochemical Processes in Industry

The significance of photochemical reactions for industrial applications is primarily that they can utilize light as a catalyst for chemical reactions, greatly enhancing efficiency and sustainability. Such reactions have been on the increase in various industries, particularly in the processes of environmental purification and energy production. For instance, semiconductor photocatalysis is revolutionizing water treatment processes by effectively decomposing organic pollutants. This process utilizes materials such as titanium dioxide to start reactions once they are exposed to light, which has brought about tremendous improvements in the cleaning of polluted water sources. The implications for industrial application are considerable because the application of such eco-friendly technologies is capable of offering solutions for life-threatening environmental issues while minimizing reliance on traditional, more harmful means of pollutant removal, ultimately resulting in a sustainable industrial climate, as illustrated in the literature reporting photochemical reactor design and operation efficiencies (Lee et al., 2004). In renewable energy, photochemical processes are increasingly valuable, particularly in the photochemical production of fuels from water. The structural and operational design of photochemical reactors for the generation of dihydrogen by photochemical water splitting is a case in point. Research into various reactor systems, such as flat-plate collectors and linear parabolic troughs, shows the unprecedented levels of energy efficiencies achievable with reported overall efficiencies of up to 11.6% under optimal conditions (Biddle et al., 2025). These processes not only reflect an advanced process of generating renewable energy but also reflect the feasibility of integrating renewable techniques into existing industrial configurations. As industries move towards being sustainable at a greater scale, photochemical techniques reflect authentic alternatives where economic viability is coordinated with environmental conservation, hence making them essential components of the energy strategies of the future. Also, the design parameters of photochemical reactors play a vital role in their ability to meet maximum functionality and operation effectiveness in industries. The type of material used in the reactor, source of light, and the photocatalyst used all influence the efficiency of photochemical reactions. For example, choice of reactor design—i.e., external illumination, annular, or circular—influences reaction kinetics and thereby the productivity of overall fuel production or overall pollutant degradation processes (Lee et al., 2004). By optimizing such design parameters, industries can enhance the sensitivity of photochemical reactors to variations in operating conditions, ultimately leading to low-cost and high-efficiency production systems. The ongoing research and technological development on reactor technologies directly highlight the increased importance of photochemical processes as central to confronting current industry challenges and promoting environmental sustainability.

8.1. PRINCIPLES OF PHOTOCHEMISTRY

Principles of photochemistry are the design and operating guidelines for photochemical reactors, i.e., design and operating guidelines for how light energy can be converted into chemical energy through various processes. Central to these principles is the light absorption by the molecule that creates excited states from which chemical transformations can be initiated. This change of energy is central to processes like water splitting for hydrogen production, in which photochemical reactors utilize sunlight to initiate endothermic reactions. Nevertheless, the efficiency of these reactions relies significantly upon the reactor system design. Through understanding the character of light-matter interactions and the wavelengths which enable maximum energy absorption, reactor performance is able to be optimized, hence overall system efficiency enhanced. In regard to considerations of different photochemical reactor designs, the choice of flat-plate collectors over linear parabolic troughs is an important consideration. Recent studies indicate that these systems have typical plant efficiencies of 10.3% and 11.6%, respectively, when optimized for the production of dihydrogen from water. The flat-plate technology is simpler and perhaps cheaper, but the trough systems superior efficiency makes it applicable to specific applications calling for greater energy conversion ratios. Lastly, the selection of reactor design must balance capital and operational costs while possessing maximum photochemical efficiencies. The ability of the engineers to handle these trade-offs is fundamental to the development of commercial viability in solar-powered hydrogen production (Biddle et al., 2025). As much as photochemical processes hold a bright future, challenge lies in obtaining the efficiencies proposed by theory calculations. Currently, the true efficiencies of systems, and particularly those designed to split water, are still short of the thermodynamic potentials, with estimates suggesting that efficiencies greater than 12 to 15% are not possible under realistic conditions. Efforts to optimize performance have focused on optimizing reactor conditions like temperature control and the use of sacrificial reagents to increase yield. Such optimization is achieved at the cost of cyclic operation and economic feasibility. With researchers looking for new solutions, there is the immediate need for a thorough understanding of operating conditions that can identify compatibility with system designs to improve efficiency and lower costs of producing renewable fuels (Biddle et al., 2025) (Figure 8.2).

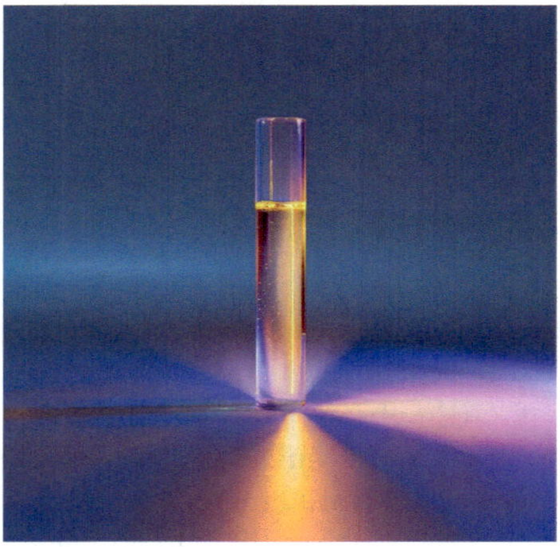

Figure 8.2. *Principles of photochemistry.*

Source: https://www.tutoring-blog.co.uk/wp-content/uploads/2024/12/photochemical-reactions.webp.

8.1.1. Fundamental Concepts of Light-Matter Interaction

Light-matter interaction is fundamental to photochemical reactor research since it uncovers how light can influence chemical reaction through excitations of electrons and energy transfers. The interaction primarily happens when photons, light quanta, interact with molecules triggering electronic transitions from ground to excited states. Owing to the fact that the excited states are predominantly transient, their properties become critical in determining reaction pathways and product formation. The efficiency of these processes relies not only on the wavelength of incident light but also on the nature of the absorbent material. The optimization of these parameters in photochemical reactors can enhance performance to enable processes such as photocatalysis. Through modulation of light intensity and wavelength, scientists can regulate reactions to their desired requirements, thereby optimizing energy efficiency and removing byproducts. A closer look at light-matter interactions reveals the role photonic structures play in speeding up reaction rates within photochemical reactors. Photonic materials can be used to generate intensified local electromagnetic fields that enhance the probability of photon absorption by reactant molecules. With the integration into reactor design, the materials enable the concentration of light energy into localized points, hence contributing to high reaction yields. This approach not only improves the general efficiency of photochemical processes, but it also allows reaction conditions to be fine-tuned for various applications. By fabricating materials with tailored optical properties, researchers can significantly optimize the interaction of light with matter so that reactor advancements become possible. This underscores the crucial contribution that nanostructured materials have in shaping the future of photochemical technologies. Together with the principles of light-matter interactions, controlled ecological life support system (CELSS) development is a prime example of the need for next-generation photochemical reactors. In the course of developing sustainable systems for long-duration spaceflight, waste treatment and effective resource recovery are important. This requires innovative photochemical systems capable of performing in changing conditions with ecological health and optimizing chemical reactions. The mechanisms of nutritionally useful compound formation and toxic chemical breakdown are contingent on effective light utilization, the cornerstone to successful CELSS activities. Workshop research suggestions emphasize the necessity for integrated methods involving nutrition and food processing, waste management, and systems modeling to facilitate meaningful advancement in CELSS. These findings highlight the integration of photochemical reactor design with general ecological and sustainability objectives, demonstrating the extensive ramifications of effective light-matter interaction in real-world application (Carden et al., 2025).

8.1.2. Types of Photochemical Reactions

Photochemical reactions feature prominently in numerous applications, primarily in the context of energy conversion and synthesis. Photochemical reactions may primarily be categorized as homogeneous and heterogeneous reactions. Homogeneous photochemical reactions occur in one phase, typically in a solution, wherein reactants are more intimately blended and interact with each other. These processes are important in fuel-forming processes where increasing higher reaction rates is a necessity. Conversely, heterogeneous reactions involve solid catalysts and gas or liquid reactants whose catalytic surface significantly influences the reaction mechanism. Both reactions have distinctive merits and demerits, most significantly with regard to efficiency and scalability, which are notable parameters to take into consideration in the design and operation of effective photochemical reactors. In the conversion of solar energy, photochemical reactions for water-splitting to produce dihydrogen as a clean energy carrier have been achieved. Such reactions' performance is based on the photochemical system design and the type of catalysts used. A theoretical maximum efficiency of 31% has been proposed in recent studies using single photosystems in an ideal system. In practical applications, this is lower, at efficiencies of 12 to 15%, which has been reported in studies investigating homogeneous and heterogeneous processes. The differences highlight challenges faced in light absorption, reaction kinetics, and optimizing product separation, which are crucial in the pursuit of sustainable energy solutions in photochemical reactor design. This is particularly relevant in items like flat-plate collector/reactors and linear parabolic troughs, where efficiency dictates economic viability (Biddle et al., 2025). Application of sacrificial reagents in cyclic photochemical processes can enhance efficiencies, but with significant qualifications. While these additions are capable of enhancing reaction rates and product yields, they divert away from the cyclic character of the process with accompanying economic and operability concerns for commercial applications. This modification also often results in a plethora multi-step reaction mechanism, requiring an even better comprehension of the involved chemistry to maximize reactor designs. In addition, advances in photoelectrochemical systems are promising, providing a more practical substitute to traditional homogeneous methodology. These hybrid systems, exploiting photochemical and electrochemical processes simultaneously, are less efficient but have greater potential with regard to cost-effective, efficient photochemical reactions. It is clear that with further research, the type of photochemical reaction employed as well as reactor design play a pivotal role towards the achievement of efficient energy generation (Biddle et al., 2025).

8.1.3. Role of Energy Transfer in Photochemical Processes

Energy transfer occupies a key role in photochemical reactions for the efficient design and operation of photochemical reactors. Photochemical reactions, which are initiated predominantly by solar energy, need efficient mechanisms of light absorption and energy conversion into chemical form. Energy transfer begins from the absorption of photons by reactant molecules, which get excited electronically. The efficiency of such a changeover, and hence the photochemical reaction in its totality, depends on the quality of the light source, reactor design, and materials used. For instance, advancements in reactor design, such as flat-plate and linear parabolic trough systems, focus on enhancing the incidence and concentration of light to provide improved energy absorption and conversion rates, thereby directly influencing the overall plant efficiency of the system, as discussed by Biddle et al. (2025).

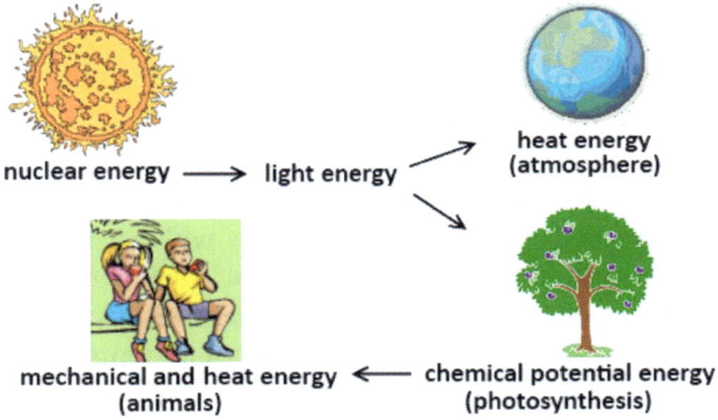

Figure 8.3. *Role of energy transfer in photochemical processes.*

Source: https://image1.slideserve.com/1562897/energy-transfer-l.jpg.

Furthermore, the thermodynamic efficiency of energy transfer processes plays a significant role in assessing the viability of photochemical processes, particularly those that are aimed at hydrogen generation through water splitting. Theoretical calculations dictate a maximum efficiency of 31% for devices operating with a single photosystem under optimal conditions. However, experimental systems tend to exhibit efficiencies significantly lower than such calculated limits. The rate of energy transfer can be optimized by using advanced materials and enhanced reactor designs to enable exciton production at higher rates and suppress energy loss. Empirical data highlights challenges in achieving theoretical energy transfer outcomes, specifically when photochemical system non-idealities and experimental realities of real-world applications are considered as described by Biddle et al. (2025). Finally, the overcoming of the inefficiencies in energy transfer associated with photochemical processes calls for innovative techniques with the ability to adapt to a wide range of operation conditions and feedstocks. The investigation of hybrid systems combining quantum and thermal conversion processes has been found to be promising towards increasing overall efficiencies. These systems seek to harness the excess energy of the photochemical reaction, providing increased scope of application and more economic value. Moreover, research on sacrificial reagent roles in cycle processes reveals opportunities to raise efficiencies, though at the expense of system cyclicity and cost feasibility. Thus, the evolution of the energy transfer methodologies and reactor systems is fundamental to raising the operational feasibility of photochemical reactors as well as their use for sustainable hydrogen production and energy conversion technology (Figure 8.3).

8.2. TYPES OF PHOTOCHEMICAL REACTORS

Under photochemical reactors, the various designs exhibit unique operational characteristics for specific applications. The most common ones are batch, continuous flow, and semicontinuous reactors. Batch reactors possess the benefit of being applied to small-scale reactions with ease of setup and ease of monitoring. Such systems find more general application in experimental work in which the reaction parameters can possibly be very strictly controlled. Conversely, continuous flow reactors allow for long-term operation and are appropriate for use in larger-scale processes, considerably increasing throughput. Semicontinuous reactors are a balance between the two, achieving flexibility in operation without compromising efficiency. The reactor type ultimately dictates the kinetics and thermodynamics of reactions being performed, dictating overall performance and scalability of photochemical processes. In the field of semiconductor photocatalysis, specific reactor geometries such as the annular and tubular ones are found to be extremely useful. Annular reactors utilize a cylindrical geometry that facilitates enhanced light penetration along with an enhanced surface area for semiconductor catalysts, which gives efficient photon absorption and photonic efficiency. Tubular reactors are typically used in those applications with intense reaction conditions, which often include the oxidation of contaminants or water purification processes. These reactors utilize the Langmuir–Hinshelwood kinetic mechanisms under steady light conditions to maximize the degradation of contaminants. Further, materials used in making these reactors, as well as their geometrical configuration, also play a critical role in determining their efficacy. Material and geometry selection are hence significant considerations during the designing process so that the reactors can achieve the desired photochemical reactions without compromising durability and performance of the structure (Lee et al., 2004). The choice between concentrating and non-concentrating solar photochemical reactors also demonstrates reactor diversity. Concentrating reactors, such as parabolic trough systems, utilize advanced collection technology to concentrate the sun onto a small surface area, maximizing energy acquisition and reaction efficiency. A case study of the design of a nominal 25,000 scmd photochemical plant, for example, illustrates clear efficiency differences between trough and flat-plate systems, highlighting their comparative working advantages (Biddle et al., 2025). Meanwhile, non-concentrating systems opt for low complexity and are more economically prudent in small-scale use. They both have varying cooling requirements based on the varied working temperatures, so an appropriate design addressing these matters is needed. Such careful study of reactor types and design parameters are needed for optimizing performance and reducing overall capital investment in the new technology of photochemistry.

8.2.1. Batch Photochemical Reactors

In the world of photochemical reactors, batch photochemical reactors are today an integral part in synthesizing and optimizing numerous chemical processes. The characteristic of streamlined operational processes is what differentiates these reactors, where reactants are subjected under controlled conditions to light for photochemical reactions without flow dynamics. One of the distinguishing features of batch reactors is their ability to accommodate a wide range of reaction conditions, such as various temperatures and light intensities, which can facilitate enhancing the efficiency of the photochemical process. Besides that, batch reactors offer high scale flexibility, permitting researchers and practitioners to investigate reaction kinetics as well as optimize conditions before scaling-up to larger systems. This flexibility is particularly invaluable for developing novel photochemical processes for water treatment, as attested by design challenges presented in recent articles, e.g., reactor glass composition and selection of light source (Lee et al., 2004). Their functioning, however, is accompanied by a number of challenges inherent in batch photochemical reactors that must be long-windedly addressed to maximize their efficacy. Among the key concerns is the potential for inhomogeneous light distribution throughout the reactor, which may result in non-uniform reaction rates and reduced overall yield. This points to the importance of accurate reactor design, particularly with reference to the light source geometry and location. Choosing the semiconductor photocatalyst is also essential to the efficiency of photochemical reactions. Different materials also have distinct absorption properties, thus affecting reaction rates and pathways. Recent research focuses on ascertaining ideal catalyst combinations and reactor shapes to optimize photonic efficiency and minimize energy needs for operation. These considerations are involved in the overall sustainability of batch photochemical technology in industrial applications (Collins et al., 2014). The future of batch photochemical reactors is bright with advancements in materials science and engineering contributing to more efficient systems. New reactor designs, such as hybrid advanced light technologies and optimized light paths, are expected to overcome some of the inherent limitations of traditional batch systems. Moreover, as the processes of photocatalytic water purification gain increasing prominence, the scalability of these reactors will be under constant testing. Practicality will rely on catalyst to starting material ratios, and thus ongoing research must seek to optimize such parameters. Understanding the kinetic principles, such as those detailed in Langmuir–Hinshelwood-type kinetics, is essential to optimizing the systems design and operation. Through relentless experimentation and ongoing improvement, batch photochemical reactors can play a revolutionary role in green chemical synthesis and clean-up methods (Lee et al., 2004) (Figure 8.4).

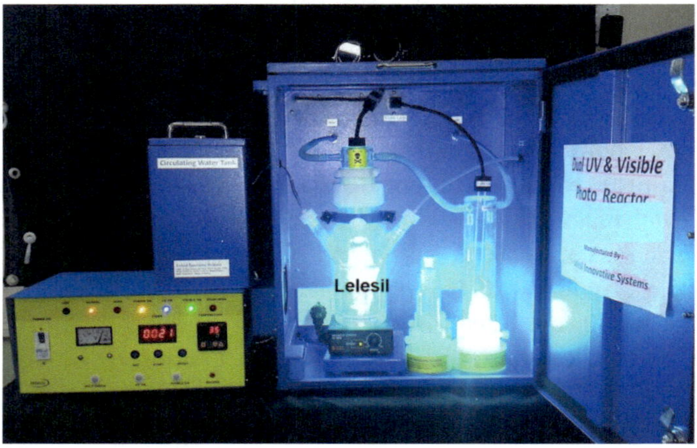

Figure 8.4. *Batch photochemical reactors.*

Source: https://5.imimg.com/data5/VW/QC/MY-2412447/photochemical-reactors-1000x1000.jpg.

8.2.2. Continuous Flow Photochemical Reactors

The structure and operation of continuous flow photochemical reactors have attracted plenty of attention due to their high efficiency and scalability in chemical processes, particularly in the purification of water. They employ a continuous flow of reactant material into a chamber where the photochemical reaction is induced by light sources. This structure facilitates improved exposure of the reactants to the light energy so that higher rates of reaction and enhanced mass transfer are achieved. In designing a continuous flow reactor, there are several factors to be remembered, such as the nature of the light source employed, reactor configuration, and flow homogeneity. The semiconductor photocatalyst employed, usually titanium dioxide, is of utmost significance as it aids in effective degradation of pollutants (Lee et al., 2004). Combining all these factors has enabled them to perform optimally and thus continuous flow reactors are a viable option for industrial application. Besides, the flexibility of continuous flow photochemical reactors does not stop at simple efficiency; they also allow for the simultaneous treatment of a wide range of organic and inorganic pollutants. With a continuous flow system, multiple stages of photochemical reactions can be carried out successively in series, optimizing conditions unique to target pollutants while minimizing the downtime associated with batch processing.

Figure 8.5. *Continuous flow photochemical reactors.*

Source: https://3.imimg.com/data3/GV/YC/MY-2412447/lab-photo-chemical-reactor-500x500.jpg.

The design of such reactors supports multiple modes of operation, such as annular and circular reactors, which support even reactant and light distribution. Worth special mention are design variations like thin falling film reactors, which enable surface area contact between catalyst and reactants, and thereby maximize the degradation potential for wastewater contaminants. As observed in the majority of research studies, application of these reactors in practice provides a robust platform for addressing pollution problems in water treatment (Lee et al., 2004). Fundamentally, the future of continuous flow photochemical reactors hinges on continuous developments in process engineering and materials science. The scientific community is directed towards an interdisciplinary method, encompassing ecological considerations alongside technological innovations in photoreactor designs. Partnerships and workshops to develop interdisciplinary guidelines can augment the deployment and efficiency of these systems. A shift towards ecological support systems not only makes these reactors more sustainable but also addresses the world's demand for effective environmental cleanup technologies. Therefore, the continuous flow photochemical reactor is not just a technological solution but a vital component of the system of sustainable methods in chemical engineering and environmental regulation (Figure 8.5).

8.2.3. Photobioreactors for Biological Applications

Photobioreactors are specially designed equipment for the cultivation of phototrophic microorganisms, particularly microalgae, for various biological applications. Their design is also vital as it influences the effectiveness of the use of light, nutrient uptake, and gas exchange, which are significant in the enhancement of biomass productivity. The integration of various operating conditions, such as light intensity, temperature, and nutrients, significantly contributes to the growth and metabolic rates of microalgae. Research emphasizes the necessity of optimizing these parameters because they play a vital role in optimizing biological productivity and efficiencies of nutrient removal. By fine-tuning such parameters to some extent, photobioreactors become a promising tool for realizing microalgae's biotechnological potential, such as bioremediation and biofuel, which is the broader potential for such nascent technologies in sustainable development (Mazzotti et al., 2015). The operational efficiency of photobioreactors is based on detailed understanding of the biochemical and physiological responses of microalgae to the environment. This is crucial not only for optimizing growth but also for maximizing the overall efficiency of biomass production processes. For instance, the effects of macronutrients, light spectrums, and temperature variations on microalgal growth illustrate how these kinds of variables dictate not just biomass yield but the quality of the yield for desired applications. In a study on a particular microalgae strain, it was found that there existed greater ability of nutrient removal under some growth conditions, which suggests that environmental manipulations strategically can significantly enhance the overall performance of photobioreactors (Mazzotti et al., 2015). Therefore, further studies on these interactions need to be conducted to optimize photobioreactor designs and operational practices. Photobioreactors in the field of sustainable biotechnology are very useful in integrating a number of applications such as food and feed, renewable energy production, and phytoremediation. Microalgae can be utilized in various biological processes, which can be effectively utilized through the technology of photobioreactors. For instance, simultaneous microalgal growth for wastewater nutrient removal and biomass yields can ensure maximum resource utilization and minimize harm to the environment. This synergy reflects the diversified application of photobioreactors to create sustainable processes in various industries (Mazzotti et al., 2015). As the quest for new and efficient biotechnologies continues, the design operation of photobioreactors will be the key to pushing the frontiers of microalgal applications in making them a cornerstone of future biological processes and sustainability initiatives.

8.3. DESIGN CONSIDERATIONS FOR PHOTOCHEMICAL REACTORS

In photochemical reactors, design considerations in such reactors are of critical value in the optimization of the efficiency and effectiveness of reactions occurring within the reactors. The design process must take into account a number of factors with utmost care, such as the type of the light source employed, reactor shape, as well as the materials used in contact with the reaction mixture. For example, the selection of light source—whether ultraviolet or visible light—may have a strong bearing on the photonic energy that is absorbed by the reactants. In addition, geometry of the reactor, such as flow patterns and the surface area available to interact with the light, is important to ensure optimal use of photons. This understanding is echoed in workshop reports specified for establishing ecological support systems, which emphasize the interdisciplinary interface that must be achieved in the development of successful photochemical systems (Carden et al., 2025). Moreover, operational parameters must be discussed simultaneously with design considerations for the purpose of harmonization between the physical characteristics of the reactor and the intended chemical processes. Important variables, including temperature, pressure, and reactant concentration, can significantly influence the yield and selectivity of photochemical reactions. For example, using the best temperatures can improve the stability of fragile compounds, and the adjustment of reactant concentrations will ensure the avoidance of side reactions that could compete with desired ones. Thus, synchronization of these operating conditions with the design framework becomes critical in order to obtain a successful photochemical reaction. Findings from workshop series to propel controlled ecological life support systems further accentuate the importance of a systemic approach to reactor design optimization (Carden et al., 2025). Finally, there is a need to devise and run photochemical reactors with consideration of safety and scalability concerns. Given that the majority of photochemical reactors are light-sensitive procedures, any design error or failure can lead to undesirable reactions or form hazardous situations. Therefore, having good safety provisions and fail-safes in place is crucial. Scalability also represents a common challenge that covers the issue of scaling up from laboratory-sized experiments to industrial-sized application. Overcoming this challenge entails the selection of materials and reaction mechanisms well adapted for scaling up into bigger systems without efficiency or safety sacrifice. The learning that is caused by the development of ecological systems underscores the need for holistic design strategies that take into account scaling issues while facilitating stable performance throughout the reactor lifetime (Carden et al., 2025) (Figure 8.6).

Figure 8.6. *Design considerations for photochemical reactors.*

Source: https://s7d1.scene7.com/is/image/CENODS/09245-scicon-Fig1a?$responsive$&qlt=90,0&resMode=sharp2&fmt=webp.

8.3.1. Material Selection for Reactor Construction

Material selection for the building of photochemical reactors is another key issue that reflects directly on the effectiveness, safety, and durability of photochemical reactors. Various issues must be considered, including chemical compatibility, thermal stability, and mechanical strength under operating conditions. In photochemical operations where ultraviolet radiation is generally employed to induce reaction, materials must be able to resist damage caused by high-energy radiation and potential photonic energies. Moreover, materials should not bring any impurities that have the potential to react with light or inhibit the desired photochemical reactions. Metal alloys, glass, and certain ceramics are most often utilized since they have good characteristics. Material-reactant interaction knowledge is also vital in obtaining maximum reaction yields and preventing process operation failure. Thus, a general analysis of the probable materials should precede the design process so that the reactor is not only in accordance with performance characteristics but also meets safety standards identified during research workshops exploring ecological systems support (Carden et al., 2025). In addition to chemical and physical characteristics, economic considerations involving material selection are significant. Material cost largely relies on availability, processing requirements, as well as durability. When designing photochemical reactors, engineers must balance high-performance materials with cost considerations in creating a viable model for production. For instance, while metals such as titanium offer improved corrosion and heat resistance, they may be too expensive to implement on a large scale. Conversely, more affordable substitutes may incur greater maintenance and reduce reactor lifespan. In addition, procurement logistics and supply chain reliability are to be considered since the availability of materials can change vastly. In conclusion, a holistic material selection approach will seek quality and performance while making sure that costs of operation remain affordable, fitting into broader sustainable production and energy efficiency strategies for photochemical reactor design (Carden et al., 2025). Finally, operating conditions play an important role in deciding suitable material for photochemical reactors. Pressure fluctuations, temperature variations, and accessibility of corrosives mean that the material behavior in specific conditions is a requirement. In all cases, materials must have properties with added attributes that are responsible for efficient energy transfer and resistance to environmental stresses. Advanced coatings and composites may be designed to enhance the performance and life of basic materials so that the reactor can run at an optimum level for extended periods. Additionally, it is important to enhance material technology to be used for designing future reactors since there is a growing need for efficient and effective photochemical processes. Hence, engineers and scientists have to continuously pursue the search for new materials, ascertain their suitability, and guarantee that they can satisfy the changing demands of contemporary photochemical reactors (Carden et al., 2025).

8.3.2. Light Source Selection and Placement

Selection of an appropriate light source is a critical factor in photochemical reactor design and operation. Efficiency of the photochemical process largely relies on wavelength and intensity of the used light as different reactions are susceptible to different spectral profiles for optimum functioning. Ultraviolet and visible light sources are commonly used because they can induce electronic transitions in the majority of photochemical compounds. For instance, application of light-emitting diodes (LEDs) has been more common due to their wavelength-tunable characteristics and energy efficiency that allows for precise control of reaction conditions. Besides, advances in luminescent solar concentrator provide a new avenue for maximizing light utilization in reactors based on the need for increased photon efficiency assessed using methods like ray-tracing simulations to improve reactor performance and economically scale up operations (Zondag et al., 2024). Positioning of light sources in a photochemical reactor has an impressive impact on light distribution and, hence, on reaction kinetics. Uniform illumination is important to ensure that all the reactants undergo comparable photonic interactions so that the yield and reproducibility of the products are enhanced. Luminance variations in the position of light sources create hot or cold spots, and these can profoundly alter the efficiency of the photochemical reaction.

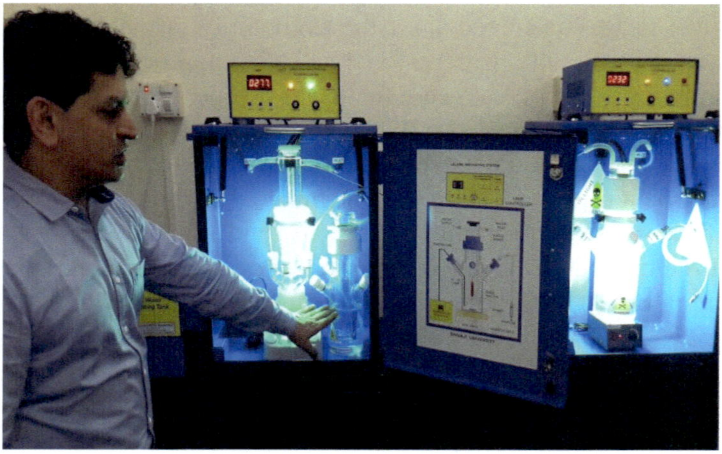

Figure 8.7. *Light source selection and placement.*

Source: https://i.ytimg.com/vi/qpDrkKtOrmQ/maxresdefault.jpg.

Therefore, proper placement of light sources is important in reactor design. Technologies such as 3D-printed reactors that are designed with optimal geometry allow for better light distribution and better interaction with streams of reactants. With the application of such novel designs, the operational problem concerning conventional reactors, such as ununiform heating or irradiation with light, can be alleviated, ultimately leading to enhanced performance in continuous-flow photochemistry (Zondag et al., 2024). In addition to selection and position, real-time measurement of light intensity within photochemical reactors is also crucial to ensure uniform reaction conditions. Fluctuations in light intensity due to equipment degradation, changes in operating conditions, or external environmental conditions can play devastating roles in reaction rates as well as product quality. Advanced photoreactor designs often consist of light intensity monitors for on-line monitoring and control systems for delivering automated compensation so that the light source is kept in the best position for the desired reactions. Furthermore, the use of feedback control systems can enhance photochemical process stability and scalability, particularly in automated systems using renewable energy sources like solar power. As photochemical usage advances, knowledge of the complex interaction between light source choice, positioning, and online monitoring will continue to be critical to maximizing reactor efficiency and widening their applied use in different areas (Zondag et al., 2024) (Figure 8.7).

8.3.3. Temperature and Pressure Control Mechanisms

Successful control of pressure and temperature is of paramount importance in the optimization of the design and operation of photochemical reactors. Photochemical reactors are intrinsically sensitive to changes in their operating conditions, which therefore impact the reaction kinetics and the efficiency of the photochemical reactions. Temperature changes can initiate alterations in the activation energy for the chemical reactions, which can reduce yield and selectivity. Equally so, pressure variations can affect reactant solubility and product and reactant diffusion rates in the reactor phase and thereby alter reaction kinetics. Therefore, having robust temperature and pressure control systems is important to maintain steady-state operation, facilitate increased conversion rates, and provide maximum throughput with minimal unwanted side reactions. The use of sophisticated control technologies guarantees continuous optimization of the operating parameters during the process. Additionally, appropriate material selection and design characteristics in photochemical reactors can optimize temperature and pressure control systems. The type of reactor materials selected should not only take thermal conductivity into consideration but also resistance against photodegradation, as well as pressure endurance at varying pressures. Reactor geometries like the use of helical flow patterns or multi-layered geometries can increase the uniformity of temperature distribution and enable efficient heat exchange. High-performance sensor technology combined with feedback control systems can closely monitor and regulate the reactor internal conditions in real time. This reduces time lag between setpoint adjustments and reactor conditions, allowing quick response to deviations. As specified in a review of operational efficiencies, complete control of such parameters can exert a strong impact on the economic feasibility and yield efficiency of photochemical processes, considering the intricacy and likelihood of inefficiency in functioning of lower performance systems (Biddle et al., 2025). Finally, studies of hybrid systems which combine photochemical processes and thermal conversion technologies open up new avenues for advancement in temperature and pressure control. These reactors can leverage the principles of thermodynamics in designing a synergistic system where processes of energy conversion are optimized. With these strategies combined, photochemical reactors can be designed to operate at high temperatures or controlled pressure conditions that significantly improve rates of reactions and yields of the desired products. Later designs, as seen through studies of advanced energy technologies, bring out the synergy between photochemical and conventional thermal processes in transcending efficiency barriers that were at some time deemed challenging (Gross et al., 2025). This is an interdisciplinary method whose progress not only enhances the reactors' operating capability but also aligns with higher sustainability goals through maximizing the power of solar and other renewable energies in fuel production.

8.4. OPERATIONAL PARAMETERS

Operational parameters in design and operation of photochemical reactors play a critical role in providing optimal reaction yields. Operational parameters consist of important factors that have a direct impact on the rate of photoreaction and target product yield, such as temperature, pressure, intensity of light, and reaction time. For instance, the choice of light source can play an important role in a photoreactor's efficiency, particularly for semiconductor photocatalyst-based systems like titanium dioxide. The conditions under which the system operates must be controlled accurately and offered in an optimal manner that is dependent upon the kinetics of the reaction, e.g., those following Langmuir–Hinshelwood-type mechanism typical of pollutant degradation. Further, the design of the reactor has to embed the process conditions along with the process demand of the photocatalysis. The kind of materials for the reactors, e.g., the glass composition, and the reactor geometries—whether external illumination systems or annular geometries—either catalyze or inhibit the sought-after photochemical reactions. Experiments have shown the operational nuances of various reactor geometries, pointing out that interactions between reactor geometry and light source impact photocatalyst activity and selectivity in a reaction directly. Integration of optimal operating parameters into these platforms not only increases the yield of reaction but also extends the functional life of photochemical reactors. More than anything else, integrating interscience knowledge cannot be overlooked when optimizing operational parameters in photochemical reactors. The suggestions that emerge from consensus workshops point to the critical need for an integrated approach in which the interface between engineering, chemistry, and environmental science is investigated with a view to maximizing reactor design. Such collaborative efforts unveil the potential for innovations in reactor design and operation, focusing on sustainability and efficiency. Such case studies demonstrating the efficacy of semiconductor photocatalysis for water purification further affirm this, indicating a growing need for operational parameters that not only favor photoreactions but also meet the laws of nature in the development of technology. These discussions are critical in the ongoing progress towards more efficient and more sustainable photochemical reactors as can be seen by the latest advancements in reactor system design and operational research studies (Lee et al., 2004; Carden et al., 2025) (Figure 8.8).

Figure 8.8. *Operational parameters.*

Source: https://www.researchgate.net/profile/Jadhawar-Prashant-2/publication/373604499/figure/fig1/AS:11431281185467814@1693656576007/UHS-operational-parameters-investigated-in-this-study.png.

8.4.1. Reaction Time and Its Impact on Yield

In the context of photochemical reactors, it is critical to comprehend the impact of reaction time on yield. The effectiveness of a photochemical process usually depends on the length of time exposed to light, as this in turn determines the degree of chemical change. Optimizing reaction time can be achieved to enhance yield, with shorter irradiation times being preferred, particularly in continuous flow reactors. They allow for improved interaction with the light source, thus enabling faster processing and enhancing the rate of turnover of the reactive species. These collaborate to provide a better yield because they raise the possibilities of successful photochemical conversions. Specifically, designed reaction times can minimize the formation of undesired byproducts, thus enhancing the overall selectivity of the process under an optimal operating setup. The impetus for enhancing photochemical reactor design is generally addressed through innovative means in managing the reaction times. Microphotoreactors, for example, enable researchers to exercise complete control over irradiation conditions for maximum reaction efficiency. One such example is a study that presents a microphotoreactor design which recorded dramatic enhancements in chemical conversion rates compared to traditional batch reactors (Aillet et al., 2013). This is because the time that reactants spend in the reactor can be minimized while at the same time the exposure given to them by light is maximized. With this, the system provides for minimum reaction time without loss of yield, ultimately transforming the means of understanding and conducting photochemical reactions in academia and industry. Optimizing such parameters makes it possible to precisely predict results and scale processes more accurately. While the chemistry community pushes ever forward to bridge theory with actual use, the development of continuous flow photoreactors becomes increasingly relevant. Not only do such equipment make operations easier, but they also allow reaction time control, which maximizes throughput. As stated by Beeler et al. (2017), continuous flow systems can achieve improvements in yield through improved control over light exposure and reactant concentration. Such overlapping of technology results in a paradigm shift where reaction parameters can be controlled in a rational manner to yield optimum results. With the focus on reaction time as a leverage point in reactor design, one can see that advancements in photochemical technology are steering the course of chemical production towards efficiency and scalability and thus overall major implications for the synthetic chemistry field in general.

8.4.2. Concentration of Reactants and Its Effects

In photochemical reactors, reactant concentration is a key aspect in managing reaction rates and product yields. At high concentrations, there may be higher rates of photochemical reactions because the likelihood of molecular collisions is high, thus enhancing the efficiency of energy absorption by the reactants. With the exception of a few photochemical reactions, the activation of light is necessary for all, so adequate reactant concentration ensures that there are sufficient molecules to act on photons and thereby realize optimal desired product formation. Concentrations of reactants that are too high also result in reaction kinetic non-linearities and saturation effects that complicate reactor design. This relationship allows engineers to develop optimal reactor conditions for specific reactions and ultimately enhance the overall efficiency of photochemical systems. In addition, reactant concentration influences the design of photochemical reactors and operating protocols. Reactor configurations, such as flow tube reactors, directly regulate reactant concentrations to obtain desired outcomes. For instance, the Caltech Photooxidation Flow Tube, where computational fluid dynamics (CFD) is applied to maximize flow profiles and ensure uniform reactant distribution. By adjusting the concentration of ozone and precursor chemicals with high accuracy, the reactor can be operated under ideal conditions for secondary organic aerosol (SOA) formation, which is vital in atmospheric research and other applications. This emphasis on reactor design for concentration dynamics reflects the significance of studying the influence of reactant concentrations on the process of reactions and the formation of products, which, in turn, improves the usability and effectiveness of photochemical processes in practical applications (Bauer et al., 2017). Also, reactant concentration levels are in direct relation with catalyst performance in photochemical reactors. In studies involving reactions such as the CO_2-CH_4 decomposition, reactant concentration level variation has much influence on catalyst efficiency and reaction rates (Davenport et al., 2025). Catalysts may exhibit varying performances at different levels of reactant concentrations, thus potentially leading to product selectivity variation and overall yields. Therefore, there is a need to define the catalyst activity over a range of the reactants' concentrations in order to maximize reactor operation. The use of a test unit with the ability to test catalyst response under varying conditions provides critical information in modifying operational factors for optimum efficiency and safety in reactor systems. Lastly, full comprehension of reaction dynamics relative to reactant concentration is imperative in order to enhance the design and operation of efficient photochemical reactors.

8.4.3. Influence of Solvent Choice on Reaction Efficiency

The choice of solvent in photochemical reactions then becomes a crucial parameter in influencing the overall reaction efficiency as it influences the solubility of reactants, the stability of intermediates, as well as the rate of energy transfer. Polar solvents stabilize charge intermediates, and this results in an enhanced reaction rate, whereas the application of non-polar solvents favors radical mechanisms by providing a favorable medium for the intermediates. Further, the dielectric constant of the solvent affects energy in excited states; hence, interaction between solvent and excited molecule can enable or hinder the reaction process. The interaction between these implies the use of an appropriate solvent to optimize reaction conditions and hence enhance efficiency.

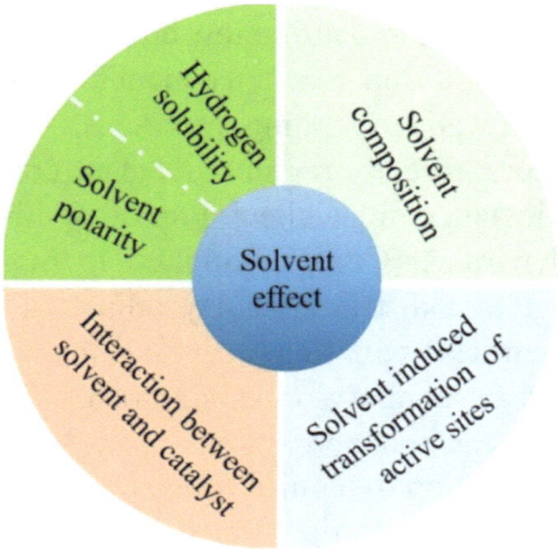

Figure 8.9. *Influence of solvent choice on reaction efficiency.*

Source: https://chemistry-europe.onlinelibrary.wiley.com/cms/asset/ef8644d9-2cb0-4e8e-bfec-677b860f7884/cctc202400120-fig-0001-m.jpg.

This aspect only increases in significance when designing reactors for industrial-scale use, where high yields and minimal costs are necessary in order to be commercially viable. Solvents also have a significant influence on the thermodynamic characteristics of photochemical reactions, beyond their role in influencing reaction kinetics. For instance, different solvents have been found to alter the activation energy for reaction to take place and thus influence the thermodynamic stability of alternative reaction pathways. Some solvents facilitate endothermic reactions more than others, and that is of significance when one is working with water-splitting in solar photochemical reactors. Current research shows efficiencies with some ionic solutions, for example, such that variations in solvent composition are demonstrated to enhance light absorption and increase quantum efficiency as they have been found to do with fuel-producing solar photochemical experiments. These results are crucial for chemists and engineers looking to create photochemical reactors not just that work at increased efficiencies but also become economically sound for practical applications, showcasing the complexity of solvent interactions in energy conversion systems (Biddle et al., 2025). Further, the interactions between solvents and reactive species must be regulated properly in order to avoid undesirable effects, e.g., solvent degradation or by-product formation, which can render the reaction less efficient. Proper design of photochemical reactors must make provision for continuous recycling and purification of solvents to maintain optimal reaction conditions. This aligns with the recent controversies regarding the development of effective and sustainable ecological support systems (Carden et al., 2025). By creating a feedback-loop system where solvent impact on reaction yield is monitored and optimized at all times, scientists might be able to optimize the whole processes of photochemical energy conversion. Hence, the effect of solvent selection is not merely limited to reaction rates of instantaneous effect, but also includes wider issues of reactor design, cost, and environmental effects, and so this subject is a key focus of photochemical research (Figure 8.9).

8.5. APPLICATIONS OF PHOTOCHEMICAL REACTORS

The applications of photochemical reactors extend far beyond theory, such as significant roles in industrial processes and the production of renewable energy. Photochemical reactors facilitate many light-initiated chemical reactions, making them indispensable in synthesizing products such as hydrogen and other hydrocarbons. A prominent application is in solar hydrogen production, in which light energy is applied to split water molecules into oxygen and dihydrogen. Studies have shown that the performance of concentrated systems like flat-plate collector/reactors and parabolic linear troughs can be 10.3% and 11.6%, respectively, depending on their design and operating conditions (Biddle et al., 2025). The economic viability of hydrogen as a source of clean energy is influenced by this performance, which underlines the reactors potential to propel clean energy alternatives. Apart from power generation, photochemical reactors also have huge potential to propel environmental remediation. These systems may be designed to facilitate the breakdown of harmful pollutants through photodegradation reactions. Employing light to activate catalysts, photochemical reactors are effective at converting toxic substances into harmless products. This application is particularly useful in wastewater treatment, where light-catalyzed reactions are initiated in the pollutants, and cleaner water is left behind that can be reused or released harmlessly into the natural system. The versatility of photochemical reactors to respond to ecological needs highlights the potential of such reactors as part of an integrated ecological support system, as advocated for by inter-disciplinary research workshops to foster an integrated knowledge base for the aforementioned technologies (Carden et al., 2025).

The design and management of the reactors therefore not only concentrate on energy sustainability but on ecological preservation. Operation design of photochemical reactors plays an essential role in achieving maximum efficiency and utilization in various industries. The engineering aspect involves optimal reactor geometry, choice of material for absorption of light, and heat exchange processes to maintain optimal temperature conditions. For instance, in the case of photochemical plants, as witnessed in the study, it is necessary to maintain the optimum temperature, especially in the case of plant configurations like the linear parabolic trough, where cooling the reactor is a must because of efficiency constraints. Additionally, as seems to be the case from economic calculations, reactor capital cost is highly dependent on plant overall efficiency, which requires strict design protocols that reconcile energy efficiency and cost factors (Biddle et al., 2025). By addressing these issues in the process within reactor design, scientists can make the use of photochemical reactors more effective in applications ranging from energy production to environmental cleanup and thereby improve technology in both fields.

8.5.1. Synthesis of Pharmaceuticals

Pharmaceutical synthesis is a key contribution to modern medicine, and the aim is to develop new and effective therapeutic molecules. Design and use of photochemical reactors have significant advantages in this field. Photochemical synthesis, using light to activate chemical reactions, is frequently able to reduce the environmental impact of pharmaceutical manufacture, since it is a preferred approach to synthesizing complicated molecules. By harnessing photonic energy, these reactors enable reactions that at thermal conditions may be slow or inefficient. This not only enhances reaction efficiency but also enables new reaction pathways that are inherently distinct from the pathways employed in traditional chemical processes. This is particularly relevant in the production of pharmaceuticals, where the construction of complex molecular architectures plays a pivotal role in the efficacy of the drug (Naresh et al., 2021). Furthermore, the integration of nanomaterials in photochemical reactor systems makes them more efficient at the production of pharmaceuticals. Nanomaterials also have new properties of high surface area and high electrical conductivity, which will be able to enhance light absorbing ability and the rate of reaction in such systems. Opportunities also lie in further tailoring these nanostructures to design reactors that are optimized for specific photochemical reactions. For instance, the application of catalysts made up of nanoparticles can lower activation energies significantly and improve selectivity in many types of chemical conversions. Thus, the use of nanotechnology to the fabrication of photochemical reactors not only makes production easier but also adheres to green chemistry principles that promote sustainable operation within the pharmaceutical industry (Baumann et al., 2020). In short, photochemical reactor synthesis of drugs is a revolutionary drug development and manufacturing technique. The revolutionary character of the application of light energy to chemical reactions and the revolutionary character of nanomaterials underscore the necessity to reconsider traditional techniques in drug synthesis. As studies continue to make advancements in this line of research, scaling up these new processes to tackle these issues will be required to their successful implementation in the business world. Lastly, the application of photochemical reactors with newer technologies holds the promise of not only increasing the sustainability and efficiency of pharmaceuticals production but also the quality and accessibility of life-saving drugs in general (Naresh et al., 2021; Baumann et al., 2020) (Figure 8.10).

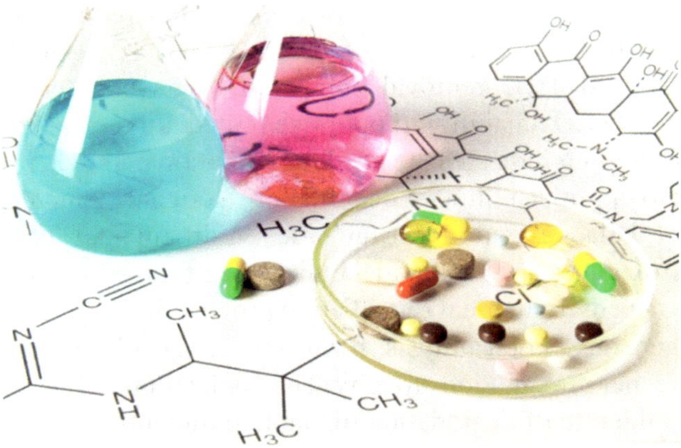

Figure 8.10. *Synthesis of pharmaceuticals.*

Source: https://external-content.duckduckgo.com/iu/?u=https%3A%2F%2Fwww.europeanpharmaceuticalreview.com%2Fwp-content%2Fuploads%2Fshutterstock_1253111212-750x500.jpg&f=1&nofb=1&ipt=8fbb0deed8a036390495a817251d4a2e807e8a3f39a64961f9dd95c9ba59e584.

8.5.2. Environmental Remediation Processes

Growing concern about environmental degradation due to industrial activity demands new approaches in environmental remediation procedures. Among some of the others, biological and photochemical (B-P) technologies have been recognized as an appropriate approach to facing pollution in aquatic ecosystems, particularly in Effluent Dominated Hydrosystems (EDHs). Such systems, notably prevalent in the Caribbean area, suffer from serious levels of contamination resulting from agricultural runoff, municipal wastewaters, and industrial effluents. The combination of biological treatments, utilizing pollutant-degrading bacteria for pre-treatment of water, with photochemical processes employing photocatalysts like Titanium Dioxide (TiO_2) can lead to a massive decline in contaminant levels. The marriage of such technologies indicates a green response as well as providing cost-effective solutions for Small Island Developing States with scarce resources and high water quality demands (Cheddie et al., 2017). Effectiveness of photochemical reactions in the degradation of a wide variety of organic pollutants also adds to their importance as part of an integrated remediation strategy. For instance, research into the photo-Fenton process indicates the effectiveness of the process for remediation of oil-contaminated soils, with significant reductions in total organic content (TOC). Experimentally, irradiation duration, hydrogen peroxide concentration, and the type of light used—visible, UV-A, or UV-C—regulate the efficiency of degradation. Interestingly enough, UV-A light has been discovered to enhance widespread TOC removal compared to its counterparts, evincing photochemical process tunability to optimize remediation efficiency (Duarte et al., 2025). This finding indicates the need for operation parameter fine-tuning in photoreactors to optimize pollutant degradation, a crucial element in recovering the ecological balance and ensuring safe environmental conditions. The incorporation of cutting-edge photochemical reactors presents opportunities and challenges in environmental remediation process design and operation. The use of heterogeneous photocatalysts, for instance, via solar energy harvesting, facilitates a greener remediation strategy. The application of these advances not only improves the rate of degradation of stable pollutants but also underscores the potential for resource recycling and recovery in environmental systems. As research continues to explore various combinations of photochemical and biological treatments, environmental cleanup in the future appears to be promising. Continuous evaluation of these technologies not only addresses today's environmental concerns but also addresses global sustainability, ultimately resulting in resilience of vulnerable ecosystems and communities threatened by pollution (Cheddie et al., 2017).

8.5.3. Production of Renewable Energy Sources

Production of renewable energy sources is at the core of efforts to find solutions to the pressing problem of climate change and energy deficiency. Perhaps one of the most promising methods of renewable energy production is utilization of photochemical reactors, with solar energy employed to trigger chemical reactions. This technology is primarily based on the transformation of vast feedstocks such as carbon dioxide into valuable chemicals and fuels through processes including photochemical and photoelectrochemical reduction.

These processes utilize sunlight, significantly reducing fossil fuel dependence and leading to a cleaner energy system. For instance, the ability to transform CO_2 directly into valuable energy is a solution to problems of greenhouse gas emission alongside offering a path towards energy security. Thus, the design and operation of photochemical reactors become significant not only to enhance efficiency but also to usher in maximum potential from renewable sources of energy into our energy equation.

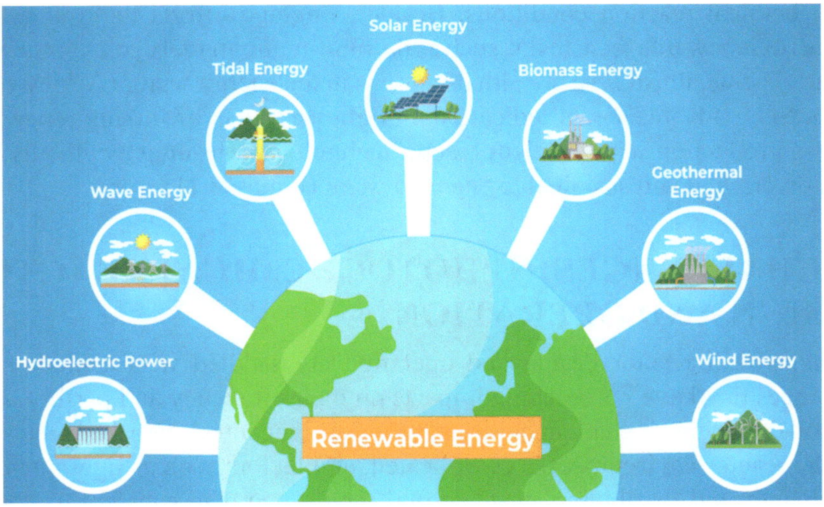

Figure 8.11. *Production of renewable energy sources.*

Source: https://www.greenesa.com/uploads/news/1632327662Renewableenergysourcestypes Greenesa.png.

Optimum performance of photochemical reactors is largely based on the selection of efficient photocatalysts. Effective catalysts must possess desirable characteristics, such as high light absorption, chemical stability, and positive charge transfer, all of which contribute to raising rates and yields of reaction. With recent advancements in the field of material science, it has become feasible to design nanostructured photocatalysts with enhanced light-harvesting capability and minimum energy losses. Also, techniques such as doping and alloying of different semiconductor materials can be employed to enhance the photocatalyst efficiency, as indicated in recent reviews on photoelectrochemical (PEC) systems for CO_2 reduction. Through a clear understanding of the base interactions in such reactors, scientists can design catalysts intentionally such that they not only increase energy conversion rates but also reduce overpotentials and ohmic losses, which in turn affect total production of renewable energy sources (Kalamaras et al., 2018). In the future, integration of state-of-the-art modeling tools and reactor designs will be the backbone to commercialize production of renewable energy by using photochemical reactors. With the optimization of PEC system design parameters and employing sophisticated characterization techniques, researchers can systematically counter the obstacles to mass deployment. Best reactor design must take into account light distribution, mass transport, and thermal management so as to maintain ideal photochemical reaction conditions. Besides, ongoing efforts to explore new sources of renewable feedstocks, such as thorium-uranium fuel cycles for nuclear power systems, illustrate a breadth of options in the energy sector (Ashley et al., 2012). Such advancements indicate the bright prospects of linking renewable process production with efficient reactor technologies, leading eventually to a more diversified and sustainable energy system (Figure 8.11).

8.6. CHALLENGES IN PHOTOCHEMICAL REACTOR DESIGN AND OPERATION

Photochemical reactor design and operation are saddled with a plethora of challenges that have far-reaching impacts on their efficiency and performance. One of the primary challenges lies in the fact that interaction between light and the reactor environment is complicated, and factors such as light intensity, wavelength, and distribution must be strictly controlled. These factors are central since they determine the reaction rates and product yields in photochemical reactions. Additionally, the very physical reactor design itself, including geometry, selection of materials, and flow patterns, can potentially have a strong influence on light penetration and reaction kinetics in general. The achievement of an optimal reactor design is therefore dependent on thorough knowledge of photonic principles and chemical engineering too, which could be an exceedingly important barrier for scientists and engineers in this field. Also, photochemical reactor process stability is an issue, particularly maintaining stable reaction conditions for long periods. Temperature, pressure, or mixture composition drift can lead to unacceptable variation in product quality and yield. Because photochemical reactions can be conducted with catalysts or photosensitive compounds, component degradation over time poses an additional problem for reactor operation. Periodic inspection and analysis become imperative to ensure the desired reactor performance on a continuous basis. Additionally, the materials chosen for the fabrication of the reactor must take into account potential photodegradation or some other chemical reaction with the reactants or products, thus further complicating the operational scenario. Consciousness of such dynamic interactions is essential to optimize photochemical reactor performance. Lastly, the scaling problem in photochemical reactor design is a serious challenge to commercial potential. Whereas it is possible to delicately control laboratory-scale reactors, large-scale operation is plagued with problems such as non-uniform light distribution and mass transfer limitations. These troubles result in lower efficiency and undermine the economic viability of photochemical processes on a commercial scale. Therefore, researchers have started probing new options in the form of combining innovative light-harvesting technologies with multi-phase reactors with improved mass transfer rates. Additionally, strategies inducing a multidisciplinary approach, especially in connecting photochemistry and cutting-edge materials science, are needed to design next-generation photochemical reactors. As outlined in future publications, these developments have the potential to break with existing barriers and therefore require continued work to move beyond the standard reactor designs (Ashley et al., 2012; Itzpatrick et al., 2015) (Figure 8.12).

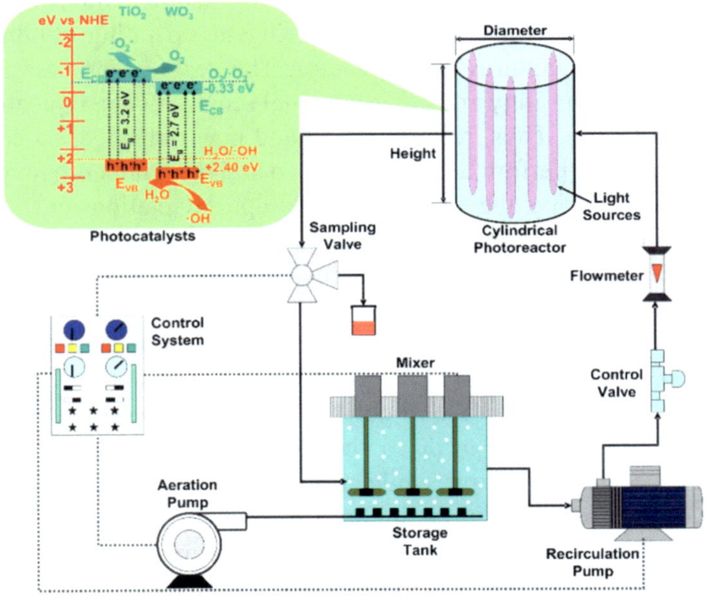

Figure 8.12. *Challenges in photochemical reactor design and operation.*

Source: https://www.mdpi.com/catalysts/catalysts-11-00556/article_deploy/html/images/catalysts-11-00556-g003.png.

8.6.1. Scale-Up Issues from Laboratory to Industrial Scale

This shift to industrially manufactured products presents a number of challenges which must be overcome for maximum design and efficiency of photochemical reactors. At the laboratory level, it is easy to maintain the reaction tightly with well-delineated parameters conducive to maximum yields, but scaling up the reactions typically reveals the mismatches owing to changes in heat and mass transfer rates affecting reaction kinetics and equilibrium. This involves the reconfiguration of reactor designs such that photonic energy distribution remains useful in larger tanks. Pilot-scale experiments are critical in identifying such scale-up issues because they enable scientists to make adjustments to operating conditions before investing in full-scale production. Knowledge of the nuances of scale-up enables engineers to develop trusted models that identify the complexity of large systems, therefore resulting in effective industrial processes. Besides, the materials and technology used in laboratory systems could be significantly different from practical applications in an industrial setting. Small-scale experiments are normally conducted in reactors with specialized materials to have efficient photon absorption and catalytic activity. Industrial reactors require materials that will survive more stringent operating conditions, for instance, higher temperatures and pressures. The scale-up process must also be in accordance with safety and environmental requirements, typically stricter at the industrial scale. Thus, reactor design and material selection can have a significant impact on performance and lifespan and must undergo much testing and qualification during the course of scale-up. Adaptation and innovation must be continuous so that reactors can be designed as optimal as possible while adhering to regulations, and this involves an integral approach to the design of the reactor. The use of renewable feedstocks also slows the scale-up of photochemical reactors. With cleaner feedstocks being eyed by industries, the scalability of the process of converting such feedstocks to useful fuel must be critically evaluated. One case in point is the novel idea of a solar refinery, which employs solar energy to drive chemical conversions previously reliant on fossil fuels. Such a transition, while promising, presents intricate challenges to reactor design, most particularly in the utilization of solar-to-fuel conversion processes. Each method, either photovoltaic integration or thermal processes, must be scrutinized for efficiency and viability at large scale, most notably in the regard to the infrastructural transformation and financing demands required. These issues need to be solved in a bid to increase economic feasibility, as well as minimize environmental impact, and the design and operation strategy of photochemical reactors being at the center of growth towards sustainable energy solutions (Tuller et al., 2017).

8.6.2. Photodegradation of Materials and Catalysts

Photodegradation is a critical aspect in the performance of photochemical reactor materials and catalysts. Photodegradation refers to material degradation due to exposure to light, particularly ultraviolet (UV) radiation. Photodegradation plays a significant role in the stability and lifetime of catalysts, which are required to catalyze many photochemical reactions. Understanding the mechanisms that govern photodegradation is essential in the design and operation of photochemical reactors as catalyst deactivation can lead to losses in efficiency, increased operational cost, and generation of secondary pollutants. By careful selection of photochemically stable materials and application of protective coating, researchers may reduce the impact of photodegradation, thereby attaining enhanced overall performance in photochemical systems (Silva et al., 2008).

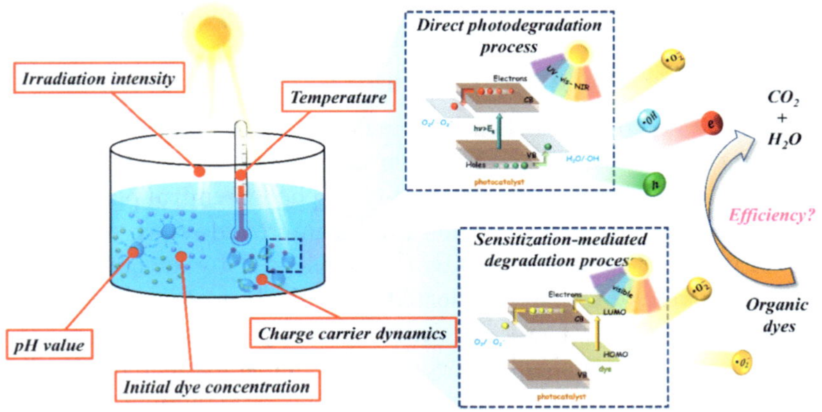

Figure 8.13. *Photodegradation of materials and catalysts.*

Source: https://www.mdpi.com/catalysts/catalysts-09-00430/article_deploy/html/images/catalysts-09-00430-sch001.png.

Solution technologies in catalyst design have also been created to address the photodegradation issues. One of them is the addition of semiconductor materials and catalysts to composite systems with greater resistance against light-induced degradation. For instance, the use of materials such as titanium dioxide (TiO_2) with metal catalysts has shown potential for advancing light absorption with improved resistance to photodegradation. Such synergy optimizes the catalytic performance and also assists in maintaining the long-term durability of the reactor components. Additionally, advanced characterization techniques allow one to investigate photoactive surfaces, and in doing so, derive valuable information that assists in developing more robust catalysts (Gulati et al., 2023). Such advancements highlight the importance of engineering material properties for specific photochemical application. Moreover, the efficiency of a photochemical reactor is most often related to operating conditions that could impact the rate of photodegradation of the materials in the system. Wavelength, light intensity, temperature, and active species presence are some factors that can influence the degradation rate of substrates and catalysts. For instance, through systematic studies it has been shown that enhancing light illumination under controlled reaction conditions will enhance the overall stability of working catalysts. Incorporating real-time monitoring techniques may also assist in assessing the status of catalysts during process progress and making the necessary adjustments to maximize performance. Last but not least, an in-depth understanding of photodegradation is required in order to promote the design and operation of photochemical reactors to ensure more efficient and sustainable energy conversion processes (Silva et al., 2008) (Figure 8.13).

8.6.3. Economic Considerations and Cost-Effectiveness

Economic viability of photochemical reactors is intrinsically linked to the general plant efficiency, which has been cited as a main indicator of cost-effectiveness. Engineering cost comparisons of photochemical plants in studies reveal that the more efficient systems exhibit improved return on investment (ROI). The performance analysis of two systems—a flat-plate collector/reactor and a linear parabolic trough—indicated general plant efficiencies of 10.3% and 11.6%, respectively. These efficiencies directly reflect in the capital and operational costs of producing hydrogen from water, arguably the most feasible solar photochemical process. Therefore, achieving maximum efficiency in such reactors not only renders them economically feasible but also meets the imperative call for renewable energy alternatives, overall furthering the case for further investment in photochemical technology advancements (Biddle et al., 2025). Apart from efficiency, the cost of capital in constructing and running photochemical reactors plays a significant role in their economic outlook. Factors like whether concentrating or non-concentrating systems should be employed decide these costs. A sensitivity analysis of the engineering parameters illustrates in detail how changes in efficiency can influence both capital expenses and cost of energy, even further underscoring the requirement for careful financial planning in the design and operation of such systems. And further, the inclusion of cooling systems improves efficiency of operation at a cost that may also have financial implications. Therefore, the economic considerations are not only of immediate costs but also include the cost dynamics of operation that affect long-term viability and cost-effectiveness (Biddle et al., 2025). While theoretical decompositions show significant efficiencies to be realized for photochemical processes, efficiencies proven in practice are often quite far below those numbers. For instance, currently available systems employed for water splitting have proved to be below 1% efficient, a rate that is not feasible economically for commercialization on a wide scale. In spite of advancement, for instance, through the utilization of sacrificial reagents, these processes render cycle efficiency difficult to attain and cost-effectiveness hard to attain. Notably, efficiencies observed in photoelectrochemical systems, though low, demonstrate greater promise compared to traditional homogeneous photochemical processes. This points to the need for ongoing research and development not only geared to optimize efficiency but also for the economic framework necessary for commercially scalable photochemical reactors (Figure 8.14).

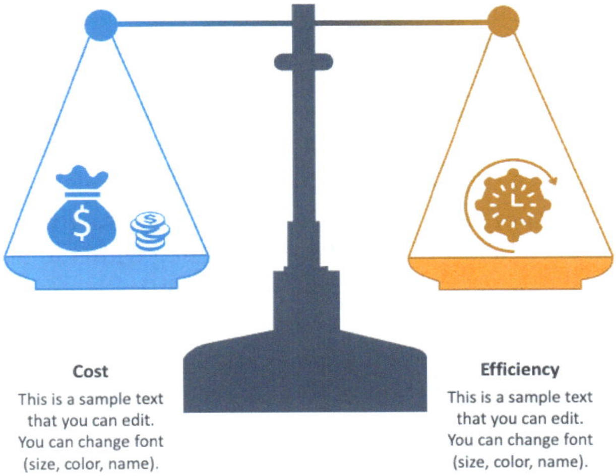

Figure 8.14. *Economic considerations and cost-effectiveness.*

Source: https://www.collidu.com/media/catalog/product/img/4/a/4a8ac0656eef77c32c5a65fde75d11b2a19267a02478260fcd07bf4cf47510f3/cost-effectiveness-slide1.png.

SUMMARY

Overall, the creation and operation of photochemical reactors are a key part in ensuring sustainable chemical production. Photochemical reactors use the energy of light to drive chemical reactions, offering an alternative that is more environmentally friendly than traditional processes driven by fossil-derived inputs. Continued advancement in photochemical technology promises increased efficiency and scalability, which will allow it to be integrated into existing industrial processes. As depicted in various case studies, photochemical reactors hold major promise for hydrogen production, which is of the highest relevance to the transition to a carbon-free energy future. The generation of hydrogen, in conjunction with future energy storage technologies, can largely neutralize the intermittent nature of renewable source energies, according to Yue et al. (2021). These developments make it imperative to innovate reactor designs in photochemistry to meet economic as well as environmental needs. Transition towards global sustainability demands high-tech reactor structures but also increased understanding of techno-economic feasibility of such systems. Photochemical reactors can be used to synthesize a wide range of useful chemicals, but their economic viability largely depends on the desired product. For instance, the transformation of carbon dioxide (CO_2) into chemical feedstocks such as carbon monoxide and formic acid appears promising under prevailing conditions, with net present values indicating economics (Jouny et al., 2018). There is a requirement for continued investment in research and development efforts to maximize the efficiencies of the processes, and to render them commercially feasible. Investigations of performance targets will allow for the identification of economically rational responses, which will result in a robust photochemical industry capable of competing with traditional processes. Ultimately, successful use of photochemical reactors relies on multidisciplinary collaboration, integrating chemical engineering, material science, and policy mechanisms. As the world faces climate change and resources dwindling, demand for non-traditional approaches has never been greater. The use of photochemical reactors in producing renewable hydrogen and other chemicals of worth will be instrumental in decoupling economic growth and carbon emission. Coordination of technological advancement with supporting economic and political institutions is, therefore, a must. The challenges of the future for reactor design and operation must prioritize better system performance through sustained advancements while ensuring scalability and cost-effectiveness. Achievement of these targets will be instrumental in defining a sustainable pattern of chemical production in relation to global energy transition visions, as characterized by current trends of hydrogen technology advancement (Yue et al., 2021).

On looking at the design and operation of photochemical reactors, several key components are crucial to the performance and longevity of these systems. Structural design choices, such as collector/reactor designs for photochemical facilities, play a critical role in overall performance. Two configurations were examined, for instance: flat-plate collector/reactors and linear parabolic troughs. Each of these designs attains different operational efficiencies with respective estimates of 10.3% for flat plates and 11.6% for trough systems from different solar photochemical efficiencies of 12.9% and 14.6% (Biddle et al., 2025). The choice between the two systems, therefore, should also consider not just efficiency but associated engineering costs and energy demands, a function that will serve in the optimization of photochemical reactor operations. Beyond structural issues, thermal management of photochemical reactors is also crucial in terms of maximizing productivity. The study underscored the way the opposing effects of concentration ratios and operating temperatures necessitate an effective cooling method, particularly with the trough system (Biddle et al., 2025). Active and passive cooling mechanisms were being scrutinized in order to prevent overheating and maximize general efficiency. This identifies a pivotal aspect of reactor design, where temperature control can be an economic determining factor for photochemical reactor efficiency. The relationship between reactor cooling regimes and operational cost implications further supports the importance of thermal management integration in the early stages of reactor design. Finally, interdisciplinary action facilitates photochemical reactor technology advancement as well as ensuring sustainable methodology. A workshop tackled the need for guidelines to promote the development of ecological support systems, convening researchers from various fields to expand knowledge of complex interactions within the emerging technology. The research partnerships will help solve the interdisciplinary problems of photochemical reactor design and operation optimization. By integrating information from disciplines such as engineering, environmental science, and economics, stakeholders can come up with improved photochemical reactors, which can pave the way towards operational applications in renewable energy conversion and industrial processing.

REFERENCES

1. Aillet, T., Dechy-Cabaret, O., Loubière, K., et al. (2013). Photochemical synthesis of a "cage" compound in a microreactor: Rigorous comparison with a batch photoreactor. https://core.ac.uk/download/18540116.pdf (accessed on 28th September 2025).
2. Ashley, S. F., Fenner, R. A., et al. (2012). Sustainability indicators for open-cycle thorium-fuelled nuclear energy. https://core.ac.uk/download/77057041.pdf (accessed on 28th September 2025).
3. Bauer, M. U., Coggon, M. M., Flagan, R. C., et al. (2017). The Caltech Photooxidation Flow Tube reactor: Design, fluid dynamics and characterization. https://core.ac.uk/download/84315702.pdf (accessed on 28th September 2025).
4. Baumann, M., Moody, T. S., Smyth, M., & Wharry, S. (2020). A perspective on continuous flow chemistry in the pharmaceutical industry. *Organic Process Research & Development, 24*, 1802–1813. https://doi.org/10.1021/acs.oprd.9b00524 (accessed on 28th September 2025).
5. Beeler, A. B., Gao, Q., Porco, J. A., et al. (2017). A photochemical flow reactor for large scale syntheses of aglain and rocaglate natural product analogues. https://open.bu.edu/bitstream/2144/32940/3/nihms888885.pdf (accessed on 28th September 2025).
6. Bezerra Duarte, M., Dantas, R., Lima Duarte, M., et al. (2025). Remediation of petroleum contaminated soil by photo-Fenton process applying black, white and germicidal light. https://core.ac.uk/download/pdf/39151999.pdf (accessed on 28th September 2025).
7. Biddle, J. R., Fujita, T., & Peterson, D. B. (2025a). Review of solar fuel-producing quantum conversion processes. https://core.ac.uk/download/pdf/42847962.pdf (accessed on 28th September 2025).
8. Biddle, J. R., Fujita, T., & Peterson, D. B. (2025b). Solar photochemical process engineering for production of fuels and chemicals. https://core.ac.uk/download/pdf/42847961.pdf (accessed on 28th September 2025).
9. Carden, J. L., & Mason, R. M. (2025). Guiding the development of a controlled ecological life support system. https://core.ac.uk/download/pdf/42867040.pdf (accessed on 28th September 2025).
10. Cheddie, D., Hills, C., & Tota-Maharaj, K. (2017). Low-cost urban wastewater infrastructure for environmental sustainability across Caribbean small island developing states (SIDS). https://core.ac.uk/download/80783931.pdf (accessed on 28th September 2025).

11. Collins, S. G., Deadman, B. J., & Maguire, A. R. (2014). Taming hazardous chemistry in flow: The continuous processing of diazo and diazonium compounds. https://core.ac.uk/download/61578559.pdf (accessed on 28th September 2025).
12. da Silva, C. S. C. G. (2008). *Synthesis, Spectroscopy and Characterization of Titanium Dioxide Based Photocatalysts for the Degradative Oxidation of Organic Pollutants*. https://core.ac.uk/download/143373410.pdf (accessed on 28th September 2025).
13. Davenport, R. J., Schubert, F. H., Shumar, J. W., et al. (2025). Evaluation and characterization of the methane-carbon dioxide decomposition reaction. https://core.ac.uk/download/pdf/42888186.pdf (accessed on 28th September 2025).
14. Fitzpatrick, D. E., Ingham, R. J., Ley, S. V., et al. (2015). Organic synthesis: March of the machines. https://core.ac.uk/download/42338797.pdf (accessed on 28th September 2025).
15. Gross, S., Millner, A., Neugebauer, M., et al. (2025). Overview study of space power technologies for the advanced energetics program. https://core.ac.uk/download/pdf/42857781.pdf (accessed on 28th September 2025).
16. Gulati, S. (2023). Investigation of light management strategies and photochemistry of Si/TiO_2 tandem microwire slurries for solar hydrogen generation. https://core.ac.uk/download/580004351.pdf (accessed on 28th September 2025).
17. Jacob-Lopes, E., & Franco, T. T. (2010). Microalgae-based systems for carbon dioxide sequestration and industrial biorefineries. https://core.ac.uk/download/322390942.pdf (accessed on 28th September 2025).
18. Jouny, M., Luc, W., & Jiao, F. (2018). General techno-economic analysis of CO_2 electrolysis systems. *Industrial & Engineering Chemistry Research, 57*, 2165–2177. https://doi.org/10.1021/acs.iecr.7b03514 (accessed on 28th September 2025).
19. Kalamaras, E., Maroto-Valer, M., Shao, M., et al. (2018). Solar carbon fuel via photo electrochemistry. https://core.ac.uk/download/288360687.pdf (accessed on 28th September 2025).
20. Lee, S. K., & Mills, A. (2004). Detoxification of water by semiconductor photocatalysis. https://core.ac.uk/download/9013846.pdf (accessed on 28th September 2025).
21. Mazzotti, M. (2015). *Physiological Studies to Optimize Algal Biomass Production in Phytoremediation Processes*. https://core.ac.uk/download/76519921.pdf (accessed on 28th September 2025).

22. Naresh, V., & Lee, N. (2021). A review on biosensors and recent development of nanostructured materials-enabled biosensors. *Sensors, 21*, 1109. https://doi.org/10.3390/s21041109 (accessed on 28th September 2025).

23. Tuller, H. L. (2017). *Solar to Fuels Conversion Technologies: A Perspective*. https://core.ac.uk/download/83232124.pdf (accessed on 28th September 2025).

24. Yue, M., Lambert, H., Pahon, E., Roche, R., Jemeï, S., & Hissel, D. (2021). Hydrogen energy systems: A critical review of technologies, applications, trends and challenges. *Renewable and Sustainable Energy Reviews, 146*, 111180. https://doi.org/10.1016/j.rser.2021.111180 (accessed on 28th September 2025).

25. Zondag, S. D. A. (2024). *Design, Characterization and Application of Continuous-Flow Photoreactors*. https://core.ac.uk/download/631133721.pdf (accessed on 28th September 2025).

CHAPTER 9

ENVIRONMENTAL, ENERGY, AND ECONOMIC CONSIDERATIONS

Contents

Introduction 353

9.1. Environmental Impact of Energy Production 357

9.2. Economic Implications of Energy Choices 362

9.3. Policy Frameworks for Sustainable Energy 367

9.4. Technological Innovations in Energy Efficiency 372

9.5. Social Considerations in Energy Transition 378

9.6. Global Perspectives on Energy and Environment 383

Summary 389

References 391

INTRODUCTION

The complexities that encompass environmental, energy, and economic considerations call for a thorough examination of sustainable alternatives that consider the dominant issues of the times. While the world is being faced with increasing dangers from climate change, the need to transition away from fossil fuels to renewable energy sources has never been more critical. The transition does not only aim to prevent environmental degradation but also foster economic resilience in an increasingly unpredictable market. Technological advancements and energy efficiency are key factors to push this change. For example, the flying sector is looking to adopt electrified planes as a way to lower its carbon emissions, such that a hopeful future of future transportation falling in line with sustainability targets may be a possibility. The use of such technologies holds great promise in dealing with a balance between economic growth and ecological health, ultimately leading to a sustainable society (Spierling et al., 2025). In building construction, the requirements to adhere to energy efficiency standards are particularly urgent. The construction sector is one of the largest consumers of natural resources, and this increases the need to integrate renewable energy systems, such as solar energy, into new developments. Solar energy has double benefits: enhancing energy efficiency in buildings and contributing to mitigation of greenhouse gas emissions. A cost-benefit study suggests the viability of various solar energy systems, such as solar heating systems and photovoltaic panels. Case studies, for example, the ones conducted in Brazil, enable stakeholders to get a precise picture of the economic and technical feasibility of such technologies. Identifying trade-offs among models is critical in facilitating informed decision-making towards sustainable construction practice (Arruda et al., 2023). The intersection of environmental, energy, and economic factors creates a very complex terrain upon which policymakers, corporations, and individuals have to navigate in order to be sustainable. To develop the productive actions that will be required, there has to be coordination to balance individual interests and broader ecological necessities. Taking advantage of breakthroughs across sectors—such as electrified flights to efficient buildings—there is an enormous potential to promote environmental stewardship while enhancing economic performance. A resolution of concern about the cost and effectiveness of renewable technology is essential to enable their extensive diffusion. Ultimately, the balance of these determinants will determine the success of endeavor toward reducing ecological stress and facilitating economic development for future generations.

Definition of Key Terms: Environment, Energy, Economy

The environment in which environmental, energy, and economic deliberations operate is premised on defining terms that guide these interconnected fields. The environment is the natural world and all that exists and lives within it, including ecosystems and biodiversity, and the physical components that sustain life on our planet. The foundation for understanding how human use impinges upon natural resources and ecosystems is the paradigm for re-evaluating activities like energy generation and waste management. The interaction with environmental parameters is critical, particularly since projects make use of alternative energy resources. The alternatives are not only measured in terms of efficiency, but also in terms of how they might impact biodiversity, climate, and other environmental indicators, as noted in studies underscored by the importance of sustainable development strategies for energy supply systems (Juan et al., 2023). Energy, the second buzzword, is the capacity to perform work and is a constituent of all aspect of life ranging from energizing homes to propelling industries. Energy analysis involves the study of its sources, utilization, and sustainability of the sources.

Figure 9.1. *Environmental, energy, and economic considerations.*

Source: https://media.greenmatters.com/brand-img/XgQVeN12M/0x0/renew-effective-enviro3-1602699720155.jpg.

The traditional sources of energy, such as fossil fuels, are dominant causes of environmental deterioration, while renewable energy technology presents a viable alternative. The transition to renewable energy entails economic consequences to be realized, including investment costs, employment within the energy sector, and public health impact. The recognition of energy as an intricate resource bridges environmental sustainability with economic viability, and thus the necessity of diversified indicators to inform energy investments and policies. Therefore, defining energy in general allows for an informed discussion on its broader ramifications in our economic and environmental spheres. Economy is the goods and services production, distribution, and consumption network of society. When understood in terms of the environment, the economy does not just refer to budgetary costs but also externalities, like environmental degradation resulting from manufacturing processes. This is where society life-cycle costing (S-LCC) enters the picture, offering a framework for analyzing the balance between economic activities and their environmental signatures. As an illustration, employing S-LCC to assess waste management programs enables policy analysts to quantify externalities that arise from emissions and environmental risks. These assessments illustrate how economic decisions can reinforce or degrade environmental sustainability, depending on how externalities are incorporated into financial analysis (Martinez-Sanchez et al., 2017). Thus, reconsidering the economy according to sustainability forces us to reconsider our conception of success, calling for models that pay as much attention to environmental well-being as to profit and growth (Figure 9.1).

Importance of the Interconnection Between These Areas

The connection between economic, energy, and environmental factors is not only favorable but also inevitable in the creation of a sustainable tomorrow. With economic development, the subsequent environmental impacts have the tendency to worsen. Such an experience is illustrated by the N-shaped shape of the Kuznets curve hypothesis that assumes economic action might aggravate environmental pollution initially but should then get better following policy changes. A deep understanding of this connection enhances policymakers' capability to design policies that minimize the adverse environmental effects while promoting economic growth. For instance, transitioning towards renewable forms of energy can not only help decrease carbon emissions but also provide economic opportunities through the generation of jobs in emerging industries. The integration of sustainable practices into economic systems is the model for the countless benefits that can be reaped from the acknowledgment of and leveraging of such interconnectedness, ultimately resulting in an equitable approach towards development with an eye toward both ecological sustainability and economic strength (Sava, 2024). A good energy policy is necessary to meet the challenges of energy security and conservation. There exists a two-way relationship between environmental degradation and energy consumption, and thus energy supplies form an important sector to take under consideration under the overall umbrella of sustainable development. Fossil fuel use has in the past inflicted massive ecological damages, which have necessitated shifts towards less environmentally destructive sources of energy. Such shifts not only address environmental concerns but also involve economic interests, particularly energy price and national security issues. Diversification of energy sources on a strategic basis is of the greatest significance, as it reduces the risks of supply disruption while also ensuring economic stability. Identification of such a correlation enables countries to develop policies that not only enhance energy security but also, concurrently, prioritize environmental integrity, thus demonstrating the linkage between such realms as necessary in driving balanced national agendas (Ibekwe et al., 2024). Secondly, the confluence of environmental, energy, and economic dimensions highlights the importance of international collaboration in preventing global challenges. As climate change worsens, the consequences of environmental degradation and energy imbalance become transboundary and require comprehensive approaches to foster cooperative solutions. Effective policies must be fashioned with a perspective towards fostering cooperative ventures among states to address shared environmental and energy issues. The complexities of global energy security and environmental protection bargaining arrangements mirror the high degree of balance that needs to be realized. Through cooperation in the research and development of clean technologies and harmonizing energy policies, nations can work together to realize common aims with respect to economic development, environmental preservation, and energy security. This international engagement is of the utmost importance, as this interdependence between these regions requires that it be done together to protect the future of both the earth and its people (Ibekwe et al., 2024).

9.1. ENVIRONMENTAL IMPACT OF ENERGY PRODUCTION

The ecological impact of energy production is a very intricate issue which warrants critical scrutiny, especially in relation to climate change and the management of natural resources. Fossil fuels, which have been the backbone of global energy infrastructure for years, are at the forefront of greenhouse gas production, whose impact is global warming and environmental degradation. Alternative forms of energy such as wind, solar energy, and biofuels can reduce these ill effects. Specifically, biofuel production has raised attention for reducing carbon emissions below that of fossil fuels and providing a sustainable option if responsibly accomplished. However, their extraction and use also raise grave concerns regarding land use change, food security, and water consumption and therefore require a cautious assessment of their environmental benefits and limitations, as highlighted by research findings from scholarly reviews of biofuel technology and sustainability (El-Araby, 2024). In recent years, the scientific community has highlighted integration of both adaptation efforts and emission reduction in restricting the effects of climate. The contribution by the Working Group II to the Sixth Assessment Report of the Intergovernmental Panel on Climate Change (IPCC) explains how human activities like energy production affect climate and biodiversity directly, hence requiring an integrated solution to managing the environment (IPCC, 2023). This research points out that climate-resilient policies are the only option to attain sustainable development by taking care of human well-being and ecological health. These policies encompass forward-looking planning and breakthrough technologies which have the potential to increase energy efficiency as well as mitigate environmental degradation. By these interlinks, policymakers and researchers can develop frameworks which merge energy generation with broader environmental goals, constructing a more sustainable energy future. In addition, the time to transition to cleaner sources means intricate economic trade-offs. Adopting renewable energy technologies does not only assure lower emissions but also stimulates economic development in terms of job creation and technological progress. Technologies used to produce biofuels, particularly, can be used to provide rural economies new opportunities for expanded growth, and the coincidence of economic and environmental objectives. Still, the economic viability of these transformations is governed by market forces, government policies, and infrastructural capacity. From what is apparent from research advances related to biofuel technologies, ongoing research is part of enhancing production efficiency and economic attractiveness without sacrificing the environmental integrity of the process. Balanced production of energy can lead to symbiosis in terms of conserving the environment, enhancing economic advancement, and sustaining community resilience (Figure 9.2).

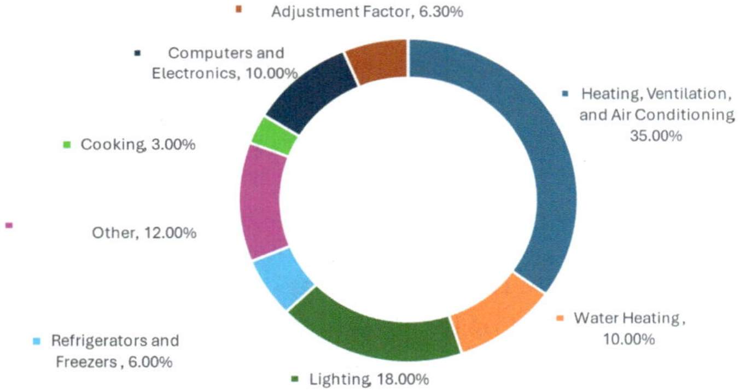

Figure 9.2. *Environmental impact of energy production.*

Source: https://www.mdpi.com/eng/eng-05-00067/article_deploy/html/images/eng-05-00067-g001.png.

9.1.1. Types of Energy Sources and Their Environmental Footprints

The quest for sources of energy reveals a diverse array of opportunities, which possess unique environmental consequences. The so-called renewable sources of solar power, wind power, and water are widely acclaimed for their minimal greenhouse emissions upon implementation. However, the overall life cycle impacts cannot be overlooked; for instance, energy is abundant in the process of producing solar panels and may result in pollution if improperly managed.

Traditional fossil fuels—coal, oil, and gas—still are the leading sources of global energy but come with substantial environmental expense, including air pollution and carbon emissions. Extraction itself tends to disrupt ecosystems, while combustion contributes significantly to key facets of climate change. An understanding of these different effects is significant for evaluating the overall sustainability of energy resources and progressing toward a cleaner energy future. The environmental impact of energy consumption extends beyond emissions to encompass resource depletion and habitat loss. For example, the extraction of fossil fuels contributes not only to greenhouse gases but also to extensive land use, which could be harmful to local ecosystems and biodiversity. Conversely, the low-emission renewable resources compete for resources and land, thus the possibility of land use conflict and increased habitat fragmentation. Bird migration routes can be affected by wind farms, and aquatic habitats can be altered by hydropower projects. Therefore, an integrated examination of the environmental consequences of every energy source should consider such trade-offs in light of informing energy policymaking. This multidimensional framework of assessing energy resources can inform actions on how to reduce negative environmental effects without compromising societal energy requirements. While moving towards sustainable energy, economic factors also come into play as an important determinant. Renewable technologies are not cheap to install, and this discourages investment despite their enduring eco-friendly as well as economic impacts. With the COVID-19 pandemic, there are probable chances of technological innovation in the energy sector, driven by increased adoption of digital platforms and remote operations. 5G and energy storage improvements are some of the technologies that can actually reduce operational footprints while increasing efficiency levels across industries. Merging this technological development with renewable energy implementation can create a more sustainable economic framework, hence finally reducing fossil fuel reliance. This calls for intensive planning and investment in infrastructure to ensure that our energy future is not only sustainable but also economically viable (Emily et al., 2021; Jiří et al., 2020).

9.1.2. Effects of Fossil Fuels on Air and Water Quality

Fossil fuel dependence has long-term effects on air and water quality, to a great extent compromising environmental integrity. Combustion of fossil fuels releases a wide range of pollutants, including particulate matter, sulfur dioxide, and nitrogen oxides, to contribute to poor air quality and human respiratory health. Urban areas with heavy traffic and industrial activity have high levels of smog and hazardous air pollutants, which pose greater threats to health such as asthma and cardiovascular illness. Additionally, fossil fuel production processes such as fracking and coal mining emit toxic chemicals into the air and the environment, leading to more public health concerns. While societies grapple with the immediate consequences of air pollution, swift action toward a transition to cleaner energy is paramount in safeguarding human and environmental health.

Figure 9.3. *Effects of fossil fuels on air and water quality.*

Source: https://green.org/wp-content/uploads/2024/01/R-2-2-e1707145162860-1200x764.png.

Even water supply is under threat from the fossil fuel industry as most extraction and production activities lead to spill events. Spills of oil, chemical poisoning, and leaching of poisonous chemicals from coal mines can devastate local ecosystems, aquatic life, and drinking water supplies. For example, significant spills such as the Deepwater Horizon spill have long-term and serious effects on marine ecosystems, leading to loss of biodiversity and damaged habitats. Moreover, the ash of fossil fuel burning, that is also likely to consist of heavy metals and other toxic elements, may settle in water bodies and pose risks to wildlife and human life through the food chain. Freshwater pollution calls for an urgent response, emphasizing the critical need for regulatory policies that prioritize a great deal of environmental protection and limit reliance on fossil fuels. Shifting to renewable power sources not only addresses air and water pollution issues but also answers economic and technological innovation needed for future sustainable development. Shifting towards cleaner energy sources, such as solar and wind, can effectively minimize air pollution and assist in lowering greenhouse gas emissions. According to recent research, the use of renewable energy technologies has been made economically viable, with the scope for significant socio-economic benefits as well as environmental protection (Gielen et al., 2019). Despite existing challenges to the achievement of the transition, such as the need for significant infrastructure and new technology, the decarbonization potential is enormous. For instance, hydrogen as clean energy offers chances for decarbonizing industries that have historically relied on fossil fuels, changing the destiny of energy production and consumption (Staffell et al., 2018). In the end, the focus on renewable energy could have implications such as enhanced environmental quality, public health, and robust economic growth (Figure 9.3).

9.1.3. Renewable Energy Sources and Their Ecological Benefits

The transition to new energy sources is essential to environmental preservation and sustainable economic growth. Renewable energy, including solar, wind, and hydro power, slashes greenhouse gas emissions significantly, which are the largest contributors to climate change. By accessing these cleaner sources, countries can avert the adverse impacts associated with the consumption of fossil fuels, including water and air pollution, loss of biodiversity, and deforestation. In addition, renewable energy technologies tend to use less water than conventional energy sources, easing the strain on precious water supplies. This transition to renewable energy is not only reducing environmental deterioration but also aligning with the world's attempts to meet climate goals and sustain ecological balance. At its most fundamental level, investment in renewable energy builds a people-friendly and environmentally sustainable setting, accentuating the need for systematic transition towards sustainable use and production of energy. To the imperative of environment sustainability, evidence shows a growing overlap between economic growth and renewable energy uptake. As explored via recent research, the use of renewable energy sources not only has environment-friendly impacts but also stronger economic performance. The study demonstrates direct causality between the use of renewable energy technologies and top-line economic indicators, such as GDP growth and employment (Chou et al., 2023). This multi-faceted development policy underscores that renewable energy is a force behind a green economy that spurs innovation and industrial productivity. Moreover, nations blessed with renewable endowment are likely to accrue enhanced investment opportunities, which implies the economic benefits of renewable energy are not only secondary but an intrinsic part of development planning in the long run. Hence, greater incorporation of renewable energy in economic policies can yield extensive environmental and economic dividends. Renewable energy's environmental benefits are accompanied by social development, ushering in climate-resilient growth that is conducive to human welfare and biodiversity. Through focus by authors in the IPCC, effective integration of renewable energy into adaptation policy produces strategies that enhance the resilience and stability of ecosystems (IPCC, 2023). This complementary relationship facilitates natural habitat conservation, reduces climate risk, and promotes sustainable land use practices that harmonize human needs with environmental needs. Renewable energy technologies are also useful to communities in a better energy security and reduced reliance on foreign fuels, and their economies are further stabilized. By responding to both the social and environmental aspects, investment in renewable energy creates a more sustainable and equitable framework for generations to come, illustrating the profound interconnections between economic prosperity and environmental well-being (Figure 9.4).

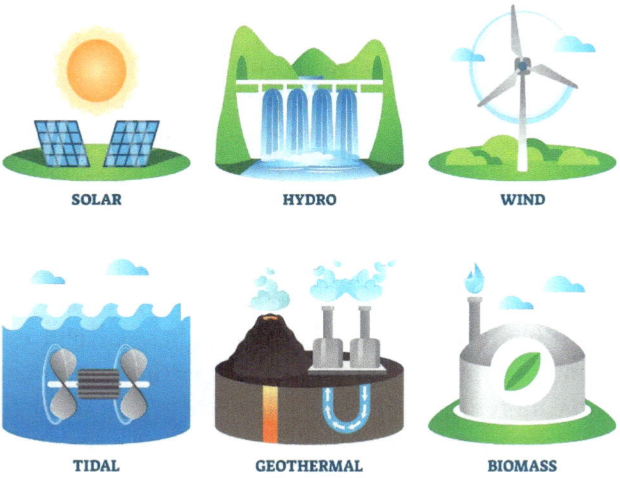

Figure 9.4. *Renewable energy sources and their ecological benefits.*

Source: https://images.prismic.io/inspirecleanenergy/types-of-renewable-energy-sources.jpeg?auto=compress,format&rect=0,0,1840,1036&w=1840&h=1036.

9.2. ECONOMIC IMPLICATIONS OF ENERGY CHOICES

The economic consequences of energy choices are multifaceted and have profound impacts on macroeconomic stability and market conduct. Central to this is the awareness that the production and consumption of energy are not technical but seminal economic activities with critical implications. Technological innovation, resource availability, and consumer choice dictate the path of energy systems and their economic consequences. As emphasized by Wang et al. (2022), energy system development relies on strategic policy choices and socioeconomic factors controlling regulatory systems as well as social acceptance. Such synergy emphasizes the necessities of policy intervention to realize sustainable energy systems in synergy with economic growth and ecological preservation. As the visibility of cryptocurrencies increases, there is a necessity to consider their economic implications alongside traditional energy sources. The green trail of cryptocurrency mining, for example, raises issues regarding energy usage and its correlation with greenhouse gas emissions. Examination reported by Wang et al. (2022) finds that increasing interest in cryptocurrencies is accompanied by market prices facing environmental scrutiny. The interaction between cryptocurrency market volatility and energy considerations reveals an immediate need for green practices for emerging technologies. Thus, with societies turning towards a digital economy, addressing the sustainability of these energy-intensive processes is imperative in a bid to assist in mitigating their adverse economic and environmental effects. Additionally, the various scenarios of energy consumption and their potential economic implications reveal the complex nature of environmental factors versus the option of energy. The Shared-Socioeconomic Pathway (SSP) methodology stipulates that energy sector development is contingent on factors from technological milestones to sociopolitical determinants (Bauer et al., 2016). The range of such projections not only reflects the future energy system uncertainty but also poses central policy implications for economic policy. Policymakers must contend with the prevailing uncertainties to provide climate-resilient energy policy that enhances economic stability without leaving large carbon footprints. Ultimately, choices made today will decide the long-term viability of energy systems and their economic bases for generations to come.

9.2.1. Cost Analysis of Renewable versus Non-Renewable Energy

The cost of renewable against non-renewable sources remains an essential factor in environmental and economic sustainability discourse. Historically, non-renewable sources such as fossil fuels have been seen as cheaper since there exist developed infrastructures for extraction and use that equate to lower upfront costs. However, these fuels are accompanied by huge externalities in terms of environmental degradation, health impacts, and the expenses of curbing climate change. These variables eliminate the perceived cost-effectiveness of fossil fuels if one looks at them in the long term. When externalities are included, one finds that while the environmental consequences of fossil fuel consumption grow more pressing, the competitive edge they possess can deteriorate.

Therefore, a well-informed cost comparison would not only consider market prices of energy production but also the socio-economic factors that accompany both sources of energy and thus indicate the complexity involved in choosing sustainable sources of energy. Solar, wind, and hydroelectricity are some of the sources of renewable energy that have come into prominence in recent years due to their long-term cost benefits and no environmental risks. While the upfront cost of renewable technology could be much higher than that of fossil fuels, ongoing innovations and economies of scale serve to reduce costs. Like in the case of lithium-ion battery technology, that efficiency of renewable energy can even be enhanced by integrating storage technologies, rendering it even more dependable.

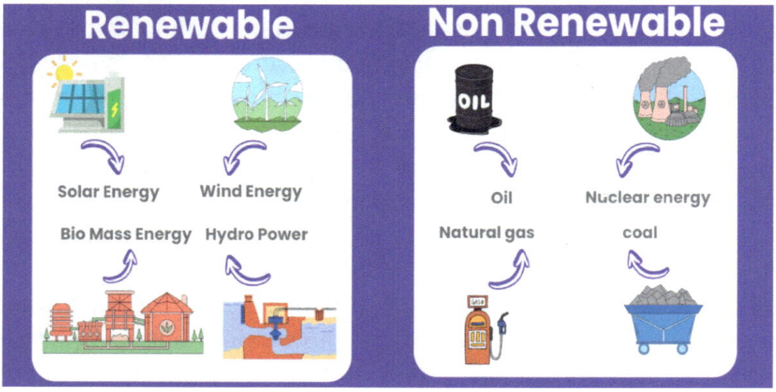

Figure 9.5. *Renewable versus non-renewable energy.*

Source: https://images.examples.com/wp-content/uploads/2024/04/Renewable-Feature-2.png.

Also, the policy landscape these days has shifted towards supporting renewable energy through subsidies, tax credits, and regulatory policies for encouraging cleaner energy production. The move away from fossil fuels has implications not only in energy security terms but also in technological changes and employment opportunities in the renewable sector, a shift that is beyond cost considerations. One significant component of such an energy shift is evaluating the potential of hydrogen technology, which can add some strength to renewable energy infrastructures. Hydrogen can be used as an energy storage solution supplementing intermittent renewable sources by enabling energy storage and re-electrification. There are various projects in the world considering different uses of hydrogen, ranging from fuel for transportation to energy storage systems. The techno-economic analysis determines that while current hydrogen technology is costly and suffers from inefficiencies, research and investment must be made to seek competitiveness. The call for improvement in performance, scaling up, and political support remains critical to enabling a cost-competitive hydrogen economy. Cumulatively, these results signify a step in the direction of a more integrative and knowledge-based appreciation of the cost factor attached to renewable energy, necessitating the integration of creative solutions into the broader framework of environmental and economic sustainability, as demonstrated in research such as that of Calvin et al. (2023) and Yue et al. (2021) (Figure 9.5).

9.2.2. Job Creation in the Renewable Energy Sector

The renewable energy industry has emerged as a key employment generation driver with long-term economic growth complemented by increased efforts to address key environmental challenges. With the transition towards clean energy sources expanding day by day, industries such as solar, wind, and green hydrogen are opening up at breakneck speed. The industries not only require a robust workforce for installation and maintenance but also create ancillary employment opportunities in manufacturing, supply chain management, and project management. To illustrate, increased utility-scale solar farms in the United States have shown this trend, which has provided tens of thousands of jobs in regional economies and increased labor involvement in cities and rural areas (Agupugo et al., 2024). Not only does increased renewable employment arrest unemployment but also presents opportunities for retraining and upskilling the workforce, leaving employees poised to meet the green economy's evolving demands. Besides, the economic impacts of employment growth in the renewable energy sector go beyond the duration of short-term jobs. Renewable energy development can also stimulate local economies through greater tax revenues and decreased dependence on imported fossil fuels. Addition of projects like green hydrogen production is also having extra possibilities for increasing employment, with studies observing the economic gains from developing this new sector (Shboul et al., 2024). These initiatives not only generate employment but also stimulate local investment, with a ripple effect positively in the community. As economies embrace renewable technology, the return is not only through jobs generated but also in energy security and reduced energy costs for consumers. All these combined underlines the urgency to promote renewable energy development as a strategy towards sustainable economic growth. Furthermore, the convergence of renewable energy work with environmental sustainability represents a paradigm shift in what economies can potentially do to advance ecological necessities. The green benefits of renewable energy technologies, such as reduced greenhouse gas emissions and improved air quality, are equaled by their ability to provide job opportunities across various populations. With mounting projects, particularly in the low-income regions, an opportunity exists to make the creation of jobs inclusive and fair. This double-pronged strategy of economic growth and environmental conservation is the need of the hour in setting up foundations for a sustainable future. Training and development programs at the workforce level will cement this nexus, preparing workers to adapt to the evolving demands of a continuously developing sector (Agupugo et al., 2024). Briefly, the renewable energy sector is a unique intersection of the environment, economy, and social development, and it occupies a central place in modern economic policy.

9.2.3. Economic Risks Associated with Climate Change

The relationship between economic stability and climate change generates serious risks that span across various sectors. As the world experiences increases in temperature, the effects of these changes have impacts that affect agriculture, infrastructure, and labor productivity. For instance, increased temperature is likely to significantly affect the production of crops, especially those regions that depend on a specific climate for agricultural sustainability. The economic implications of reduced agricultural productivity are widespread, affecting food supply but also raising prices and inflation. Volatility can also enhance socioeconomic gaps, leaving poorer segments of society more vulnerable to food insecurity and economic distress. The potential for massive job and livelihood transformations contributes further to the need to deal with these risks in the context of overall economic planning strategies based on climate-induced disruption resilience. In evaluating the economic risk represented by climate change, consideration should be given to the prospect of widespread financial loss across a variety of industries, particularly natural-resource-based ones. The energy sector is particularly vulnerable, with investments in fossil fuels exposed to increasing questioning and potential devaluation in a changing economy. As renewable energy technologies develop and gain increasing popularity, investment in traditional energy firms dwindles, with monetary losses for them being enormous. More importantly, the transition to green energy is risky business, from stranded assets in fossil fuel investments to demands of titanic amounts of capital in renewable infrastructure. Policymakers are forced to deal with these uncertainties in a bid to provide an atmosphere that would be conducive to innovation while limiting economic shocks that could stifle growth. Thus, the economic system must evolve alongside environmental considerations to provide viable opportunities for sustainability. Adaptation and preparatory measures cannot be overstressed as one works with the challenges posed by climate changes economic impacts. The transition to renewable energy sources is both an opportunity to minimize environmental impacts and a challenge in terms of economic expense and infrastructure growth. As established in current literature, the speeding up of the transition towards renewable energy technology can achieve massive economic benefits, from creating jobs to increased energy security. But this transformation must be met with a strategic policy strategy to make regulatory climates conducive to fostering growth in the clean tech and energy efficiency sectors. The finance in clean technologies as well as energy efficiency must be prompted by a higher-level plan that considers the chance for economic dislocation. Through this, stakeholders can work together in a manner that maximizes synergies of economic growth and environmental sustainability, thus turning the economic climate change threats into opportunity for resilience and innovation (Emily et al., 2021; Gielen et al., 2019).

9.3. POLICY FRAMEWORKS FOR SUSTAINABLE ENERGY

The formulation of robust policy guidelines to support sustainable energy is at the center of the response to the intertwined challenges of environmental degradation and economic progress. These guidelines often revolve around the twin objectives of enhancing energy efficiency and promoting the adoption of renewable energy resources. Technological advancement becomes a pillar of such policies due to the fact that it not only supports the development of green technology but also necessitates a regulatory environment in favor of green practices. The integration of technology and economic policies, particularly through channels such as environmental taxation, can promote investment in renewable resources and reduce fossil fuel reliance. Renewable energy policies necessitate a comprehensive knowledge base of how innovation, when combined with taxation policy, can push back against environmental impacts and at the same time establish resilience in economies, confirming the reasonability of an equitable strategy as posited by recent evidence (Mohania et al., 2025). Sustainable institutional structures are essential in the successful enforcement of renewable energy policies. Institutional quality is significant in the transition to renewable energy because it captures the level of governance, transparency, and legal accountability required for promoting environmental sustainability. This concept is largely aligned with the United Nations Sustainable Development Goals (SDGs), i.e., SDG-7 that concerns affordable and clean energy, and SDG-13 that concerns climate action. Strong institutions facilitate the transition to renewable energy technologies through regulatory implementation and innovation support, enforcing policy systems' relationship with ecological traces. Further, through making it explicit and establishing mechanisms of accountability, institutional systems can curtail adverse ecological effects and spur investment in sustainable energy options, thus facilitating a favorable environment for economic development while remaining in environmental compliance (Li et al., 2025). The integration of economic incentives into policy frameworks can propel revolutionary change towards supporting sustainable energy initiatives. Governments can use monetary tools like subsidies, tax rebates, and concessional credits to initiate investments in renewable energy technology and infrastructure. Furthermore, tying these economic tools to institution-building enhances regulatory standards and reinforces accountability culture in the energy sector. As these policies become more widespread, they both support the utilization of renewable energy and cultivate an atmosphere of stakeholder engagement, the point of departure on environmental and economic sustainability. Thus, the interplay between institutional quality and certain financial incentives is essential in a bid to have policy systems not only encourage sustainable behavior but also reduce the dependence on nonrenewable resources, enhancing the overall environmental and economic framework (Li et al., 2025) (Figure 9.6).

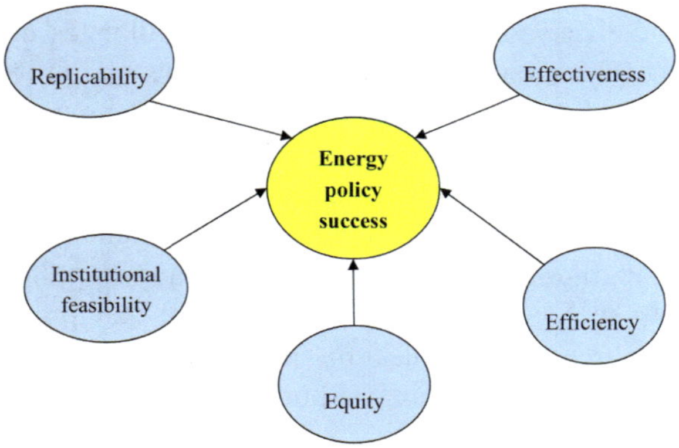

Figure 9.6. *Policy frameworks for sustainable energy.*

Source: https://pub.mdpi-res.com/sustainability/sustainability-12-05078/article_deploy/html/images/sustainability-12-05078-g006.png?1592906883.

9.3.1. Overview of International Agreements (e.g., Paris Agreement)

Global agreements are of crucial significance in establishing global reactions to environmental concerns, namely climate change. Of the agreements, the Paris Agreement is a landmark document constructed under the United Nations Framework Convention on Climate Change (UNFCCC). Adopted in 2015 with the aim of limiting the world's temperature way below 2°C over pre-industrial levels and making efforts towards containing the rise in temperature at 1.5°C, it has been instrumental in setting a departure from the business-as-usual scenario. The Paris Agreement also calls for signatory nations to provide and submit their long-term low greenhouse gas emission development plans and check on their progress from time to time. By promoting a collective effort, the Paris Agreement emphasizes the importance of contribution by developing as well as developed nations towards unified climate action and thus towards the environmental as well as socio-economic aspects of climate change. Besides the Paris Agreement, other international agreements target particular elements of environmental management. The Kyoto Protocol, passed in 1997, for example, specifically addressed industrialized countries and introduced legally binding reduction commitments for greenhouse gas emissions. The dynamics of climate governance, however, have since changed, resulting in a more integrated system that takes into account the role of sub-national actors. This evolution has been encapsulated in what since 2016 has been referred to as urban pragmatism, seeking to redefine cities as major players in climate policy. Cities' entry into the international climate agenda stems from the awareness that cities are critical in enacting local practices with immediate impacts on global climate results, influencing both energy use and economic activity (Cast et al., 2020). The relationship between economic conditions and environmental regulations is more visible in the light of global agreements. These collaborative actions do not only aim to mitigate climate impacts but also aim to correct the follies brought about by conventional climate policies. Climate resilient development issues a call for double thinking wherein coping with climate change has to be achieved while, on the other hand, sustainable economic development is being encouraged. Reports, such as the ones of the IPCC, affirm this fact by requiring coordinated approaches that consider ecological, social, and economic aspects (IPCC, 2023). Lastly, the enforcement of international agreements like the Paris Agreement relies on their ability to create functional frameworks that are balanced and fair and strengthen environmental governance, and this will enable societies to address the obstacles of climate change in an effective manner.

9.3.2. National Policies Promoting Renewable Energy

In recent years, national renewable energy policies have emerged as critical instruments for addressing environmental, energy, and economic issues. By placing emphasis on the development of renewable energy, the government aims to lower greenhouse gas emissions, mitigate the impact of climate change, and enhance energy security. These policies entail various mechanisms, from incentives to renewable technology, tax credits, to regulatory systems that facilitate integration of clean energy sources into existing energy infrastructure. It is not just environmental imperative to wean away from fossil fuels to renewables; it is also an economic stimulus by way of new jobs in newer industries. In an effort to join their energy policies and climate goals, policies of sustainable development place emphasis on the necessity for joint action for a robust and flexible energy system that benefits present and future generations (Ahmed et al., 2024). Additionally, national policies to promote renewable energy play a significant role in helping nations meet growing energy demands while protecting public health and the environment.

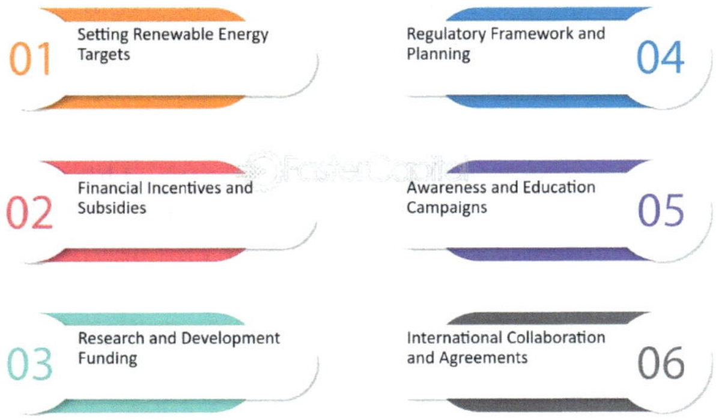

Figure 9.7. *National policies promoting renewable energy.*

Source: https://fastercapital.co/i/Navigating-Government-Policies-and-Regulations-in-the-Renewable-Energy-Industry--Understanding-the-Role-of-Government-in-Promoting-Renewable-Energy.webp.

Energy diversification is made feasible through key technology advances, along with the support of governments, while reducing dependence on fossil fuels. Where exposure to severe weather events is high, for instance, the use of renewable energy in the mix increases resilience but reduces vulnerability to climate-related health hazards. Policy strategies that consider the health effects and promote the use of renewable energies can make significant improvements in public health. These types of projects, though having energy security and climate mitigation functions, have a synergistic impact that ultimately enhances national economies in terms of further investments and technology advancement (Zhang et al., 2023). Furthermore, the successful implementation of national policies favoring renewable energy involves interdisciplinarity collaboration among government branches and levels. This includes the engagement of actors from the academic sector, enterprises, and civil society to open a holistic discussion on energy transition policies. Sound policymaking will need to take into account socio-economic realities, ground-level realities, and technological innovation to ensure policies are responsive and localized. Increasingly, as nations continue to mature and enhance energy policy, active consideration of health and public engagement will play a key role in driving meaningful change. The ongoing evaluation of existing policies and research focus on renewable technology use can be a significant factor in the achievement of low-carbon future with sustainable energy systems, thus reconciling environmental and economic viability alongside enhanced human welfare (Ahmed et al., 2024) (Figure 9.7).

9.3.3. Local Initiatives and Community-Based Energy Solutions

Local action and community-based energy solutions are increasingly viewed as the necessary elements of addressing the interdisciplinary challenges of energy sustainability and environmental responsibility. Grassroots action gives the local community the ability to use their unique resources to build localized energy creation that minimizes the use of non-renewable resources. Community solar programs, for instance, allow neighborhoods to collectively invest in solar power cooperatively, while benefiting economic resilience with reduced greenhouse gas emissions. Involvement of local government in such schemes normally ensures that the voices of the population in the community are represented in decision-making, which translates into equitable energy distribution. Moreover, by attending to local needs and making sense of the local context of energy demand, such schemes promote the principles of environmental justice in that marginalized communities receive a fair share of provision of clean energy, thereby tackling historically entrenched disparities in energy access. Incorporation of community-based energy solutions extends even further than mere energy generation; it involves awareness and education programs that enlighten the residents about the benefits of renewable energy. Local institutions typically take the lead in providing workshops and training sessions in order to promote energy efficiency and sustainability practices. Education through this approach brings with it a feeling of collaboration among residents, leading to innovative solutions tailored to benefit the specific problems of the community. By increased awareness, the people of the community are not only better placed to make optimally informed choices in the use of energy but also to advocate for policies that enhance sustainable use of energy. By empowering public perception of energy dynamics, such community-led initiatives contribute significantly to advocacy of broader structural changes in energy policy arrangements. Moreover, the economic effects of local energy initiatives cannot be overstated. Participation in community-based energy programs supports the creation of employment, particularly in the renewable energy sector. Green cooperatives and small enterprises with green initiatives not only provide employment opportunities but also stimulate local economies through the circulation of energy expenditure within the local context. Through partnerships with institutions emphasizing corporate social responsibility, communities are enabled to construct robust infrastructure that fosters innovation in energy. As illustrated by some case studies, as reported by Widianti et al. (2025), and Chauhan et al. (2025), communities with successful integration of local energy projects into economic planning are likely to have enhanced resistance towards external economic shocks, leading to long-term sustainable development. Thus, local energy projects play a significant role in a balanced system of environmental, energy, and economic parameters and stress the importance of localized measures for resolving global issues.

9.4. TECHNOLOGICAL INNOVATIONS IN ENERGY EFFICIENCY

The rapidly evolving energy efficiency technology is a keystone factor in the transition to renewable energy systems. The development in this field not only maximizes the performance of the energy systems but also reduces environmental impacts significantly. For instance, the application of artificial intelligence (AI) in energy management has become the trend transforming how energy systems are handled to enable further optimization of the utilization of renewable energy sources. AI can forecast energy consumption patterns, optimize energy distribution and storage, and even manage the operation of renewable energy facilities with eerie precision. All these directly help to minimize waste and increase the reliability of the energy supply, a crucial necessity in creating a sustainable energy future. By their implementation, stakeholders can attain enormous economic benefits while upholding environmental stewardship, eventually realizing societal dreams for a cleaner energy mix (Ohalete et al., 2023). Moreover, the complementarity between energy efficiency and turbocharged growth in renewable energy technologies cannot be overemphasized. As underlined in recent research, renewable energy sources, primarily sunlight and wind, can supply more than a substantial portion of the energy requirements of the world by 2050, by substantially lowering greenhouse gas emissions and thus countering climate change. Synergies resulting from efforts in energy efficiency complement this transition and make technology function better and more efficiently. These technologies such as smart grid technology, energy-efficient appliances, and advanced insulation materials allow consumers and industries to decrease their energy consumption. Such a two-pronged focus on renewable production and increased energy efficiency not only assists in the achievement of environmental objectives but also stimulates economic development through the creation of jobs in new green industries and technologies (Gielen et al., 2019). Finally, the success of these technological advances hinges on the presence of facilitating regulations and policies. To get consumers and businesses to adopt energy-efficient practices and invest in renewable technologies, one needs to provide an environment that has incentives and financing mechanisms in place. Governments play a critical role here in facilitating this transition, funding research and development, and enhancing public-private partnerships. With upcoming infrastructures such as electric vehicle charging points and decentralized energy production plants, the need for integrated planning and coordination of policies is increasingly imperative. Through energy efficiency and the adoption of state-of-the-art technologies, society can lay the groundwork for a safe energy future with a good effect on the environment and economy as well (Gielen et al., 2019) (Figure 9.8).

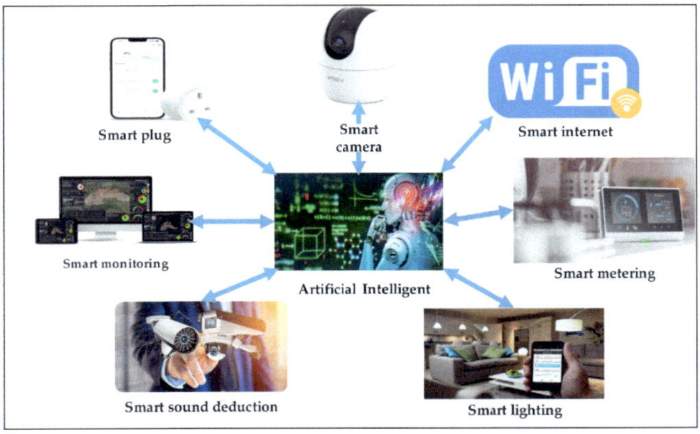

Figure 9.8. *Technological innovations in energy efficiency.*

Source: https://pub.mdpi-res.com/smartcities/smartcities-05-00071/article_deploy/html/images/smartcities-05-00071-g001.png?1666580987.

9.4.1. Advances in Renewable Energy Technologies

The innovation of renewable energy technologies is the key milestone in the mitigation of climate change as it concurrently responds to rising energy demands. Not only do they present a future direction in reducing greenhouse gas emissions, but also the solution to energy security and price uncertainty issues. Despite the seeming merits, the massive deployment of renewable energy technologies is typically hampered by structural and economic barriers.

Market competitiveness is typically uneven with traditional energy sources having the capability of enjoying favorable treatment in present frameworks. Nevertheless, research indicates that such barriers are normally evaded through strategic deployment and strengthened support for R&D. Although it is clear that many renewable technologies can make substantial contributions to energy generation, the issue continues to be how these answers can obtain the best market share they need to make a meaningful difference in energy systems worldwide (Neuhoff et al., 2006). Again, highlighting the significance of interlinked markets, technological progress in the field of renewable energy also depends on international policy and larger international frameworks. Migration to international markets is a challenge as well as an opportunity for technology developers.

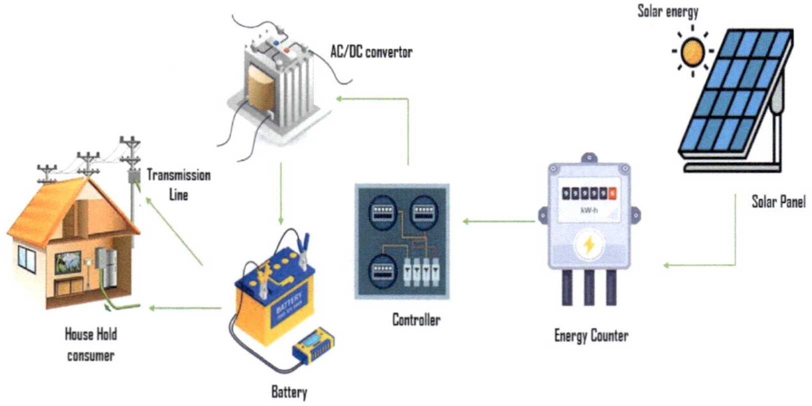

Figure 9.9. *Advances in renewable energy technologies.*

Source: https://pub.mdpi-res.com/energies/energies-16-06579/article_deploy/html/images/energies-16-06579-g001.png?1694578489.

When they are backed by governments, it not only enables the potential for dramatic technological growth but also increases the confidence of developers in the business opportunities' feasibility. The increased confidence then leads to access to financial resources, permitting higher R&D spending and better production processes. Furthermore, larger markets encourage international competition, enabling the best technology that is available to be utilized. This is, however, a price that accompanies international advancement, where global effort does not enable national governments to take advantage without significantly contributing themselves. This, thus, requires global collaboration to design the renewable energy system economically while embracing new technologies (Neuhoff et al., 2006). In addressing the interconnected problems of environmental impact, energy diversity, and economic sustainability, technological innovation in the field of renewable energy is to be tackled through a multi-dimensional lens. With countries transitioning from fossil fuels to greener energy sources, governments also have an important part to play. Policymakers must create incentives that promote R&D and reduce market entry barriers for renewable technology. Apart from it, establishing the regulatory structure to aid equitable competition will make such innovations easier to integrate into existing energy systems. In terms of a substantial decrease in environmental effects combined with energy security, intentional and strategic integration of government support, market readiness, and international collaboration is seen as an essential component of sustainable energy in the long term. The promise of taking on renewables can lead to revolutionary changes that are not only economically feasible but also environmentally necessary (Figure 9.9).

9.4.2. Smart Grid Technology and Its Economic Benefits

Smart grid technology is a revolutionary advancement in electricity transmission and distribution. By injecting the power system with new communication and information technologies, smart grids enhance efficiency and reliability of operation. Smart grids enable real-time monitoring of data, enabling utilities to manage demand more effectively and reduce operating costs. Therefore, smart grids allow for the integration of renewable energy sources (RES) and thereby promote a greener energy world. The economic impact of such a change is immense; with more efficient energy production and consumption optimization, smart grids can lead to lower electricity prices for the consumers and improved profitability for the energy suppliers. Further, enhanced grid resilience lessens the economic impacts of outages and disruptions and thus highlights economic benefits accrued from the adoption of smart technologies. Integration of dynamic line rating (DLR) and energy storage systems (ESS) in the smart grid infrastructure is another key development that yields substantial economic benefits. The application of DLR technology facilitates real-time monitoring of the capacities of transmission lines and maximizes their efficiency and limits congestion costs. The integration of ESS presents a solution to store excess energy produced by RES, thus preventing the generation and consumption variability. Integration not only provides support for the stability of the grid but also the economic viability of RES through constant provision of electricity. According to recent studies, DLR and ESS installation in modern power systems has been proven to significantly reduce greenhouse gas emissions while reducing overall operating costs. Such diverse advantages shed light on how smart grid technologies can make energy systems more financially feasible, with financial goals aligned with environmental requirements (Talebi et al., 2024). The financial benefits of smart grid technology are not restricted to utility companies but extend to consumers themselves through enhanced energy efficiency and savings. With real-time feedback on their energy consumption, consumers can behave smart and change their energy usage patterns so as to optimize consumption. This demand-side management approach encourages energy conservation habits and reduces total electricity bills. In addition, since smart grids allow distributed generation (DG) and clean energy technologies (CET) to be integrated into the power grid, residential consumers can participate in this decentralized power system. This opens a window for consumers to become part of energy generation through rooftop solar systems or other renewable systems, thereby enhancing their monetary returns in terms of potential savings and incentives. Convergence of such technologies not only ensures economic stability in power usage but also empowers consumers, substantiating the argument that smart grid innovation provides multilateral economic benefits for standalone homes as well as the overall energy sector (Mange et al., 2024) (Figure 9.10).

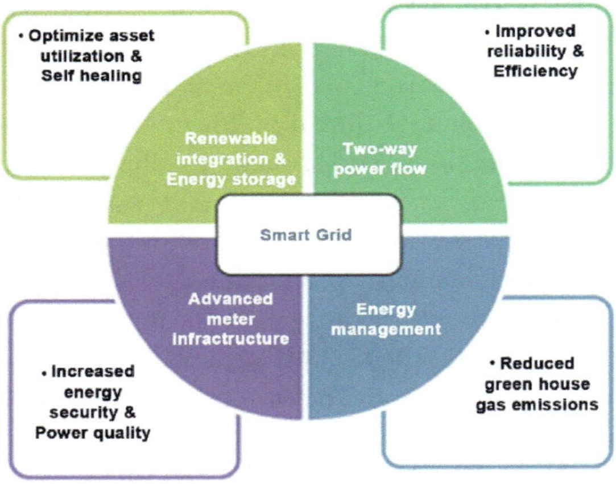

Figure 9.10. *Smart grid technology and its economic benefits.*

Source: https://www.researchgate.net/profile/Harrison-Bhatti/publication/325122283/figure/fig2/AS:626171481972738@1526302323616/Smart-grid-benefits-25.png.

9.4.3. Energy Storage Solutions and Their Environmental Impact

The increasing use of renewable energy technologies has put the focus on the central role of energy storage technologies in the modern energy system. As solar and wind power is weather- and time-dependent, energy storage reduces the inherent intermittency of these resources. Lithium-ion batteries pumped hydro storage, and emerging technologies in the form of solid-state batteries are being developed to store excess energy generated during peak production phases. These technologies, along with contributing to a secure supply of energy, will also lower the dependency on fossil fuel backup, which is one of the primary sources of greenhouse gas emissions. However, the environmental impacts of energy storage technologies cannot be overlooked because the extraction and processing of minerals used in batteries have possible life-cycle environmental degradation and pollution impacts, which necessitate that life cycle impacts of such technologies as well as their operational benefit be considered. In addition, the manufacturing processes involved in energy storage devices, particularly batteries, are environmental concerns. The manufacture of lithium, cobalt, and nickel necessary to produce batteries has also been concerning in terms of loss of habitat and water management, especially in areas where the mines are situated. These elements have a propensity for being mined in nations where there is little environmental protection measure, thus having large ecological footprints. In addition, the use of high energy in producing batteries can also result in increased carbon emission, particularly if the energy source used is not green. Addressing these issues necessitates the shift towards more sustainable sources and the manufacture of less mineral-intensive technologies. Alternatives such as battery recycling and the development of alternative materials are key in cutting down the general ecological footprint of energy storage products as the world shifts toward a cleaner, more sustainable energy future. Furthermore, energy storage systems need to be integrated into overall energy management systems in order to maximize their environmental benefit. In a time when the transition to a circular economy is gaining momentum by the day, effective utilization of energy storage solutions can enhance energy efficiency and curb wastage. For instance, integrating storage with smart grids enables the most efficient allocation and utilization of energy, hence lowering overall energy demand from conventional sources. However, unless executed intelligently and planned carefully, the disadvantages of increasing energy consumption through new technology might be greater than the benefits for the environment. As during the COVID-19 pandemic, reliance on new technology has changed consumer behavior and energy demand drastically, their long-term effects need to be studied (Jiří et al., 2020). Thus, in the future, research must be directed at studying how energy storage, demand-side management, and socioeconomic aspects are interlinked so that there is a balanced and sustainable strategy for energy storage systems within the larger context of environmental considerations (Emily et al., 2021).

9.5. SOCIAL CONSIDERATIONS IN ENERGY TRANSITION

The shift to renewable energy sources has a variety of social causes which have significant impacts on societies and communities. Society shifts from fossil fuels to renewable energy, and its concerns vary from losing jobs in traditional energy sectors. The majority of workers in coal, oil, and gas sectors have their livelihood uncertain as they may not be skilled in the new solar and wind power technologies. This change necessitates a whole-of-society strategy that entails retraining programs meant to equip workers with the necessary skills. Secondly, engaging the local community in the energy transition process enhances ownership and mitigates resistance to change. Local inclusion-focused participation activities can decimate social costs associated with energy transitions, and hence community resilience as well as acceptance of new energy systems can be enhanced. It is necessary to view these developments not only as technological changes but also as profound shifts in society. Moreover, equitable sharing of gains realized from renewable energy schemes is a crucial social factor during the transition to clean energy. It has been established in a research that poor communities and marginalized people are most harmed by the degradation of the environment but deprived of enjoying its economic benefits. As such, policymakers must make sure poor groups are prioritized when renewable energy schemes are planned and put in place. Equitable policies might include employment generation schemes in affected communities or economic benefits for households for access to renewable energy technology. Additionally, incorporating social justice planning mechanisms into the process may be able to negate structural injustices that previously disadvantaged some groups. Such forward-looking action reconciles economic advantage with social justice, creating an energy sector that is more socially inclusive, upholding the principle of justice and fighting climate change. This is necessary to make a just transition in which the vulnerability of various groups of people is taken into account as part of the general environmental agenda. Lastly, transition to energy and social forces must also consider cultural forces and local values since this help determine local attitude toward change. Different communities may have unique historical relationships with historic energy sources, which can make it challenging for them to prepare for alternatives. As highlighted in the IPCC's Sixth Assessment Report, climate resilient development is not solely technological advancement but also an understanding of the intricate relationship between environmental preservation and local cultures (IPCC, 2023). It is precisely these cultural factors that can assist a firm in determining a successful communication strategy in adopting new energy solutions. Secondly, engagement with local leaders ensures cultural sensitivities are respected, thereby paving the way for smoother transitions. By valuing and respecting the social and cultural fabric on which energy transitions happen, actors can build more integrated and sustainable transitions that resonate with local identity and forge lasting positive change (Figure 9.11).

Figure 9.11. *Social considerations in energy transition.*

Source: https://www.officialenergyasia.com/wp-content/uploads/2023/06/The-Social-Impacts-of-a-People-Centric-Energy-Transition_FB-IG_1080x1080.jpg.

9.5.1. Public Perception and Acceptance of Renewable Energy

Public perception plays a crucial role in the use of renewable energy technologies and has a direct impact on implementation and viability. An enlightened public can contribute significantly to the transition from dependence on fossil fuels to clean energy. Studies indicate that there are awareness of climate change and its impact, including massive economic losses in natural disasters, but there tends to be lack of understanding in the public regarding the benefits of renewable energy. Specifically, events like wildfires, droughts, and floods demonstrate costly climate-related expenditures that highlight the necessity for sustainable actions. When individuals understand the environmental necessity and potential economic advantages of renewables, like energy autonomy and job creation, their likelihood of favoring such actions increases directly with how much they understand about the danger of doing nothing (Fawzy et al., 2020). Moreover, the use of renewable energy technology also relies on public attitudes shaped by education and public communication. It was discovered that poor awareness of the functionality and availability of renewable energies is a great setback to its usage.

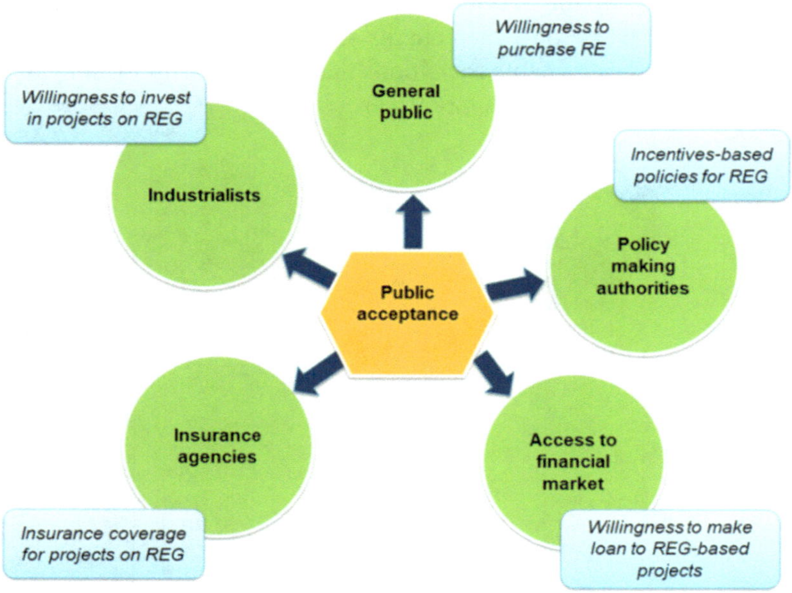

Figure 9.12. *Public perception and acceptance of renewable energy.*

Source: https://www.researchgate.net/profile/Nousheen-Fatima-2/publication/351588956/figure/fig2/AS:1023596096921613@1621055735991/Public-acceptance-of-renewable-energy-generation-Source-authors-elaboration.png.

The majority of the residents lack clear information on how solar, wind, and biomass energy help in limiting greenhouse emissions due to the reluctance strengthened by the ongoing dominance of fossil fuels in power production. Though renewables produced only 26.5% of electricity globally in 2017, the shift is gradually approaching with alternative sources regarded as contributory alternatives towards the energy crisis (Qazi et al., 2019). Creative communication and informational campaigns with the aim of disseminating information on the technological advancement and economic benefit of renewable energy can foster a better climate, thus enhancing public acceptance of these technologies. Social media is also an influential platform for shaping the image of renewable energy among the public. Modern communication channels enable instantaneous interaction, whereby individuals can express support or opposition towards renewable activities. A systematic study of public opinion on social media platforms like Twitter has made this dynamic visible, with traces of public trend that can be utilized by policymakers and scholars alike. In particular, direct involvement in such forums enables proponents to correct misunderstandings and address the advantages of renewable energy.

Since online conversations continue evolving, application of public opinion on social media can prove to be an important element in propagating a setting for more acceptance of renewable energy, which ultimately contributes to the larger objective of mitigating climate change and promoting sustainable procedures (Qazi et al., 2019) (Figure 9.12).

9.5.2. Equity Issues in Energy Access and Distribution

Equity issues in energy access and distribution are of utmost importance when environmental, energy, and economic interests come into discussion. Social inequality is further boosted by inequalities in access to energy, largely impacting vulnerable groups globally. Unequal access to energy exacerbates the existing vulnerabilities, trapping poorer communities in a vicious poverty trap, as well as limiting them in investment in alternative sustainable energy (IPCC, 2023). Ensuring fair access to energy not only enables inclusive economic growth but also increases resilience to climate shocks. Closing these inequalities requires an integrated strategy with social, economic, and infrastructure interventions for accessing energy by marginalized groups and thus meeting social justice and sustainable development. Energy resource patterns have a tendency to reflect larger systemic injustices and, as such, create geographic and demographic injustices. Rural locations, for instance, tend to experience underinvestment in energy infrastructure in comparison to urban counterparts. This disproportionate share not only impedes economic prospects but also entrenches health inequality since populations with no access to stable energy outlets experience substandard living conditions. The Global Burden of Diseases study reveals the health impacts of inadequate energy access, demonstrating that individuals whose energy comes from unsustainable sources are more prone to non-communicable diseases and injuries (Spencer et al., 2018). Reducing these inequalities could assist policymakers in improving individuals' health and, simultaneously, supporting a more sustainable energy system to serve the entire society. Improving energy access equity strategies must also address the intersection of various social determinants like race, gender, and socioeconomic status. Women, for instance, bear disproportionately the adverse effects of inadequate energy access, which compromise their access to education and opportunities for work. The inclusion of gender-responsive policies in energy planning is likely to promote equitable access and empower the marginalized and women. Here, IPCC-fostered climate-resilient development policies focus on inclusive strategies that are sensitive to the subtleties of energy inequality (IPCC, 2023). Through the equitable supply of energy, societies become capable of establishing a solid basis for sustainable development, which brings economic growth while ensuring social cohesion and environmental resilience.

9.5.3. Role of Education in Promoting Sustainable Practices

Learning is at the heart of making sustainable practice possible through skills, knowledge, and awareness necessary to solve environmental problems. Formal education structures offer the chance for pupils to study concepts of sustainability and their implication on the environment, economy, and society. Curricula that include sustainability in the practice of science, economics, and social studies foster critical thinking and instill a sense of accountability for the earth. For instance, incorporating green technologies into study can push innovations forward in reducing carbon prints and improving energy efficiency in the future. Further, schools can be a role model by emulating sustainability in themselves as a living laboratory where students and the community at large can see and participate in environmentally friendly activities. These experiences not only create the significance of sustainability but also prepare students to advocate responsible practices in professional and personal lives, moving towards permanent behavioral change. Besides formal education, informal education and community life are accountable for raising awareness towards sustainable practices. Community initiatives, workshops, and public campaigns can communicate effectively regarding information about sustainable living, protection of resources, and alternative energy sources. These projects typically target a range of audiences such that all members in the community, irrespective of age or background, can be reached and educated. Community education in an empowering and accessible manner not only builds a culture of sustainability but also enhances the ability of individuals to contribute towards broader environmental goals. By targeting vulnerable populations and applying human rights-based approaches, learning programs can tackle local issues while promoting inclusiveness and social justice. Providing access to marginalized populations guarantees that environmental issues are addressed on a collective basis and are reflective of the needs of the people, demonstrating that education can be a powerful force for social justice and sustainable development (Priya, 2025). Education is also key to instilling a sense of environmental stewardship that will endure throughout the lives of people. This duty is important in bringing up future leaders to prioritize sustainable development in various sectors, including business and politics. As professionals, qualified people will most likely implement sustainable approaches to their line of work, either by advocating for energy-efficient practices of operation, promoting sustainable purchasing, or guiding policy development that fosters conservation. The addition of sustainability to post-secondary program curricula, particularly in business and environmental studies, can create new thought that addresses critical global challenges. Also, promoting individuals into sustainability careers, like renewable energy and environmental activism, has the ability to push transformation at the systems level in industry sectors. By designing an education system that prioritizes sustainability as a guiding philosophy, we pave the way for a healthier world and a distributive economic future (Chandrasekaran et al., 2025) (Figure 9.13).

Figure 9.13. *Role of education in promoting sustainable practices.*

Source: https://fastercapital.co/i/Environmental-awareness-and-education-Sustainable-Solutions--How-Environmental-Awareness-Drives-Innovation-in-Business--The-Role-of-Education-in-Promoting-Sustainability.webp.

9.6. GLOBAL PERSPECTIVES ON ENERGY AND ENVIRONMENT

The interconnectivity of the world's energy systems and environmental sustainability are increasingly being identified in the context of climate change and its effects. With nations looking towards energy security, attention towards a transition to low-carbon energy sources has gathered momentum. A prime example is the decision of nations such as the United Kingdom to achieve net zero emissions by the year 2050, calling for an accelerated transition to renewable energy technology. But the change is defined by problems that are economic and social, necessitating an equilibrium path towards sustainability. An objective evaluation of the energy picture reveals complementarity across various generation technologies, like nuclear power, which offers huge low-carbon energy potential. Therefore, it is imperative to develop a positive discussion regarding the strengths and weaknesses of every energy source in order to provide pathways to an eco-friendly energy future and resolving environmental concerns comprehensively (Taylor et al., 2024). In addition to education about conventional sources of energy, advanced technologies, particularly in the area of photovoltaic (PV) technologies, play a revolutionary role in preventing environmental deterioration.

The rapid expansion of the PV industry is a stark indicator of a heightened global shift towards clean energy technologies. As illustrated by its dynamics in China and the USA, the PV sector is guided by a high-level interaction between government policies, economic incentives, and geopolitical factors. China's emergence as a leading player in PV technology is energized by supportive supply-side policies that improve production and export capabilities. Conversely, the United States gives emphasis to demand-side policies grounded in market mechanisms, which are a different approach entailing protection policies for domestic industries. The understanding of these different directions emphasizes the energy transition complexity and illustrates the importance of cooperative global measures to realize greater environmental sustainability while maintaining competitive global markets (Zhang, 2024). Lastly, seeking sustainable energy sources necessitates multiplicity of strategy in implementing various energy technologies and cooperative measures. This holistic approach is key in addressing today's pressing energy and environmental challenges. With nations confronting the reality of global warming, the role of all possible sources of energy in improving matters becomes critical. For instance, nuclear energy emerges as a significant player in the low-carbon energy range, capable of meeting huge energy demands while confining greenhouse gas emissions to moderate levels. However, its adoption is condemned owing to the safety and waste management problems, calling for the educationally driven public debate whose exigency is particularly critical. In the same way, the development of photovoltaic technology presents a promising avenue for diversifying energy mix portfolios, triggering economic development as it advances environmental sustainability. Therefore, to navigate the intricacies of global energy policies means to address and balance environmental, economic, and social implications towards sustainable ends (Taylor et al., 2024; Zhang, 2024).

9.6.1. Comparison of Energy Policies in Developed versus Developing Countries

The contrast in energy policy between the developed countries and developing countries is a very typical example of the complexity of harmonizing environmental, energy, and economic concerns. Developed economies have a sophisticated legal system and financial capacity to invest heavily in renewable energy initiatives, advanced technology, and extensive research in sustainable processes. For instance, European Union countries have implemented widespread policies that aim to reduce carbon emissions and ensure a shift towards renewable energy. On the other hand, though, developing nations are usually constrained by immense obstacles such as poor financial abilities and technical competency, which more frequently than not lead to their reliance on fossil fuels. The imbalance is representative of the need for policy frameworks with modifications that consider the unique challenges for developing nations while promoting a transition towards clean energy alternatives (Lestari et al., 2024).

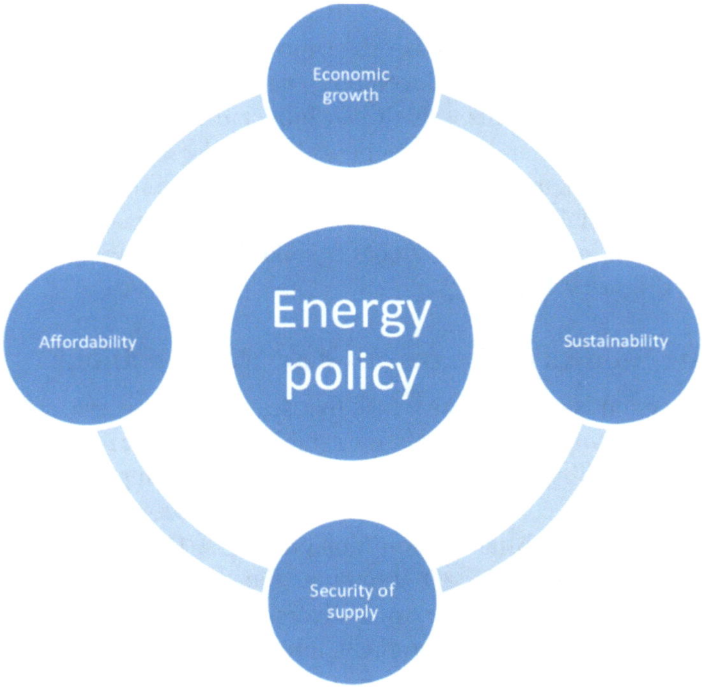

Figure 9.14. *Comparison of energy policies in developed.*

Source: https://www.researchgate.net/publication/347803534/figure/fig1/AS:1022185539923969@1620719432344/The-objectives-of-energy-policy.png.

In the majority of developing countries, the application of renewable energy sources is gaining popularity as a principal method of sustainable development and greenhouse gas mitigation. Solar photovoltaics, wind energy, and biogas are gaining recognition as they contribute to carbon footprint reduction and improved access to energy. However, to tap the full potential of such technologies, one requires strategic policy initiatives that transcend barriers such as exorbitant up-front capital costs and restricted finance. Initiating inter-country cooperation between developed and developing nations can facilitate transfer of technology as well as capacity building, ultimately resulting in actualization of the potential for effective renewable energy deployment (Kamau, 2024). Such collaborations not only make the environment more sustainable but also yield socio-economic benefits in terms of employment creation and energy security in the Third World. Good regulation is also key to influencing energy policies that promote sustainable behavior. The developed world is likely to have cleaner regulatory spaces that promote investments in renewable energy and innovation. These structures consist of investment incentives for the private sector, effective permitting processes, and provisions supporting the grid integration of renewables. On the contrary, developing nations are likely to have such integrated regulatory systems, which can hinder the pace towards renewable energy. The governance and policy support have to be enhanced in developing nations to facilitate enabling environments to encourage investments and low-carbon transition. Making sure that those policies are properly implemented is of utmost concern in shaping the global agenda for sustainable energy (Lestari et al., 2024) (Figure 9.14).

9.6.2. Case Studies of Successful Energy Transitions

Energy conversion is an imperative task that several regions are undertaking in order to achieve environmental goals and establish economic resilience. A strong example is that of Yibin, a Chinese city located in Sichuan province, which has been renowned for its coal and traditional baijiu production of the hard liquor, and each has powered substantial pollution problems. The city's transformation is particularly noteworthy since it surmounts the structural barriers of moving from a resource-based economy to a diversified and sustainable industrial complex. Through strategic government intervention, Yibin has managed to establish a thriving power battery industry, which is a significant about-turn towards cleaner energy solutions. This groundbreaking process showcases the potential of dedicated policy frameworks and research and development investments in assisting historically dependency-based cities to envision new economic futures as well as address critical environmental concerns (Li, 2025). A similar thought-provoking case study is found in various global practices where sustainable finance proves to be the turning point in changing energy. New financing instruments, such as green bonds and sustainability-linked financial instruments, have become pivotal for funding the renewable energy sector. They not only provide funds to renewable projects but also initiate the incentives for the financial markets to have their activities aligned with climate action goals. By using these financing tools, the governments are actually able to reduce the investment barriers for clean energy technologies. This symbiotic relationship of public policy and private investment is necessary since it brings room for renewable energy to thrive, leading eventually to a balancing of economic growth and environmental sustainability (Onabowale, 2025). Analysis of effective global transitions recognizes that disadvantages accompany superior progress. While cities like Yibin demonstrate the efficacy of government intervention in transitioning towards greener economies, they still possess long-term sustainability challenges. The need for continuous innovation and policy adaptation cannot be overstated; it is through constant fine-tuning that such cities can stay away from the pitfalls of stagnation and market hubris. In addition, the partnerships between governments, corporations, and financial institutions must be encouraged for the aims of supporting sustainable growth in the renewable sector. A coordinated approach that honors the sustainability of existing industries and fosters innovation presents a path towards future-proofing economies against environmental strain. It is, therefore, essential that knowledge gleaned from these case studies be employed in informing broader strategies for sustainable energy transformations at the global level (Li, 2025; Onabowale, 2025).

9.6.3. International Cooperation in Addressing Environmental Challenges

International cooperation is key to effectively addressing the intricate environmental challenges that cross borders. As climate change, biodiversity loss, and pollution intensify, cooperative frameworks become more imperative by the day. States realize that individual action alone is insufficient; international agreements provide much-needed frameworks for setting common goals and specifications. Such systems, like the Paris Agreement, have mechanisms for communication between developing and developed countries that enable knowledge and technology transfer. When countries collaborate on environmental concerns, they can pool their resources and share expertise, thus providing resilience against climate change impacts while, at the same time, promoting sustainable action. The dependency reflects the truth that environmental health is proportional to economic prosperity and stability, as well as calling for global collective action. The foreign direct investment (FDI) function stands out as the most prominent mechanism in global cooperation to address environmental challenges. FDI has been discovered to boost growth and technological development, particularly in the field of renewable energy, under greater economic unpredictability like the COVID-19 pandemic (Chrystella et al., 2025).

Through attracting investment in green technologies and green behavior, countries can stimulate local economic benefits while converging with global environmental goals. Moreover, smart partnerships can improve the quality of infrastructure, improve upskilling of staff, and improve the regulatory frameworks to provide a good platform for sustainable investment. This cooperation is particularly vital as nations navigate the ramifications of international climate agreements, knowing that local treaties make such investments more efficient in attaining environmental and economic goals. Seeing the specific steps of major economies such as Russia and China reveals more the dynamic nature of international cooperation in environmental contexts. Both countries have proved to have different methods of incorporating climate initiatives into their economic frameworks, as indicated in recent studies (Astapov et al., 2025). Russia's incremental integration of climate needs into its legacy energy-based economy differs from Chinas aggressive transit to green energy solutions. These two directions point towards the complexities involved in balancing economic growth and environmental sustainability and call for active partnerships, particularly with developing nations. Also, international climate accords demand not only compliance but also leadership from top players, underlining the need for policies that encourage green investments. The trend points to the need to establish shared economic incentives to facilitate transitions towards sustainability, eventually leading to a shared global effort to solve environmental problems (Figure 9.15).

Figure 9.15. *International cooperation in addressing environmental challenges.*

Source: https://www.ielts.net/wp-content/uploads/2024/08/international-cooperation-environmental-issues-66af5f.jpg.

SUMMARY

In conclusion, the intricate interdependence of environmental, energy, and economic considerations calls for the implementation of holistically integrated approaches to sustainability. Highlighted in recent literature, the balance between economic progress and care for the environment is now a core ambition for nations worldwide. The implementation of sustainable development practices is not merely an ethical requirement; it is a business necessity to stay ahead in a continuously evolving global environment. The data coming from various studies confirms that regions with sustainable practice adherence can experience a myriad of benefits, ranging from improved environmental conditions to improved economic performance, as is the case with the state of Ceará, which strategically positions itself on the international energy narrative through quality practices (Barbosa et al., 2025). Furthermore, carbon emissions trading scheme (ETS) analysis reveals that schemes of policy are the drivers of advocating energy transitions. The empirical evidence confirms that pilot provinces in ETS schemes have recorded significantly higher energy transition rates than their non-pilot counterparts. The results confirm the significance of policy instruments in transforming local government behaviors towards sustainability. The economic incentives established under the ETS express how policy regulation can spur environmental progress and economic resilience. When cities are empowered to make their priorities more focused on green initiatives, as observed in various research studies, the result is a more resilient and robust energy profile (Tang et al., 2025). Upon consideration of these observations, it appears only logical that an integrated approach towards sustainability would be the panacea for future prosperity. The integration of environmental considerations into economic models not only makes the planet healthier but also fuels innovation and growth in various sectors. Stakeholders are committed to prioritizing more collaborative action to ensure that the environment, energy, and economic objectives are all sought after simultaneously. With such alignment, there can be a vision for a world where sustainable actions are the norm rather than the norm, thereby avoiding adverse environmental impacts while making economic sustainability all the more feasible. Finally, the future is reliant upon an unshakeable commitment to sustainable development that acknowledges the interdependence of our world system (Barbosa et al., 2025; Tang et al., 2025). One of the emerging areas of investigation is the intersection of environmental, energy, and economic interests and, more particularly, their interconnection to climate change. Climate change has significant implications for ecosystems, human health, and the economy, and this necessitates an interdisciplinary approach to mitigation and adaptation efforts. The contribution of Working Group II to the Sixth Assessment Report of the IPCC identifies key information regarding the imperative for climate resilient development, which brings together human endeavor and environmental sustainability. This model not only aims to curtail greenhouse gas emissions but also incorporates adaptive actions that enhance greater societal resilience, which in turn leads to a sustainable future for human beings as well as for biodiversity (IPCC, 2023). While the Earth continues to warm up, the world observes growing effects of inaction, and therefore efforts must be directed towards combined solutions. The discussion of energy transitions is also central in this respect; renewable energy sources must be at the forefront to achieve sustainability goals. Clean technology breakthroughs are indicated in reaction to the excessive reliance on fossil fuels, which drive climate change and economic instability. Hydrogen has been described as a game-changer and has been discussed as an impressive option for decarbonizing the world's energy system. Its use across power generation, heating, and transport underscores its involvement in the creation of decarbonization (Staffell et al., 2018). Transition to hydrogen as a carrier of energy can reduce greenhouse gas emissions while fostering economic growth in emerging industries. The economic implications are staggering, spanning job creation in green technologies to improving the efficiency of industry. Lastly, the implications extend to policy framework and economic investments in sustainable methods. Policy makers are realizing increasingly that the coincidence of economic growth with ecological stewardship is a key to lasting prosperity. The incorporation of the lessons of the IPCC report signifies a multilateral endeavor by numerous stakeholders such as government, enterprise, and civil society to step up to meeting these shared goals. Such collaboration will enhance public awareness and motivate individual commitment to sustainability. Therefore, understanding and integration of environmental, energy, and economic considerations appear as not only desired but necessary components for the creation of a sustainable future that benefits all aspects of society (IPCC, 2023). The call for action is immediate, suggesting that the time for substantive change is at hand.

REFERENCES

1. Agupugo, C. P., Ajayi, A. O., Salihu, O. S., & Barrie, I. (2024). Large scale utility solar installation in the USA: Environmental impact and job creation. *Global Journal of Engineering and Technology Advances*. https://www.semanticscholar.org/paper/c3004070679fbcdd5ba59ef69b78e4886167e906 (accessed on 28th September 2025).
2. Astapov, K., & Shen, C. (2025). Influence of international climate regulation on economic development strategies of Russian Federation and People's Republic of China. *Russian Journal of Industrial Economics*. https://www.semanticscholar.org/paper/983c70acd2ce78cf825b644fde0610a975a2bdbf (accessed on 28th September 2025).
3. Barbosa, M. C., & Gomes, R. (2025). Sustainable development and socio-environmental and corporate responsibility as a competitive advantage in the energy transition in the state of Ceará. *IOSR Journal of Business and Management*. https://www.semanticscholar.org/paper/98e1bdb8cd990e7acee756ef60a60ae1b6683337 (accessed on 28th September 2025).
4. Bauer, N., Calvin, K., Emmerling, J., Fricko, O., Fujimori, S., Hilaire, J., Eom, J., et al. (2016). Shared socio-economic pathways of the energy sector–Quantifying the narratives. *Global Environmental Change, 42*, 316–330. https://doi.org/10.1016/j.gloenvcha.2016.07.006 (accessed on 28th September 2025).
5. Bender, E. M., Gebru, T., McMillan-Major, A., & Shmitchell, S. (2021). On the dangers of stochastic parrots. https://doi.org/10.1145/3442188.3445922 (accessed on 28th September 2025).
6. Calvin, K., Dasgupta, D., Krinner, G., Mukherji, A., Thorne, P., Trisos, C. H., Romero, J., et al. (2023). IPCC, 2023: Climate change 2023: Synthesis report. Contribution of Working Groups I, II and III to the Sixth Assessment Report of the Intergovernmental Panel on Climate Change [Core Writing Team, H. Lee and J. Romero (Eds.)]. *IPCC, Geneva, Switzerland*. https://doi.org/10.59327/ipcc/ar6-9789291691647 (accessed on 28th September 2025).
7. Castán Broto, V., & Westman, L. (2020). Ten years after Copenhagen: Reimagining climate change governance in urban areas. *Wiley Interdisciplinary Reviews Climate Change, 11*. https://doi.org/10.1002/wcc.643 (accessed on 28th September 2025).
8. Chandrasekaran, S., & Narayanan, M. (2025). Technological enhancement of banking sector: A special reference to green banking initiatives. *Asian Journal of Management*. https://www.semanticscholar.org/paper/54f71cf2439b2ac68f5dcc0f8e4322836b218066 (accessed on 28th September 2025).

9. Chauhan, S., & Sharma, P. (2025). The influence of CSR on sustainable urban development. *International Journal of Scientific Research in Engineering and Management.* https://www.semanticscholar.org/paper/11df2e6e176fcdbb5b2b14f81d68eafd64496383 (accessed on 28th September 2025).
10. Chou, C. H., Ngo, S. L., & Tran, P. P. (2023). Renewable energy integration for sustainable economic growth: Insights and challenges via bibliometric analysis. *Sustainability, 15*, 15030. https://doi.org/10.3390/su152015030 (accessed on 28th September 2025).
11. Chrystella, D. F., Ibrahim, S., & Lihua, Y. (2025). Maximizing the impact of foreign direct investment: Trends, determinants, and strategies for sustainable economic and social development. *IOSR Journal of Economics and Finance.* https://www.semanticscholar.org/paper/17e6d9ac9460e7f859dace396f0104168c3ac22d (accessed on 28th September 2025).
12. De Arruda, R. N., Figueiredo, K. V., Vasco, D., Haddad, A., & Najjar, M. K. (2023). Cost-benefit analysis of solar energy integration in buildings: A case study of affordable housing in Brazil. *Frontiers in Built Environment.* https://www.semanticscholar.org/paper/7576b8a907f8354f66be486e7b42e797d6e6cd26 (accessed on 28th September 2025).
13. El-Araby, R. (2024). Biofuel production: Exploring renewable energy solutions for a greener future. *Biotechnology for Biofuels and Bioproducts, 17.* https://doi.org/10.1186/s13068-024-02571-9 (accessed on 28th September 2025).
14. Elkhatat, A. M., & Al-Muhtaseb, S. A. (2024). Climate change and energy security: A comparative analysis of the role of energy policies in advancing environmental sustainability. *Energies, 17*, 3179. https://doi.org/10.3390/en17133179 (accessed on 28th September 2025).
15. Fawzy, S., Osman, A. I., Doran, J., & Rooney, D. W. (2020). Strategies for mitigation of climate change: A review. *Environmental Chemistry Letters, 18*, 2069–2094. https://doi.org/10.1007/s10311-020-01059-w (accessed on 28th September 2025).
16. Gielen, D., Boshell, F., Saygin, D., Bazilian, M., Wagner, N., & Gorini, R. (2019). The role of renewable energy in the global energy transformation. *Energy Strategy Reviews, 24*, 38–50. https://doi.org/10.1016/j.esr.2019.01.006 (accessed on 28th September 2025).
17. Ibekwe, K. I., Etukudoh, E. A., Nwokediegwu, Z. Q. S., Umoh, A. A., Adefemi, A., & Ilojianya, V. I. (2024). Energy security in the global context: A comprehensive review of geopolitical dynamics and policies. *Engineering Science & Technology Journal.* https://www.semanticscholar.

org/paper/c59c9a292740a9f582f0ea72d1aa14119f7f37db (accessed on 28th September 2025).

18. James, S. L., Abate, D., Abate, K. H., Abay, S. M., Abbafati, C., Abbasi, N., Abbastabar, H., et al. (2018). Global, regional, and national incidence, prevalence, and years lived with disability for 354 diseases and injuries for 195 countries and territories, 1990–2017: A systematic analysis for the Global Burden of Disease Study 2017. *The Lancet, 392*, 1789–1858. https://doi.org/10.1016/s0140-6736(18)32279-7 (accessed on 28th September 2025).
19. Kamau, J. (2024). Efficiency of renewable energy sources in reducing carbon footprint in developing countries. *International Journal of Environmental Sciences*. https://www.semanticscholar.org/paper/c2f081c869040ce8735d76802dcb1013ee899bdc (accessed on 28th September 2025).
20. Klemeš, J. J., Fan, Y. V., & Jiang, P. (2020). COVID-19 pandemic facilitating energy transition opportunities. *International Journal of Energy Research, 45*, 3457–3463. https://doi.org/10.1002/er.6007 (accessed on 28th September 2025).
21. Lakshmi Priya, S. (2025). Topic: Strengthens and weaknesses in human rights-based approaches to developing vulnerable communities in society. *International Journal of Advances in Engineering and Management*. https://www.semanticscholar.org/paper/7290ca0ea4558f8927f108b95d8a1f9e114aa1bc (accessed on 28th September 2025).
22. Lestari, S. D., Leon, F. M., Riyadi, S., Qodariah, Q., & Putra, A. H. P. K. (2024). Comparison and implementation of environmental law policies in handling climate change in ASEAN countries: A comparative study of Indonesia, Malaysia, and Thailand. *International Journal of Energy Economics and Policy*. https://www.semanticscholar.org/paper/205e23762e838cb2bc9d30c5b5a8228e25509d2c (accessed on 28th September 2025).
23. Li, Q., & Zhang, S. (2025). Impact of globalization and industrialization on ecological footprint: Do institutional quality and renewable energy matter? *Frontiers in Environmental Science*. https://www.semanticscholar.org/paper/dec0ca0d28b235f799210c1543284f4e79a9e445 (accessed on 28th September 2025).
24. Li, X. (2025). Sustainability in the economic development of China's industrial cities: A case study with new energy industries in Yibin. *Journal of High School Science*. https://www.semanticscholar.org/paper/9545f006138cebcf5bbedbd6691788e517f274a1 (accessed on 28th September 2025).

25. Mange, P., Savant, R., Singh, K., & Lule, A. (2024). Smart grid optimization dimensioning methods for residential solar cell dimensioning. *2024 2nd International Conference on Device Intelligence, Computing and Communication Technologies (DICCT)*. https://www.semanticscholar.org/paper/bc55e4ae6e35508a9deadafeab07d0a8efd833da (accessed on 28th September 2025).

26. Martinez-Sanchez, V., Levis, J., Damgaard, A., DeCarolis, J., Barlaz, M., & Astrup, T. (2017). Evaluation of externality costs in life-cycle optimization of municipal solid waste management systems. *Environmental Science & Technology, 51*(6), 3119–3127. https://www.semanticscholar.org/paper/159559a274c1b2c8f7f21ca29b05bff94b48fd00 (accessed on 28th September 2025).

27. Mohania, S., Khare, J., Vikram, A., Samal, A., Sandhya, N., & Kotehal, P. U. (2025). Technological advancements and their impact on sustainable development: A policy review on green innovation and environmental taxation. *International Journal of Computational and Experimental Science and Engineering*. https://www.semanticscholar.org/paper/9f90edf98fd48e0545dd1270f2c74f3990925910 (accessed on 28th September 2025).

28. Montalvo-Navarrete, J. M., & Palacios, A. (2023). Selection of environmental criteria for multi-criteria evaluation of energy supply alternatives in non-interconnected zones of Colombia. *2023 Advances in Science and Engineering Technology International Conferences (ASET)*. https://www.semanticscholar.org/paper/be528d055cc4220f1a0f4a141f339b472ae0614d (accessed on 28th September 2025).

29. N/A. (2018). *Kyoto Protocol to the United Nations Framework Convention on Climate Change. Cambridge University Press eBooks*. https://doi.org/10.1017/9781316577226.067 (accessed on 28th September 2025).

30. N/A. (2023). *Climate Change 2022–Impacts, Adaptation and Vulnerability*. https://doi.org/10.1017/9781009325844 (accessed on 28th September 2025).

31. Neuhoff, K., & Sellers, R. (2006). *Mainstreaming New Renewable Energy Technologies*. https://core.ac.uk/download/1248990.pdf (accessed on 28th September 2025).

32. Ohalete, N. C., Aderibigbe, A. O., Ani, E. C., Ohenhen, P. E., Daraojimba, D. O., & Odulaja, B. A. (2023). AI-driven solutions in renewable energy: A review of data science applications in solar and wind energy optimization. *World Journal of Advanced Research and Reviews, 20*, 401–417. https://doi.org/10.30574/wjarr.2023.20.3.2433 (accessed on 28th September 2025).

33. Onabowale, O. (2025). Energy policy and sustainable finance: Navigating the future of renewable energy and energy markets. *World Journal of Advanced Research and Reviews*. https://www.semanticscholar.org/paper/d64389a38843fc627c1a66697099eb65d69f7e43 (accessed on 28th September 2025).
34. Qazi, A., Fayaz, H., Abd Rahim, N., Hardaker, G., Alghazzawi, D., Shaban, K., & Haruna, K. (2019). Towards sustainable energy: A systematic review of renewable energy sources, technologies, and public opinions. *IEEE Access, 7*, 63837–63851. https://doi.org/10.1109/access.2019.2906402 (accessed on 28th September 2025).
35. Savaş, D. A. (2024). Renewable energy and CO_2 emissions: A case study of Turkey (1985–2022). *Akademik Yaklaşımlar Dergisi*. https://www.semanticscholar.org/paper/784ea32c4d3df68c5e64657dc1e3b9264ad49aef (accessed on 28th September 2025).
36. Shboul, B., Zayed, M. E., Marashdeh, H. F., Al-Smadi, S. N., Al-Bourini, A. A., Amer, B. J., Qtashat, Z. W., et al. (2024). Comprehensive assessment of green hydrogen potential in Jordan: Economic, environmental and social perspectives. *International Journal of Energy Sector Management*. https://www.semanticscholar.org/paper/58cab9a38c97689bdf6a3f3347a2d921addf640f (accessed on 28th September 2025).
37. Spierling, T., Lynch, A., & Bradley, T. (2025). Battery optimization for sustainable aviation: A comprehensive analysis for battery electric aircraft. *Journal of Aircraft*. https://www.semanticscholar.org/paper/a90107271a9c8fec7a3fc718d6659873b245445c (accessed on 28th September 2025).
38. Staffell, I., Scamman, D., Velazquez Abad, A., Balcombe, P., Dodds, P. E., Ekins, P., Shah, N., et al. (2018). The role of hydrogen and fuel cells in the global energy system. *Energy & Environmental Science, 12*, 463–491. https://doi.org/10.1039/c8ee01157e (accessed on 28th September 2025).
39. Talebi, A., Agabalaye-Rahvar, M., Zare, K., & Gozel, T. (2024). Economic-environmental analysis of smart power system in the presence of dynamic line rating and energy storage system. *2024 9th International Conference on Technology and Energy Management (ICTEM)*. https://www.semanticscholar.org/paper/86a96fdf1bfa9a744b796618166e94422e41c8d8 (accessed on 28th September 2025).
40. Tang, Y., Li, S., & Wu, F. (2025). How does carbon emissions trading impact energy transition? A perspective based on local government behavior. *Sustainability*. https://www.semanticscholar.org/paper/4657c8fbaf92d4094c54b3a9eeae2e8709c8e366 (accessed on 28th September 2025).

41. Taylor, R., Bodel, W., Banford, A., Butler, G., & Livens, F. R. (2024). Sustainability of nuclear energy—A critical review from a UK perspective. *Sustainability*. https://www.semanticscholar.org/paper/7577b70df290278 10fe28695ce878dfd0eb38006 (accessed on 28th September 2025).
42. Wang, Y., Lucey, B. M., Vigne, S. A., & Yarovaya, L. (2022). An index of cryptocurrency environmental attention (ICEA). *China Finance Review International, 12*, 378–414. https://doi.org/10.1108/cfri-09-2021-0191 (accessed on 28th September 2025).
43. Widianti, F., & Aptasari. (2025). Rent-seeking theory in the asymmetry of blue economy benefit distribution: Economic and sustainability accounting perspectives. *SAR (Soedirman Accounting Review): Journal of Accounting and Business*. https://www.semanticscholar.org/paper/2f71 bacde934b3e4f15c533c51aa1036a72561a2 (accessed on 28th September 2025).
44. Yue, M., Lambert, H., Pahon, E., Roche, R., Jemeï, S., & Hissel, D. (2021). Hydrogen energy systems: A critical review of technologies, applications, trends and challenges. *Renewable and Sustainable Energy Reviews, 146*, 111180. https://doi.org/10.1016/j.rser.2021.111180 (accessed on 28th September 2025).
45. Zhang, S., Zhang, C., Cai, W., Bai, Y., Callaghan, M., Chang, N., et al. (2023). The 2023 China report of the Lancet Countdown on health and climate change: Taking stock for a thriving future. *The Lancet Public Health, 8*, e978–e995. https://doi.org/10.1016/s2468-2667(23)00245-1 (accessed on 28th September 2025).
46. Zhang, Z. (2024). Powering the future under geopolitical pressure—PV industry analysis for China and USA. *Transactions on Economics, Business and Management Research*. https://www.semanticscholar.org/paper/a0b1e07d543b88f9f7f58e15b9491012b81542da (accessed on 28th September 2025).

CHAPTER 10

FUTURE PROSPECTS, INNOVATIONS, AND CASE STUDIES

Contents

Introduction 397
10.1. Fundamental Principles of Photochemistry 402
10.2. Innovations in Photochemical Techniques 408
10.3. Applications in Environmental Science 413
10.4. Photochemistry in Medicine 418
10.5. Case Studies of Successful Photochemical Applications 423
10.6. Future Prospects in Photochemistry 427
Summary 432
References 435

INTRODUCTION

Photochemistry has been extensively advanced, particularly with regard to understanding excited electronic states and energy transfer phenomena, both of which were extensively explored during the 1980s. These fundamentals have proven to be fertile ground for novel applications, such as the construction of covalent organic frameworks (COFs). COFs represent a new type of crystalline porous material, and their remarkable structural tunability and enormous surface area are the distinguishing features. The capability of COFs to act as visible-light photocatalysts gains importance in a broad variety of research fields, including organic transformations. As photocatalysts derived from COFs based on photonic matter continue to appear in ongoing research, their mechanisms of catalysis and the impact of dimension on performance are crucial for ongoing photochemical application development (Alberto et al., 2022).

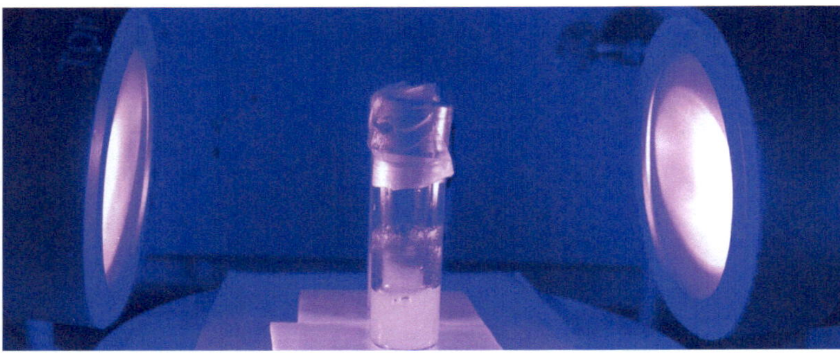

Figure 10.1. *Photochemistry.*

Source: https://pitrelab.okstate.edu/images/djmediatools/1-home-page-slideshow/Reactions_Website.jpg.

This underlines the necessity for photochemical research, which stands at the crossroads of numerous practical applications. In the past few years, intersection of photochemistry and biotechnology has drawn much interest, especially in microalgal biotechnology. Microalgae's ability to utilize sunlight to generate useful bioproducts such as drugs, lipids, and biofuels is one of the possibilities of integrating photochemical processes with green production routes. Genetic engineering technology is advancing fast as groundbreaking tools to enhance the productivity and efficiency of microalgae in the lab. These advancements, coupled with improvements in photochemical techniques, are critical to overcome existing barriers in the field (Samir et al., 2022). The possibility of applying genetically modified microalgae for industrial purposes creates a window in which photochemistry could be a massive contributor to sustainable growth and environmental conservation. As the world grapples with energy sustainability and environmental destruction, photochemistry in the future offers unimaginable opportunities. Innovations such as the development of COF photocatalysts and genetic engineering of microalgae hold the secret to totally revolutionizing not only chemical reactions but also the overall field of sustainable practice across many industries. The research landscape is fraught with potential, where the efficiency of light-driven processes can be tuned to address world energy needs as well as promote ecologic balance. There are some difficult problems, however, including scalability and industrialization of laboratory showcase stories. Therefore, constant discussion between researchers, industry participants, and policy makers is important in order to harness the full potential of photochemistry in generating a sustainable and innovative future (Alberto et al., 2022; Samir et al., 2022) (Figure 10.1).

Definition and Scope of Photochemistry

Photochemistry, as a branch of chemistry that deals with the phenomenon of the interaction between light and chemical compounds, is an integral component of different scientific disciplines such as environmental science, medicine, and material science. Photochemistry relies on the premise that certain chemical reactions can be initiated or caused to proceed through the uptake of light. Essentially, photons transfer energy that can result in breaking of chemical bonds, electron transitions, and a variety of chemical conversions. Photochemistry is comprised of various mechanisms like photodissociation, photoisomerization, and several light-induced processes. Due to evolving industries, the necessity for efficient photochemical reactions is on the rise, pointing to its role in photolithography, solar energy conversion, and photonic device design. The scope of photochemistry is quite vast, extending into the development of new applications that address contemporary challenges. A prime example of this would be the development of optical sensors, which is a pragmatic realization of photochemistry principles, in which optics can be used to enhance selectivity and sensitivity of chemical sensing systems. These sensors have vital applications in medical diagnostics, environmental monitoring, and industrial processes, thereby showcasing the very necessary interdisciplinarity of photochemistry. As highlighted by recent studies, technological advancements in optical sensor technologies have enabled the creation of miniaturized and cost-effective sensing solutions based on photochemical effects. This bridging of the gap between photochemistry and sensor technology not only indicates its inherent value toward the enhancement of present systems but also the potential for new innovations that could be generated from present studies in this ever-evolving field (Mário et al., 2023). Also, the photographic environment is constantly evolving with research unveiling new opportunity for light-activated chemicals' health benefits and further phytonutrients and nutraceuticals from foods. Some of these nutraceuticals and phytonutrients possess therapeutic activity when combined with photochemical processes. For instance, carotenoids are fat-soluble compounds that have been proven to be involved in disease prevention and control. The convergence of nutrition and photochemistry underscores the importance of understanding how light-activated activities might influence the bioavailability and efficacy of such nutrients. With continuing scientific studies that deconstruct such interactions, opportunities are opened for the development of functional food ingredients that optimize health gain, positioning photochemistry at the center of food science and more (Rao et al., 2023) (Figure 10.2).

Figure 10.2. *Scope of photochemistry.*

Source: https://d2cbg94ubxgsnp.cloudfront.net/Pictures/380x253/9/2/3/537923_setupcopy_376534_crop.jpg.

Historical Background and Development

Photochemistry has an inherent historical background that emphasizes its development and significance in scientific advancement. As a stand-alone science in the mid-nineteenth century, photochemistry examines the reaction between light and chemicals. One of the landmarks in the history of this science was Heinrich Hertz's discovery of the photoelectric effect, which opened the door to understanding the dual nature of light as a wave and as a particle. This basic understanding further led to understanding the role of light in chemical reactions. Most significantly, the evolution of spectrophotometry as a tool for the analysis of light-absorbing compounds further contributed to advancing photochemical research by enabling scientists to investigate molecular changes created by light irradiation. Such advancements led to the development of using photochemistry in an extensive range of applications from materials science to biological systems and now serves as a cornerstone of contemporary chemical investigation. Later developments in photochemistry are defined by the merging of emerging technologies and methods that have led the field to new frontiers. In the first half of the 20th century, there were breakthroughs in quantum mechanics and photochemical kinetics that allowed scientists to describe light-driven processes at a molecular level better. These developments were important in shaping the understanding of photochemical reactions, particularly in artificial photosynthesis and conversion of solar energy. Furthermore, greater environmental sustainability concern led to the quest for photochemical applications in waste treatment and sequestration of carbon, which demonstrates the ability of the field to change to meet the needs of society. Modern advances in optical sensors, as described by Mário et al. (2023), attest to ongoing advancement in the field with tests of how photochemistry remains addressing issues in different industries, including the environment and medicine. The history of photochemistry also describes it as being multidisciplinary, where it brings chemistry together with other sciences and industries. Notably, the intersection of photochemical process and reaction engineering has spurred the development of new materials, as well as efficient energy conversion systems. According to the findings of Bollini et al. (2023), the evolution of reaction engineering focuses on the ability to understand more than one reaction mechanism and the operations of scaling, which form the basis for the optimization of photochemical utilization. This cross-disciplinary approach has not only enhanced the efficiency of photochemical reactions but also given rise to potential for sustainable chemical industry processes, such as a shift toward continuous processing and electrical operation of chemical reactors. With its basis in history and ongoing developments, it is evident that photochemistry is poised to lead future developments and applications toward global sustainability goals.

Importance in Modern Science and Technology

The pertinence of photochemical progress comes clearly into focus in the field of existing scientific study and technological development. As the eternal push for efficiency and sustainability continues, photochemistry is a tool that is absolutely vital in the development of new materials and processes. For instance, the study of Metal-Organic Frameworks (MOFs) has enabled one to reimagine the sensing of gases and chemicals, which are of significant importance for environmental monitoring and the industrial sector. According to recent studies, MOF-derived nanocomposites exhibit enhanced sensitivity, selectivity, and stability, ensuring reproducible performance in miniaturized gas sensors. Such a capability is essential for addressing such pressing issues as air quality and industrial security, emphasizing the importance of photochemical advancements towards developing an adaptive and resilient technological ecosystem (Nandi et al., 2024). No less vital is the role played by optical sensors, which have undergone tremendous evolution in design and function. These devices, at the heart of modern diagnostic processes, utilize the photochemistry mechanisms to deliver unprecedented sensitivity and selectivity. Optical sensors are used in an enormous variety of fields, from medical diagnosis and environmental monitoring to defense, as per detailed studies. The development of cost-efficient, miniaturized optical sensing technologies is a reflection of the progress of photochemical applications. The technologies not only accelerate data gathering but enhance the analytical capability required for solving complex real-world issues. The blending of photochemistry with optical technologies is therefore a revolutionary power, creating new sensor paradigms and elevating the level of scientific measurement (Mário et al., 2023). Future prospects in photochemistry are to further cement its pertinence to contemporary civilization and science. Materials science breakthroughs, specifically the development of novel photochemical methods and nanomaterials, will be capable of creating devices with the property of working under hostile conditions, which is critical to the expansion of sensor applications. The work aims at combining advanced materials and creative optical designs in order to overcome present performance and endurance limits. This convergence will open the way for new-generation sensors able to track a broad range of parameters with high efficiency. As the world is facing environmental and health problems, the relevance of photochemistry in developing future solutions cannot be exaggerated, so it remains at the center of scientific research and technological innovation (Figure 10.3).

Figure 10.3. *Importance in modern science and technology.*

Source: https://www.cest.org.uk/wp-content/uploads/2021/07/cest-importance-of-science-business.jpg.

10.1. FUNDAMENTAL PRINCIPLES OF PHOTOCHEMISTRY

The fundamental principles of photochemistry are the light-matter interactions, which explain how photons initiate chemical reactions. At the heart of these interactions is excited electronic states in which molecules take up light energy to excite electrons into higher energies. This uptake of energy propels a lot of processes, such as photochemical reactions, whereby molecular configuration of a compound becomes irreversibly altered. An understanding of such principles not only illuminates the reaction mechanisms but also underpins emergent applications such as photocatalysis. By manipulating light to trigger chemical reactions, researchers are able to design sustainable chemical processes that save waste and minimize the use of energy, in accordance with needs for green technologies that are current.

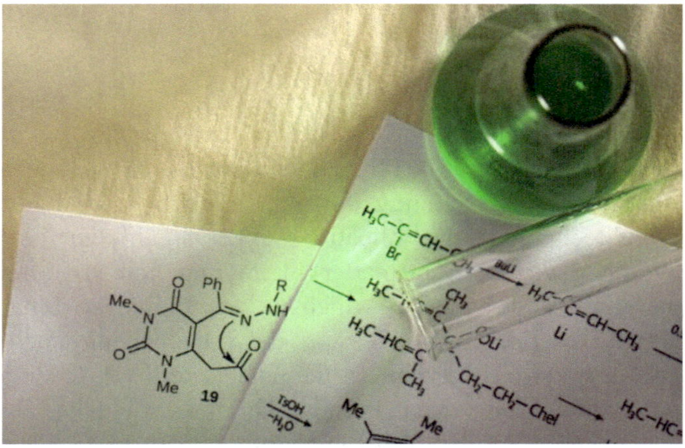

Figure 10.4. *Fundamental principles of photochemistry.*

Source: https://microbiozindia.com/wp-content/uploads/2023/11/photochemistry.jpg.

Additionally, the ability to modulate molecular attributes through structural modifications has formed a platform for covalent organic frameworks (COFs) as upcoming photocatalysts in organic reactions. COFs are characterized by their instructive tunability, with which researchers can design their structure and functionalize them to enhance light absorption and reactivity. According to recent organic photochemical studies, insights gained from the understanding of excited state and electron transfer phenomena in simpler molecular systems are being used in developing these advanced materials. This finding is a milestone change in photochemistry where principles developed by traditional photochemical science drive breakthroughs in material science, particularly in COF-based photocatalysts (Alberto et al., 2022). As the science of photochemistry develops, much work is still to be done, notably towards the optimization of the photocatalytic activity of COF-based photocatalysts for real-world applications. Scientists are working towards understanding the effects of dimensionality and crystallinity on their efficiency, seeking to develop the catalytic reaction mechanisms further. The effective implementation of these concepts in practical applications holds the potential to revolutionize renewable energy solutions and sustainable chemical processes. Apart from this, the innovations will make an extended contribution in fighting urgent global challenges that range from climate change and energy shortages. The future of photochemistry is thus tied to the advancement in COF technology and presents a window for further research that links basic science to practical use (Luna, 2022) (Figure 10.4).

10.1.1. Light–Matter Interaction

At the heart of photochemistry lies the intricate process of light-matter interaction, an event that describes the phenomena of material behavior when exposed to photons. Understanding the interaction is crucial in the design of a number of technological applications, most notably in the fabrication of new photonic devices. Light-matter interaction comes in the form of absorption, emission, and scattering, which are pivotal in describing the optical characteristics of materials. As researchers delve deeper into such processes, they find ways of manipulating light at the nanoscale level, leading to energy efficiency innovations and new materials. This new field has the potential to transform industries ranging from telecommunications to energy, putting light center stage of future technologies. Recent progress in material science, particularly concerning flexible conductive materials, illustrates the importance of light-matter interaction towards the development of novel applications. The materials have been of great interest due to their flexibility and potential integration into flexible energy storage devices, touch panels, and sensors. The behavior between light and matter is best observed when examining how these materials may be used to enable performance in devices through mechanisms of energy transfer and light absorption. Through work on the sensitivity of these interactions, researchers try to achieve the highest efficiency and functionality of technologies reliant upon such materials. The unveiling of light-matter interactions here not only adds to the cognitive advancement of knowledge but also holds the promise of the formation of beneficial applications that can address current challenges in energy and electronics (Dutta et al., 2024). Optical sensors are another extremely significant application field influenced by light-matter interactions, bearing witness to their profound influence on technological advancement both industrial and scientifically. These sensors exploit the inherent properties of light to reach record sensitivity and selectivity, and they cover a broad spectrum of applications from environmental monitoring to biomedical diagnostics. The design and process of creating beneficial optical sensors involve a multidisciplinary grasp of the optical, material, and environmental characteristics that determine their performance. Future-generation sensors, including laser and plasmonic sensors, display the practical applications of light-matter interaction with studies developed to meet the demands of different applications. The utilization of novel materials that have enhanced light absorption characteristics can potentially support the creation of future sensing technologies in the form of efficient, cost-effective, and compact technology, opening the door towards a new era of optical sensing technology (Mário et al., 2023) (Figure 10.5).

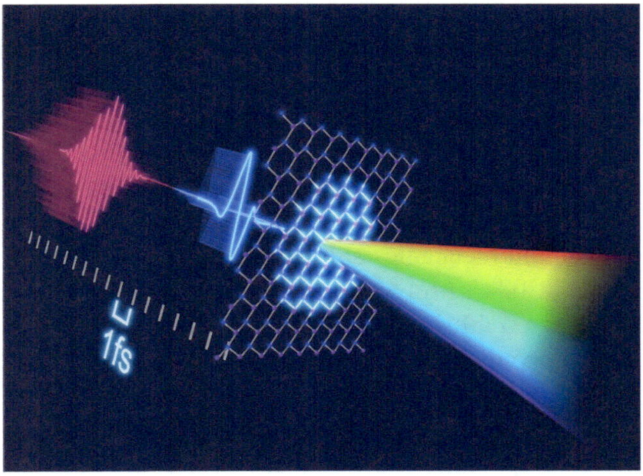

Figure 10.5. *Light–matter interaction.*

Source: https://www.photonics.com/images/Web/Articles/2022/7/6/thumbnail_681 66.jpg.

10.1.2. Quantum Mechanics in Photochemical Reactions

The quantum mechanics subtlety forms the core of photochemistry, governing in a major way the molecular dynamics in light-catalyzed chemical reactions. At the quantum level, matter-photon interaction generates effects like excitation, electron transfer, and energy migration. These are the fundamental processes that result in the initiation of chemical reactions upon photon absorption. Quantum mechanics provides for the framework necessary to study molecular orbitals, which are critical to anticipating reaction outcome as state transitions of energy. As researchers learn more about these interactions, they apply quantum mechanical theory to develop innovative ways of manipulating reaction pathways, optimizing yields, and developing new materials with desired characteristics. This understanding not only enhances the effectiveness of photochemical reactions but also serves as the basis for novel applications across multiple disciplines, such as energy conversion and pollution cleanup. Moreover, the integration of quantum mechanics with advanced materials science presents promising directions for the development of new photochemical systems. New advancements in materials, particularly in the integration of polymers with porous organic frameworks through covalent approaches, have been highly promising in terms of transcending the limits set by tradition. Materials like metal-organic frameworks (MOFs), covalent organic frameworks (COFs), and hydrogen-bonded organic frameworks (HOFs) can be designed with enhanced porosity and enhanced stability. Their ability to function well in photochemical applications is based on their unique capability to alter electronic properties through molecular design. By taking advantage of the quantum mechanics for tailoring these properties, researchers can open up new avenues of applications, and these hybrid structures are used in catalysis and gas separation, thus establishing the interdependent relationship of quantum mechanics and material innovation in photochemistry (Md. Hossain et al., 2024). Moreover, the intersection of quantum mechanics with nanotechnology establishes the direction for research in photochemistry. The last 7th workshop/conference hosted by the Ladoke Akintola University of Technology (LAUTECH) Nanotechnology Research Group focused on the relevance of nanotechnology in improving photochemical reactions. Through these developments, the attendees talked about the capability of nanostructures to improve the absorption of light and energy transfer, resulting in enhanced reaction efficiencies. This discussion points to the promise of using nanotechnology in controlling quantum effects at the nanoscale, which would transform energy storage and conversion processes. The convergence of several institutions and research centers is an assiduous move to harness the power of quantum mechanics in advancing the field of photochemistry further, pointing towards the necessity of future interdisciplinary research (N/A, 2023).

10.1.3. Energy Transfer Mechanisms

The energy transfer mechanism is a unifying topic in the study of processes that lead to photochemical reactions. Photochemistry involves matter-light interaction that leads to energy exchange that induces chemical change. At the core of the process are various mechanisms of energy transfer, including singlet and triplet energy transfer, Förster resonance energy transfer (FRET), and Dexter energy transfer. Each of these mechanisms possess distinct properties and efficiencies that have been used to create new applications, such as photovoltaic cells and photodynamic therapy. This exploration of how energy can be effectively transferred and utilized not only provides further insight into fundamental chemical principles but also positions us for the development of better solar energy harvesting and targeted medicine therapies, illustrating the potential for future development in photochemistry. Improvements in the understanding of energy transfer mechanisms have triggered some extremely worthwhile contributions to materials science, particularly with the advent of new photonic materials.

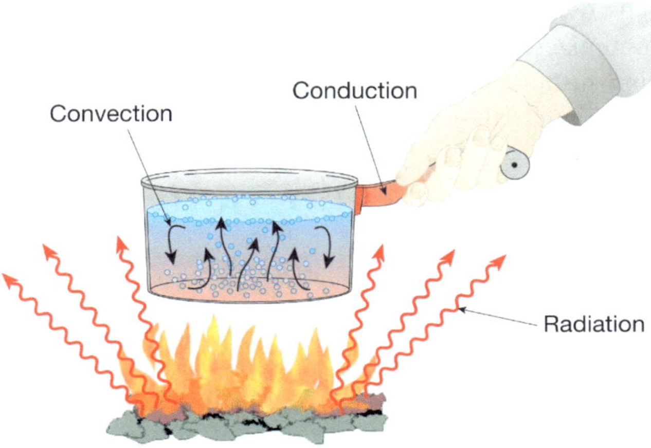

Figure 10.6. *Energy transfer mechanisms.*

Source: https://www.wattco.com/wp-content/uploads/2024/05/Understanding-Thermal-Energy-Transfer.jpg.

By designing materials that are optimized to enhance energy transfer, researchers have made it possible to fabricate systems with improved efficiency for light harvesting and energy conversion. For instance, introduction of nanostructures can optimize the efficiency of light-harvesting complexes to accelerate energy migration and, in consequence, improve the yield of photovoltaic and photosynthetic systems. Moreover, as seen in redox mechanism studies in various applications, energy transfer optimization can significantly improve the electrochemical efficiency of energy storage devices, such as zinc-based batteries, indicating the relationship between photochemistry and universal energy solutions (Tang et al., 2024). This marriage of photochemical research and materials design confirms the revolutionary potential of novel energy transfer ideas. Collaborative efforts by industry and academia have also pushed the investigation of energy transfer processes into reaction engineering. The desire for photochemical systems encourages multi-disciplinary efforts involving chemical engineering, materials science, and photophysics. By identifying impediments to development and advocating for the line of study in response engineering, experts highlight the importance of integrating mechanisms of energy transfer with a number of industrial processes, such as pharmaceuticals and specialty chemicals (Bollini et al., 2023). This general vision not only identifies new paths for enhancing chemical processes but also targets sustainable operations, such as electrification of chemical reactors and converting traditional batch processes into more effective continuous ones. With continued advancements in the field, these efforts envision a promising future where optimized energy transfer mechanisms can contribute significantly toward enabling sustainable development and technological advancements in photochemistry (Figure 10.6).

10.2. INNOVATIONS IN PHOTOCHEMICAL TECHNIQUES

Photochemical technologies are changing technologies in numerous various domains, particularly agriculture and environmental protection. With increasing food demands from the world population as a result of climate change, efficient mechanisms of crop production have been a key need. Fresh nanotechnology integrated with photochemical processes offers an innovative solution for improving agricultural productivity and sustainability. Through targeted application of agrochemicals in the form of nanomaterials, these techniques provide for the precise delivery of pesticides and nutrients, improving their effectiveness with minimal wastage. Targeted and therefore more efficient application not only increases agricultural yields but minimizes the environmental burden of standard agricultural application, proving that new photochemical methods are central to solving future food security challenges (Shang et al., 2019). In materials science too, photochemistry is advancing to a great extent through advanced polymerization mechanisms. The shift to light-catalyzed polymerization is offering a level of control that is not possible in most scenarios through classical thermal or chemical systems. Photopolymerization networks employ light as the energy source, enabling rapid reaction and the ability to tailor molecular structures with high accuracy. This innovation is instrumental in the creation of advanced materials that can be designed for specific applications, from drug delivery systems to high-performance coatings. The adaptability provided by these processes is central in advancing material science, demonstrating that photochemical innovations are still a cornerstone in the future of producing innovative synthetic materials (Corrigan et al., 2018). Besides, the extrapolation of photochemical techniques to inter-disciplinary applications symbolizes a wider trend within science for integrated solutions. Between molecular detection and photonic devices, light's application for facilitating chemical reaction incites innovation in as broad a range of disciplines as medicine, electronics, and pollution monitoring. The innovations facilitate real-time data capture and analysis, which enhances our ability to react to environmental issues. Through continued exploration of the intersection of photochemistry with material science and nanotechnology, researchers can access new techniques that not only improve existing systems but also allow for entirely novel paradigms of scientific inquiry and industrial use. The potential in the future of photochemical processes offers great promise for both scientific advancement and societal needs.

10.2.1. Advances in Photocatalysis

With the ongoing evolution of the field of photocatalysis, synthesis and design of materials have progressed significantly, transforming its efficiency as well as applications. The key break-through is the discovery of new semiconductor materials with enhanced photocatalytic activity in visible light. The materials, like altered titanium dioxide and perovskite-structured oxides, possess better electron-hole generation properties, and hence they are capable of facilitating a broader range of chemical reactions. The introduction of heterojunctions, or the combination of different semiconductors, has further increased charge separation efficiency, yielding higher rates in reactions like water splitting and pollutant degradation. All these advancements are essential for fostering sustainability solutions and resolving environmental problems because they drive the accessibility of photocatalysis in practical applications. As described by Bollini et al. (2023), the emphasis on the development of such high-tech materials is crucial in overcoming obstacles and improving photocatalytic technology. Besides, the integration of photocatalysis with other technologies holds considerable promise in extending its efficiency as well as range of application. For instance, the coupling of photocatalytic reactions with electrochemical systems is an available avenue towards the improvement of the reaction kinetics and selectivity. This hybridization allows for greater efficiency in converting solar energy into chemical energy, thereby opening up new applications like carbon dioxide reduction and hydrogen production. In addition, employing machine learning algorithms to optimize photocatalytic reaction conditions is likely to enhance these processes further, leading to more efficient designs and faster scaling up to industrial scales. The research of such multi-dimensional approaches is crucial in the development of a top-level framework that encompasses both elementary and practical aspects of photocatalysis, as noted in the current discourse in reaction engineering by Bollini et al. (2023). Considering the future, the prospects of photocatalysis in addressing world challenges, most notably in environment remediation as well as sustainable energy production, is in the forefront. The ability of photocatalysts to increase the degradation of lasting organic pollutants offers a green solution while reducing reliance on harmful chemical treatment. Concurrently, as pointed out in, advancements in biomaterials, and particularly algae, are demonstrating potential synergistic effects if blended with photocatalytic systems for application to cancer treatments, which also signifies the extent and breadth of photocatalysis outside traditional application. The intersection of health with environmental science thus highlights the broader importance of photocatalytic innovation. By continued innovation and expansion of the use of photocatalysis, scientists can make a meaningful contribution to sustainable development goals, ensuring that this technology remains at the forefront of scientific progress in years to come (Figure 10.7).

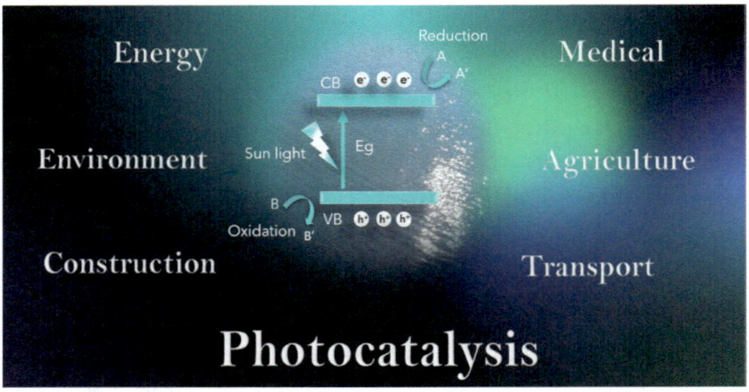

Figure 10.7. *Advances in photocatalysis.*

Source: https://cicenergigune.com/media/uploads/mediacenter/photocatalysis-cicenergigune1.png.

10.2.2. Development of Photochemical Sensors

Photochemical sensor development is a revolutionary approach to sensing technologies that employs the principles of photochemistry for better detection. The sensors apply light for the activation or catalysis of chemical reactions that produce measurable signals, thereby providing information critical to many applications from environmental monitoring to medical diagnosis. Optimum optimization of such chemical reactions previously required significant amounts of manual experimentation, a process fraught with inefficiency and constraints. But recent advances in machine learning and laboratory automation have revolutionized the optimization process. Algorithmic approaches propel high-velocity prototyping, and an ability to modify multiple reaction parameters at once reduces the time and human effort needed to determine optimal conditions by orders of magnitude. This paradigm demands the ability of photochemical sensors to quickly evolve and adapt in parallel with developments in computational approaches (Velasco et al., 2025). Moreover, the variety of applications of photochemical sensors in medicine, defense, and environmental science is pointed out as an example of the versatility of such sensors. The sensors are designed to operate in diverse spectral regions with advanced photonic materials and nanostructures for enhancing selectivity and sensitivity. For example, laser sensors and plasmonic sensors are currently under extensive investigation to increase the detection limits as well as response times. The subtle interaction between sensor operation and material properties calls for a deep understanding of optical principles, plus the ambient environmental factors which potentially impact results (Mário et al., 2023). So, researchers are working together on both advances in sensor technology and materials science to further broaden the capability of photochemical sensing systems. The future outlook of photochemical sensors appears promising, with the hope of future technological advances enhancing their performance and uses. The availability of nanotechnology and materials engineering can facilitate the development of even smaller and cheaper sensors, which can be used for diverse applications like smart environmental monitoring systems and *in vivo* biomedical diagnostics. Because interdisciplinary research enables the combination of optics, chemistry, and data science, putting highly advanced analytical methods together will enable the creation of highly specialized sensors for specific applications. The history of photochemical sensors is an ideal demonstration of dynamic collaboration between engineering practice and science advancement, showcasing the way progress in one direction will be a stimulus for realization in another that will further increase our awareness and interaction with the world around us (Velasco et al., 2025; Mário et al., 2023).

10.2.3. Innovations in Solar Energy Conversion

The evolution of solar energy conversion technology has undergone groundbreaking innovation aimed at enhancing efficiency and making it more affordable. Among the most promising of these is the use of new materials in photovoltaic devices. Recent advances in perovskite solar cells have demonstrated record-breaking efficiencies of over 25% while lowering their cost of production compared to traditional silicon-based cells. This revolution not only opens up the possibilities for large-scale deployment, but it also opens up opportunities for small-scale, off-grid solar systems. Research indicates that hybrid schemes with multifunctional materials, i.e., nanostructures, can further enhance charge carrier transport as well as light absorption, thereby making solar energy conversion more efficient under various environmental conditions. In this way, the coupling of high-performance materials with emerging manufacturing technologies is positioning the solar industry for groundbreaking growth during the forthcoming decades. Another hopeful field in solar energy research and development concerns the use of flexible conductive material, which has been of interest given multiple applications. This type of material is crucial for the creation of lightweight, transportable solar panels that can be easily embedded in different surfaces and frameworks.

Figure 10.8. *Innovations in solar energy conversion.*

Source: https://d2fryjlmubyuuh.cloudfront.net/wp-content/uploads/2024/12/18045335/The-Future-of-Solar-Energy-Emerging-Trends-and-Innovations.webp.

The integration of flexible solar cells into wearable technology and building-integrated photovoltaics is a step towards more flexible energy options. By harnessing technologies that manage the electronic properties of such conductive materials, scientists are developing solar systems that not only offer energy generation but also contribute to the energy storage solution. These advances supplement the agenda of creating sustainable energy architectures to initiate an age where the capture of energy becomes seamless and ubiquitous across life, thereby greatly enhancing energy usability and availability. Moreover, the role played by photochemical processes in the invention of solar energy conversion technologies cannot be overstated. Photochemistry-led technologies such as artificial photosynthesis are at the forefront of addressing the urgent need for carbon-free energy solutions. Systems that mimic closely natural photosynthesis are being developed to convert sunlight directly into energy-rich fuels such as hydrogen or methanol, which can be stored and utilized with great efficiency. These reactions require sophisticated catalysts and design reaction pathways to enhance efficiency, with promise in closing the gap between intermittent solar energy capture and continuous fuel output. Systematic exploration of these redox processes is crucial in developing solar technologies like Zn-based batteries, which are central to energy storage (Tang et al., 2024). Continuing research in this field not only provides the potential for unlocking pragmatic energy solutions but also serves to further important environmental objectives of reducing carbon footprints and greenhouse gas emissions (Figure 10.8).

10.3. APPLICATIONS IN ENVIRONMENTAL SCIENCE

The intersection of photochemistry and environmental science is fast becoming a subject of attraction for new applications to address the pressing ecological problems. Photochemistry employs the power of sunlight to drive chemical reactions, offering green alternatives to water treatment, air purification, and waste disposal. The use of titanium dioxide (TiO_2) photocatalysis is a case in point, as it has emerged as a leading technology to degrade toxic pollutants such as organic pollutants and heavy metals from wastewater. The ability of photocatalytic systems to operate under visible sunlight increases their effectiveness in a broad spectrum of environmental circumstances. Such applications do not only assist in pollution control but also contribute to the development of a circular economy, whereby by-products of the processes can be recycled or safely disposed of from ecosystems, hence promoting environmental rehabilitation and resource efficiency. In renewable energy potential, photochemistry is also vital in the production of sustainable energy systems, particularly through the conversion of solar energy. Photochemical processes are accountable for the performance of solar cells and artificial photosynthesis, phenomena that attempt to duplicate natural processes to harness solar energy and convert it into chemical fuels. Significant advances in this area can create more efficient solar-to-fuel systems that tap and store sunlight energy in chemical bonds, such as hydrogen or carbohydrates. As these advancements solidify, they can transform the generation of energy, maintaining fossil fuel consumption substantially lower and reducing climate change impacts. By successful integration of these technologies into existing energy infrastructure, environmental scientists are able to make a transition viable towards cleaner, more environmentally friendly sources of energy, simultaneously addressing issues of energy security and environmental degradation. Furthermore, addressing the complexity of climate change issues using photochemical innovations also extends to agricultural applications. Photochemical technologies can optimize crop growth by increasing fertilizer efficiency and reducing chemical pesticide use through targeted delivery systems. For instance, advanced photochemical techniques can be used to formulate smart agrochemicals that respond to environmental stimuli and disperse nutrients only when necessary. This not only provides the maximum yield from the crops but also minimizes the environmental footprint of agriculture, protecting biodiversity and soil fertility. Further, accessing these technologies also enables more public and private investment in sustainable agriculture. By creating collaborations between environmental scientists and chemists, the agriculture sector will benefit a great deal from these photochemical advances in terms of food security as well as climate resilience (Tang et al., 2024; Bollini et al., 2023).

10.3.1. Photochemical Processes in Water Treatment

Photochemical processes are a promising way to develop water treatment technologies, particularly in the elimination of emerging contaminants such as pharmaceuticals and personal care products. Utilizing light energy to drive chemical processes, photochemical reactions offer a sustainable solution compared to traditional treatment techniques, which have a tendency to use heat or chemicals. For instance, photolysis, where ultraviolet (UV) light is used to degrade pollutants, has been effective at degrading complex organic molecules that can slip through traditional filtration. The technology not only achieves optimal treatment efficiency but is aligned with global sustainability goals as it reduces reliance on harmful chemicals and waste. By utilizing natural sunlight or individual wavelengths of artificial light, photochemical reactions can be integrated into traditional water treatment facilities to enhance their contaminant removal capacity, a step towards smart, lean utilities as highlighted by recent case studies. The energy usage and greenhouse gas emissions associated with advanced treatments are issues for the operational effectiveness of modern-day water treatment plants. Studies show that methods like reverse osmosis and UV disinfection are very energy-intensive, raising the issue of their viability with increased energy costs and potential effects of climate change (Eleweuwa et al., 2025). Strategically used photochemical processes have a long way to alleviate such pressures. For example, the use of renewable energy sources such as sunlight in UV disinfection systems can make energy use more efficient and bring water treatment habits more in line with broader ecological processes. New ideas, such as using hybrid systems incorporating photochemical and traditional processes, open the door to optimized energy use and maintaining water quality parameters. The research on major utilities confirms that there can be great advantages to the use of integrated photochemical solutions. Moreover, the destiny of photochemical treatment of water is being shaped by ongoing research and advancements. Recent advances in nanotechnology, particularly the construction of photocatalysts, have enhanced the efficiency and rate of photochemical reactions, which are more viable on a large scale. These catalysts, generally composed of titanium dioxide, facilitate the photo degradation of water contaminants under light, thus extending the scope of application of photochemical treatments beyond their fundamental use (Manyoundou, 2025). Case studies demonstrate fruitful applications of such technologies in various environments, where they effectively attack the contaminants while recording lower operating costs than conventional practices. While researchers continue to investigate novel combinations of materials and light sources, the prospects for these advanced photochemical processes that are transforming water treatment are brighter than ever before. Integrating these technologies holds not just the promise of improved treatment but also to ensure sustainable water management practices (Figure 10.9).

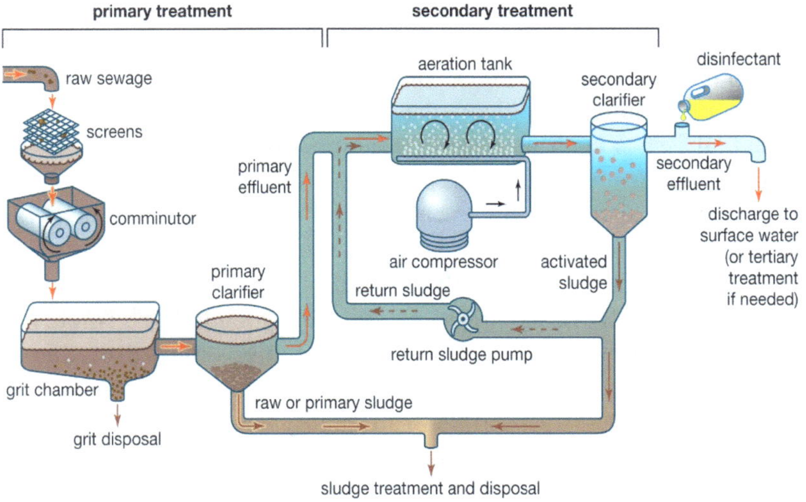

Figure 10.9. *Photochemical processes in water treatment.*

Source: https://cdn.britannica.com/55/23955-050-A2939E03/treatment-sewage-sludge-process.jpg.

10.3.2. Role in Air Pollution Control

Municipalization of air pollution through new strategies is the way to go in achieving cleaner environments, particularly in urban centers whose levels of pollution consistently exceed safe limits. Photochemistry has proved to be one of the most significant methods of controlling air pollution, where sunlight is employed to energize chemical reactions that can neutralize toxic emissions. This case is especially pertinent for volatile organic compounds (VOCs), which are major culprits behind ozone formation. Through application of photochemical processes, researchers have developed incremental reactivity (IR) scales that are environment-specific, such as in megacities such as Guangzhou. Scales help in approximating the ability of VOCs to form ozone, a significant parameter for effective control strategies for air quality. As evident in current studies, site-specific IR scales can be used to guide emissions policy and ensure interventions target the most impactful pollutants to maximize overall air quality management (Zhang et al., 2021). The health impacts of air pollution need to be addressed through an integrated response using scientific progress, policy framework, and societal involvement. The American Heart Association cites the longstanding link between fine particulate matter and cardiovascular risk, necessitating broad policy action to lower emissions and increase air quality. Insertion of photochemical innovation can play a significant role in such policies, particularly as air cleaning technologies based on photocatalytic reactions to break down pollutants. These advances not only address the near-term effects of pollution but also enhance policies addressing more profound environmental issues such as climate change and socioeconomic disparities. Lastly, the implementation of these new technologies in policy can facilitate greater access to cleaner air for everyone and thus advance public health on a number of different levels (Joel et al., 2020). As photochemistry advances, its role in air pollution control will expand. New technologies, such as new photocatalysts and solar-based processes, have the potential to be more efficient and effective in pollutant removal. Future innovations may also address how these emerging technologies can be made compatible with existing infrastructure, and so cities become less susceptible to air quality issues. A collaborative effort by governments, researchers, and the private sector will be critical in taking advantage of these advances. Public-private partnerships, in this instance, might speed the deployment of scalable solutions combining photochemical methods with the inclusion of community in the issue of air quality. Thus, the prospects for photochemistry in air pollution control are promising, with the potential to reduce environmental and health impacts significantly while simultaneously providing socioeconomic benefits in the form of cleaner technologies as well as fair policy interventions.

10.3.3. Contributions to Green Chemistry

The advances in green chemistry mark a milestone in the strategy towards chemical processes, particularly environmental sustainability. Reducing waste and minimizing chemical production's impact on the environment is one of the central principles of green chemistry, keeping pace with the greater demand for sustainable development. This is in line with photochemistry development, where photochemical reactions are increasingly being studied to reduce toxic by-products. For example, the use of nanomaterials in chemical processes can increase the efficiency of energy transfer and light absorption, thus fewer resources are used, and less waste is produced. Not only do these practices help in creating a sustainable environment but also bring economic benefits as they can lower the cost of production and enhance market competitiveness. Hence, it becomes crucial to incorporate these innovations into modern chemical practices for creating a more sustainable industrial environment.

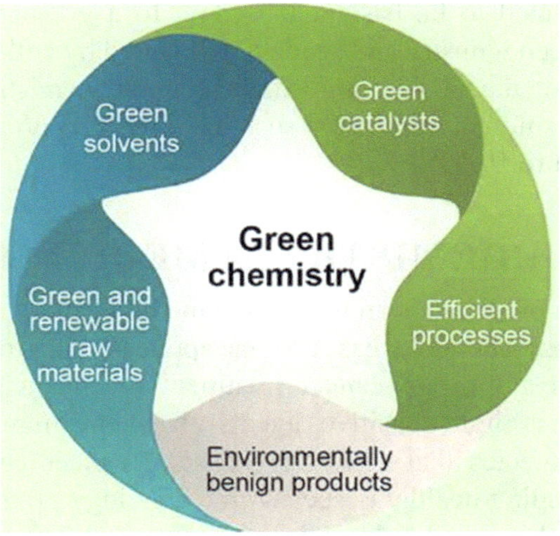

Figure 10.10. *Contributions to green chemistry.*

Source: https://chempedia.in/wp-content/uploads/2024/06/green-chemistry-1.jpg.

The use of nanomaterials in photochemistry illustrates the synergy between materials chemistry and green chemistry. It has been found in recent studies that nanomaterials can significantly improve the efficiency of photochemistry. Using nanostructured catalysts, scientists have been able to demonstrate enhanced rates and selectivity of reactions for various chemical conversions, reducing the need for traditional, more harmful reagents. This contribution is also highlighted in liver cancer therapies, as discussed in recent bibliometric research examining nanoscale innovations in the last 20 years. The increased emphasis on targeted drug delivery systems using photothermal therapy is just one of the general implications of coupling photochemical principles with medical applications. Through such critical health issues addressed through sustainable means, the influence of green chemistry extends well beyond the industry to attain public health and welfare as main objectives. In addition, the future of green chemistry is intimately related to the continued development of reaction engineering principles placing substantial emphasis on sustainability. As emphasized in recent expert opinion, the ability to address the challenges of plastics circularity and decarbonization is central in shaping the chemical industry. Advances in reactor design, catalysis, and process intensification are instrumental for this transformation. For instance, electrochemical devices and non-conventional types of reactors are still untapped fields with the potential to enhance energy efficiency and reduce emissions. With these advanced technologies set to be integrated into industrial procedures, the effect on green chemistry is expected to be hastened, leading to a greener era. Continuous engagement between industry and academia, directed by collective knowledge, is sure to be charting the path for innovations in photochemistry as well as green chemistry, prioritizing environmental sustainability in approaches to chemical development (Figure 10.10).

10.4. PHOTOCHEMISTRY IN MEDICINE

Photochemistry in medicine is an innovative method that uses light to advance therapeutic actions. The field has various applications, from photodynamic therapy (PDT) as a form of cancer treatment to enhancing drug delivery systems. PDT uses light-sensitive agents that, on illumination, generate reactive oxygen species that selectively destroy cancer cells with minimal damage to surrounding healthy tissue. With technology progressing, advances in formulation of drugs and light delivery systems are in progress to increase efficacy and reduce side effects. Additionally, insertion of nanotechnology into photodynamic agents has emerged on the horizon, which could improve the pharmacokinetics of such treatment as well as open up lines for combination therapies. These breakthroughs underscore the pivotal role of photochemistry in reinventing therapeutic paradigms in medical practice (Adamovich et al., 2022). Besides, the rising demand for electron–donor–acceptor (EDA) complexes is an indicator of another most significant facet of the application of photochemistry in medicine. Through photoexcitation of EDA complexes with radiation, researchers are capable of triggering electron transfer reactions that result in radicals that can subsequently elicit therapeutic reactions. This process is highly favored due to the fact that it is performed at mild conditions, reducing the use of strong chemical catalysts and adhering to the principles of green chemistry. With the increasing demand for sustainable and green methods of drug synthesis, the EDA process not only increases reaction efficiency but also has the potential to decrease the environmental impact of pharmaceutical production. As these techniques are progressed, they have potential to become a major component in more cost-effective methods of drug synthesis and the development of novel therapeutics (Yang et al., 2021). Future applications of photochemistry in medicine are not only at the procedural levels but also at the diagnostic level. Technologies such as photonic biosensors utilize the interaction between light and biological molecules to enable fast and sensitive detection of disease markers. This technology has enormous potential in early detection and disease monitoring, particularly in cancer and infectious diseases. Advanced imaging procedures, leveraging the use of specific photochemical principles, allow real-time observation of therapeutic effectiveness and patient response. With advancement in technology, the integration of artificial intelligence and machine learning into photochemical procedures can further optimize these diagnostic tools, leading eventually to strategies for personalized medicine. The intersection of light and biochemistry reveals an innovative path to advancement in diagnostics and therapeutics in medicine, demonstrating the vast potential of this interdisciplinary research (Adamovich et al., 2022).

10.4.1. Photodynamic Therapy for Cancer Treatment

Photodynamic therapy (PDT) is now a pivotal advancement in the treatment of cancer with the application of synergism between photosensitizers, light, and oxygen for localized tumor destruction. The action of PDT is through the action of a photosensitizer drug selectively accumulating within the tumor cells. Exposure of the drug to light of a given wavelength activates the drug to generate reactive oxygen species that lead to cell killing primarily through necrosis and apoptosis. This new modality is distinct from traditional modalities such as chemotherapy and radiation in that it is more specific and causes fewer systemic side effects. The use of PDT is most effective in the treatment of skin, head, and neck cancers, where it has been shown to be encouraging in results and provides patients with a less invasive option compared to traditional surgical interventions, hence marking the beginning of the evolution towards more patient-focused therapies in oncology. Recent improvements in laser technology have greatly enhanced the effectiveness of PDT, making its use more feasible for wider applications in cancer therapy. In recent decades, research has demonstrated the optionality of laser sources capable of generating singlet oxygen, an active compound in the photodynamic process, *in situ* in biological tissues. The technology was achieved based on quantum-dot laser diodes that emit in the near-infrared (NIR) range, which enables deeper tissue penetration with minimal damage to adjacent healthy tissue. Direct singlet oxygen production also avoids earlier weaknesses characteristic of traditional photosensitizers, such as limited light penetration and heightened sensitivity to systemic toxicity. As recent work has highlighted, such as advancements in PDTs cerebrovascular dimensions, utilization of next-generation laser technologies presents promising strategies towards enhanced clinical efficacy and expanding the therapeutic window of PDT for cancer treatment (Bragina et al., 2017). Together with technological developments, advancements in photosensitizing agents have fueled the evolution of PDT. A good case in point is the second-generation photosensitizer 5-aminolevulinic acid (5-ALA), which possesses a superior safety profile compared to previous generations. 5-ALA is a protoporphyrin IX (PpIX) precursor, causing selective trapping in cancer cells and fewer side effects. Experiments on the combinatorial effect of 5-ALA mediated PDT and γ-irradiation have indicated potential additive cytotoxic activity without synergism. Moreover, the understanding of DNA cell cycle kinetics dynamic following treatment intervention can optimize PDT treatment in a manner that therapy is designed based on tumor type and response to treatment. Combined, these advances form a solid basis for a future in which photodynamic therapy continues to revolutionize cancer treatment modalities (Allman et al., 2025) (Figure 10.11).

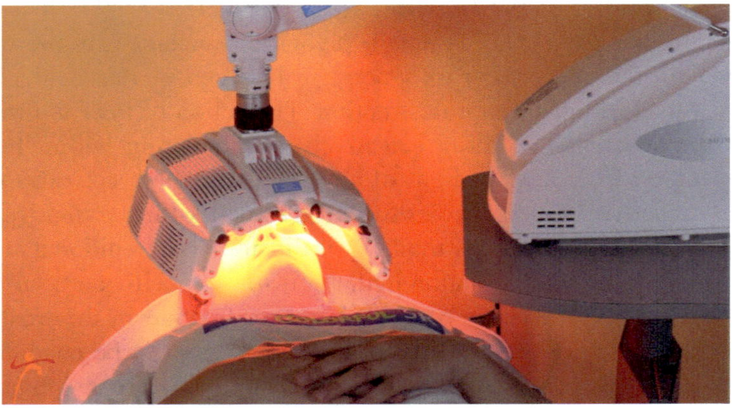

Figure 10.11. *Photodynamic therapy for cancer treatment.*

Source: https://www.skinclinicgoldcoast.com.au/wp-content/uploads/2020/10/photodynamic-therapy-brisbane-pdt.jpg.

10.4.2. Use of Photochemical Reactions in Drug Delivery

Employment of photochemical reactions in drug delivery systems is a significant step for targeted therapy and personalized medicine. By taking advantage of the selectivity of light-driven processes, scientists are able to prepare drugs that are dormant until irradiation at the site of action. This strategy enhances therapeutic efficacy with fewer systemic side effects. Photochemical reactions are employed to activate drugs upon exposure to certain wavelengths of light in order to achieve precise control over the release profiles. A very good example is employing photoresponsive polymers that exhibit a change in conformation upon illumination, thus controlling the rate of release of enclosed drugs. These developments are a paradigm shift in medication delivery, and they bring to the fore the significance of local treatment paradigms over the traditional systemic forms. As the technologies keep being researched by scientists, there is more potential for more effective therapies. Another very promising area of research is the engineering of nanoparticles possessing photochemical capability for drug delivery. These nanoparticles can be engineered to uptake specific wavelengths of light, thereby causing drug release, typically referred to as "light-induced drug delivery." Nanomaterials are so adaptive that they can be made responsive in terms of size, shape, and functionalization, potentially enhancing the absorption and penetration of the therapeutic drug. Nanoparticles can also be targeted with targeting ligands that improve specificity towards the target tissues, for instance, tumors. Development of such multi-functional drug delivery systems is on the rise because it has the potential to enhance the pharmacokinetics and biodistribution of the therapeutic agents. Additionally, as discussed in systematic reviews of nanocomposites, continuous advancement in the field is constantly addressing the fundamental needs of drug delivery systems, including sensitivity, stability, and cost-effectiveness (Nandi et al., 2024). New photochemical-based technologies not only enhance drug delivery systems but also potentially track the therapeutic efficacy in real-time. When sensing capability is integrated into delivery devices, researchers can use feedback loops to regulate drug release automatically from pre-defined biological signals. For instance, photochemical-based systems integrated with chemical sensors can detect changes in the tumor microenvironment and release the drug when the cancer cells are most in need. This synergistic approach allows clinicians to tailor treatments according to patient reactions, thereby setting the bar higher for personalized medicine. However, there remain hurdles, like compatibility of these systems with biological systems and their long-term stability. Thus, as photochemical innovation continues, overcoming these barriers will be the key to making photochemical reactions available for their complete potential in transforming drug delivery as a therapeutic modality.

10.4.3. Innovations in Diagnostic Imaging

Diagnostic imaging technology has rapidly changed the history of medical diagnostics and ensured that the detection of diseases has become more accurate and efficient. Advancements in technologies such as magnetic resonance imaging (MRI), computed tomography (CT), and molecular imaging have allowed medical professionals to gain previously untold amounts of knowledge about the human body, and this has resulted in earlier and more precise diagnoses. Not only do these technologies improve the visualization of anatomical structures, but they also enable the identification of biochemical changes at the cellular level, opening doors to more individualized medicine. The application of nanomaterials in imaging techniques, particularly their use for hepatocellular carcinoma, has been an exemplary innovation. Studies prove that such materials can enhance contrast in imaging and allow for targeted drug delivery, offering the potential for more effective treatment protocols in oncology, particularly for primary liver cancer, as revealed in the in-depth assessment of the research trend on this subject (Fan et al., 2024).

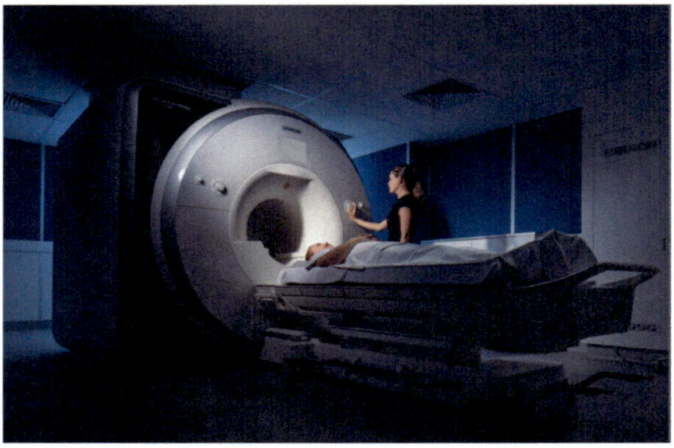

Figure 10.12. *Innovations in diagnostic imaging.*

Source: https://intranet.tri.edu.au/filething/get/58812/20170801AR_207V2.jpg.

Together with advancements in imaging modalities, the uses of photochemistry and nanochemistry have also instigated new advances in diagnostic imaging. The use of reversible thermochromic polymers in combination with traditional imaging methods provides a novel way to include heat detection capability, where new fields of research can involve energy loss in insulation within buildings and potential lapses in medical settings. Through the use of their color-switching properties, these materials have the capacity to show signs of heat fluctuation presence, thereby enhancing diagnosis in thermal imaging applications. Despite this, reversible thermochromic nanocomposites are still encountering obstacles in their application in clinical diagnostics, such as their limited color palette and reliance on external stimuli. Mitigating these challenges is essential to realizing their maximum capability and potential in medical imaging and other areas, which has a critical bearing on patient treatment

(Bakar et al., 2024). With increasing diagnostic imaging, interdisciplinarity—merging photochemistry understanding, nanotechnology advancements, and medical science discoveries—will be imperative for creating new innovations. Future directions include not only refining existing technologies but also looking for synergistic effects of merging diverse imaging modalities with new imaging agents. This may lead to improved specificity and sensitivity of disease detection, allowing early interventions. Further, the fusion of machine learning and artificial intelligence algorithms in imaging analysis can transform the diagnostic process, leading to more accurate interpretations of complex imaging datasets. With greater interdisciplinary interaction and focus on breaking through existing bottlenecks, the future of diagnostic imaging has the potential to harbor miraculous breakthroughs, finally translating into improved patient outcomes and enhanced healthcare delivery systems (Fan et al., 2024; Bakar et al., 2024) (Figure 10.12).

10.5. CASE STUDIES OF SUCCESSFUL PHOTOCHEMICAL APPLICATIONS

Utilization of photochemistry has paved the way for staggering advancements in various fields, particularly favoring the presentation of its successful uses. Among them is the synthesis and use of photocatalytic processes in conjunction with covalent organic frameworks (COFs). COFs have become favorably known due to their significant structural characteristics, including high porosity and stability, favoring the applicability in photocatalytic use. It has been established based on research that such materials are able to utilize visible light effectively for organic transformations and therefore reduce their reliance on traditional energy-consumptive processes. Rational investigation and design strategies towards COFs not only reflect the versatility of photochemical applications but also offer a sound understanding that can function towards developing future breakthroughs in the field, as explained by recent work committed to their applications in various catalytic reactions (Tang et al., 2024). In the past few years, photochemical use has been an emerging branch of investigation in energy storage devices. Notably, zinc-based batteries have received the focus of researchers due to their preferable properties, which include safety as well as ecocompatibility. The photochemical mechanisms behind enhanced electrochemical activity of these batteries reveal the profound relationship between light-activated reactions and efficiency in energy. Case studies of redox reaction optimization of Zn-based batteries show the means by which innovative design principles can lead to revolutionary improvement of functionality. These developments not only provide outstanding models of the practical applications of photochemistry towards renewable energy technology but also lead the way for further breakthroughs that can revolutionize energy storage solutions (Tang et al., 2024). Another striking instance of successful photochemical application is in the area of environmental remediation. Photocatalysis is found to be an imperative technology in mitigating pollution, and its aim is the photolytic decomposition of harmful organic pollutants with sunlight activation. Case studies emphasize the efficacy of semiconductor-based catalysts in breaking down intricate molecules into less harmful forms. These studies have led to the development of new materials that can store solar energy efficiently as chemical energy, thereby speeding up pollutant degradation rates. This use showcases the key position of photochemistry towards the achievement of sustainability goals, illustrating its capability to redefine conventional methods of managing environmental challenges. With the creation of newer materials, future prospects for advanced photocatalytic systems also appear bright, with a rosy future in the pipeline for this vital area (Alberto et al., 2022).

10.5.1. Case Study: Photocatalytic Degradation of Pollutants

Environmental pollutant degradation via photocatalysis is a promising approach for overcoming the profound challenges of industrial waste and chemical pollution. Photocatalytic reactions utilize semiconductor materials that harness the sun's energy, thereby triggering catalytic reactions that break down harmful substances into less toxic products. In a specific case study, titanium dioxide (TiO_2) has been studied exhaustively as a photocatalyst due to its stability, non-toxicity, and effectiveness in the degradation of organic pollutants under ultraviolet light exposure. Its ability is complemented further through metal doping or coupling with other materials such that it is capable of broadening light absorption beyond that in the UV regime. The progress in photocatalytic technology means that with further research, TiO_2 and other materials can become key elements in sustainable environmental management strategies. Photocatalytic degradation technologies transcend experimental applications in the laboratory to practical uses in real-world situations, emphasizing the utilitarian importance of new materials. New studies have also proposed the application of covalent organic frameworks (COFs) as new generation photocatalysts, which possess superb structural tunability and engineered photocatalytic characteristics. The ability to synthesize COFs with desired functionalities allows researchers to tailor their efficacy towards degrading pollutants, particularly in visible light sources. Notably, synthesis of hybrid COFs—containing various functional groups—allows for greater contact with pollutants, improving reaction rates and selectivity. The approach is supported by an in-depth knowledge of the catalytic reactions involved in COF-mediated degradation, which points to the vast potential of such materials toward developing photocatalysis as a viable mechanism for pollution abatement (Alberto et al., 2022). Optoelectronic sensors are also essential in photocatalytic systems in monitoring pollutant degradation process and system performance. Optical sensors based on advanced technology can give real-time feedback about pollutant concentration and efficiency of photocatalytic reaction. Such capability makes treatment processes more responsive so that adjustments can be made to optimize the rates of degradation and ensure the safety and efficacy of the process. While use of optical sensors in photocatalysis is still in the research stage, the convergence of various disciplines—engineering and materials science—is signaling a revolution in monitoring and regulating photocatalytic degradation in environmental terms. The problem in sensor design, including sensitivity and selective detection, is a sign of the conversation between application and innovation within the photochemical domain (Mário et al., 2023).

10.5.2. Case Study: Solar Fuels Production

The emerging field of solar fuel synthesis presents a cogent case study of the intersection of sustainable energy alternatives and advanced photochemical techniques. As solar energy utilization processes continue to enhance, researchers have increasingly directed efforts toward converting sun radiation into chemical fuels such as hydrogen or hydrocarbons. This phenomenon is not only an efficient utilization of renewable materials but also addresses acute requirements concerning carbon emissions and energy deficits. Current work in this area examines a variety of methodologies, including photochemical water splitting and photocatalytic utilization, which facilitate the conversion of solar energy into storable forms of chemicals. The positive results of these strategies point to the revolutionary capability of solar fuels in revolutionizing the energy industry, thereby paving the way for a more environmentally friendly future. The use of flexible conducting materials in the creation of solar fuel systems points to a great technological breakthrough. These materials are witnessed to be making headlines with their multifaceted applications and capabilities of utilizing them in energy storage systems, sensors, and even solar cells. Through the use of flexible conductive structures, solar fuel generation systems can be made more efficient and effective, allowing them to be used in expanded applications and in conjunction with other technologies. The physical flexibility of the materials themselves can also lead to innovative designs that optimize space and performance. As discussed, the development of such materials is needed in order to obtain a more efficient energy conversion process and improve the overall viability of solar fuels as a significant source of substantial energy in the future (Dutta et al., 2024). Furthermore, waste product upcycling has been an intriguing process within the solar fuels manufacturing process. Through the use of plastic waste, specifically polyethylene terephthalate (PET), researchers are beginning to demonstrate how these can be efficient feedstocks to produce hydrogen or other high-value fine chemicals. Not only does this address the environmental issues of plastic waste, but it also aids the circular economy as it derives useful energy sources from waste materials. With the augmenting demand for green energy solutions, upcycling coupled with solar fuel production may become a symbiotic relationship between waste management and energy generation, thus opening a novel path towards a greener energy future (Jia et al., 2023) (Figure 10.13).

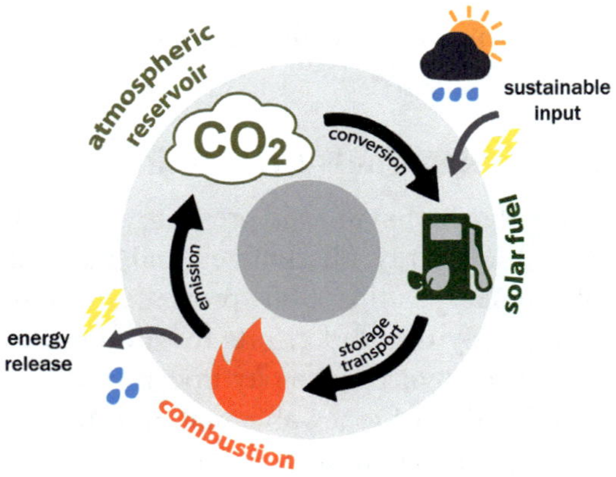

Figure 10.13. *Solar fuels production.*

Source: https://external-content.duckduckgo.com/iu/?u=https%3A%2F%2Ftse4.mm.bing.net%2Fth%2Fid%2FOIP.t-PXowvj-EnN2VmkkviPIgHaF-%3Fpid%3DApi&f=1&ipt=642ec82a90580658f00f5ed9d2f7635eaadef0bbf97ce77cd55c7d80eb663aef.

10.5.3. Case Study: Photochemical Synthesis of Pharmaceuticals

The photochemical synthesis drug development case study is a classic example of the greater application of advanced technologies to produce medicines. Light is employed to drive chemical modification through photochemical reactions, a more environmentally friendly method than the typical processes of synthesis that can involve drastic conditions and toxic reagents. This new technique not only minimizes wastage but also increases overall efficacy of synthesizing complex molecules, which is important for the pharmaceutical industry in the case of producing APIs. In this work, photochemical reactions have been used effectively to create numerous drug molecules, testifying to the practical benefits of this technique over traditional synthesis. As the pharmaceutical sector steps in the direction of more environment-friendly methods, exploring the advancement in photochemistry is increasingly significant in defining how the future of pharmaceutical manufacturing can progress in meeting regulatory and environmental demands. Artificial intelligence (AI) plays a crucial role in enhancing the efficiency of photochemical synthesis, as recent developments in the field have illustrated. Through the application of machine learning algorithms, researchers can predict reaction outcomes, control experimental conditions for optimization, and propose possible synthetic pathways tailor-made for individual pharmaceutical compounds. This integration of AI in photochemical reactions represents a paradigm shift towards intelligent pharma synthesis, transforming drug discovery and drug-making processes. Apart from this, learning models that combine theories and applications of AI help to build a skilled workforce that is well-versed in emerging technologies (Sun et al., 2025). With the combination of AI and photochemical synthesis, the pharmaceutical sector not only maximizes efficiency but also provides an environment of innovation, empowering students and professionals with the needs of the imminent challenges in drug making. Prospects for photochemical synthesis in pharmaceuticals are bright, with continuing innovation and multidisciplinary collaboration. Researchers increasingly recognize the potential of combining photochemistry with other areas of technology development, such as blockchain to handle supply chain management and enhanced data analytics to streamline processes. These systems approach not only accelerates the development of new drugs but also ensures compliance with rigorous safety and quality standards. The application of advanced learning tools, including virtual labs and interactive simulations, serves to bridge the theory-application gap of photochemical methodologies in real-world scenarios (Uddin et al., 2025). With the growth of this research area, consideration of ethical principles and green practices will be crucial in determining the future of pharmaceutical manufacturing, eventually coming up with a better, more innovative sector.

10.6. FUTURE PROSPECTS IN PHOTOCHEMISTRY

With the development of the field of photochemistry, one of the brightest prospects for future development lies in integrating the field with emerging technologies. With the ongoing evolution of digital realms, particularly in immersive technologies such as the Metaverse, the future research and education patterns in photochemistry are beginning to be redefined. With virtual reality and augmented reality entering the scientific fraternity, scientists are able to simulate complex photochemical reactions in real-time to facilitate improved visualization and interaction. Not only do such experiences promote deep insight into esoteric ideas, but also collaborative science across researchers globally. By using such advanced platforms, schools are able to educate young chemists with the proper knowledge to navigate a rapidly changing scientific world, ultimately leading to groundbreaking findings in photochemistry (Buragohain et al., 2023). In addition to advances in pedagogy, advancements in experimental approaches are also critical to the future of photochemistry. Techniques such as advanced laser and advanced spectroscopic methods are widening the scope of researchers to explore the dynamics of photochemical reactions with unparalleled accuracy.

Figure 10.14. *Future prospects in photochemistry.*

Source: https://www.metatechinsights.com/images/blog/trends-in-photochemistry-harnessing-light-for-innovation.png.

For instance, improved control of wavelength and pulse length can help researchers to identify transient phases of matter and generate extremely reactive intermediates that play a central role in unraveling complex mechanisms. Further, the use of artificial intelligence and machine learning in data analysis is revolutionizing analysis of advanced data sets to enable chemists to predict more accurately. Besides enhancing the degree of research integrity in photochemistry, these developments are facilitating optimization of materials for specific applications such as conversion of solar energy and drug delivery (Fenstermacher et al., 2021). Lastly, the future of photochemistry is highly interlinked with its potential applications to clean energy and remediation of the environment. While global energy needs continue to rise, photochemical processes promise to utilize solar energy efficiently in systems like photocatalysts for using sunlight to drive chemical reactions. Research on photochemical-based photovoltaic materials can be helpful in the creation of greener solar cells with improved efficiency in capturing and converting energy. In addition, photochemical processes for detoxifying pollutants through advanced oxidation processes are one of the most significant areas of research in environmental chemistry to solve pressing ecological issues. Focusing on these emerging applications, photochemistry can significantly contribute to energy sustainability and environmental health so that the future generation can be left with a cleaner and more sustainable planet (Buragohain et al., 2023) (Figure 10.14).

10.6.1. Emerging Technologies and Research Trends

Emerging technologies in photochemistry are demonstrating vast promise, not least in the development of new materials such as reversible thermochromic polymers. These polymers are designed to alter color with changes in temperature, and application is possible across a range of disciplines including building insulation and thermal sensors. The potential of the material to signal temperature change offers a separate opportunity for enhancing energy efficiency within buildings. But the application of these nanocomposites is hampered by limitations like a narrow color gamut and dependence on external stimuli, which may deter their extensive application. Overcoming these limitations through continuing research is critical to their full potential. By expanding the color palette and diminishing reliance on exterior power supplies, future technologies can further enable the use of these materials in real-world applications and thereby contribute positively to sustainable construction and architectural practices (Bakar et al., 2024). In parallel with progress in thermochromic polymers, research into optical sensors is increasingly determining the field of photochemical science. With increasingly ubiquitous roles of sensors in contemporary applications, ranging from medical diagnostics to environmental monitoring, the demand for highly sensitive and selective optical sensors has increased. Of particular concern to research is the concern of an understanding of the interaction between optical behavior and the environment for sensor design and functionality. Specifically, advances in micro-engineered and plasmonic sensors demonstrate the potential of enhanced detection capacity over a range of spectral regions from visible, infrared, and terahertz light. This diversity of sensor technology enables more compact and more robust solutions that are application-specific, thus addressing the specific challenges posed by various operating conditions (Mário et al., 2023). The intersection between these emerging technologies reflects broader trends in photochemistry science towards innovation and sustainability. As researchers explore photonic systems and intelligent material applications, it is increasingly clear that interfacing disciplines such as materials science, biology, and environmental engineering is unavoidable. Interdisciplinary research encourages the development of multifunctional materials not only responsive to environmental changes but also contributing to energy conservation and resource optimization. The current work on optical sensors and the development of responsive materials like thermochromic polymers indicate a clear direction towards combining chemistry and technology to achieve more benefits for society. Ongoing research and contribution towards this field can be expected to yield revolutionary innovations, overcoming boundaries to what is possible now in photochemistry, and redefining the rulebook of various industries (Bakar et al., 2024; Mário et al., 2023) (Figure 10.15).

Figure 10.15. *Emerging technologies and research trends.*

Source: https://blogs.bcm.edu/wp-content/uploads/2023/04/FUTURE-OF-MEDICINE-image.png.

10.6.2. Potential for Sustainable Development

The prospect of sustainable development in the field of photochemistry is experiencing a significant boost, with a stimulus coming from new approaches based on harnessing solar power for a variety of purposes. Harnessing renewable energy using photochemical reactions not only assists in addressing the urgent challenge of climate change, but it also aids in complementing the principles of sustainable development. Solar power, being a sustainable source of energy, can be transferred as chemical energy using photochemical reactions, giving rise to carbon-free fuels and reducing dependence on fossil fuels. Work is currently centered on the development of efficient photocatalysts and solar-to-fuel conversion processes, which would revolutionize energy storage and utilization, providing a feasible way to achieve energy sustainability. These innovations not only enhance energy efficiency but also contribute to minimizing greenhouse gas emissions, thereby assisting in the promotion of the principles of sustainable development. Photochemical innovations will also assist in positively impacting waste management and recycling, which is in conformity with the principles of sustainable development. Photocatalytic systems may be engineered to decompose toxic pollutants and poisonous materials, enabling environmental cleansing while simultaneously making it possible to recover resources. For instance, photochemical processes used to break down plastics provide a successful solution for tackling plastic waste globally. By recycling waste into a valuable chemical feedstock, these processes reduce environmental stress as well as create a reusing resources circular economy, minimizing virgin materials' need. The innovative application of photochemistry to address the waste problem is an exemplary example of its revolutionary ability, hence creating a platform for sustainable practices that can be used across all industries, and thus enhancing environmental resilience. Furthermore, the interaction of photochemistry with the principles outlined by Bollini et al. (2023) has a promising future for sustainable industrial processes. The chemical industry, which has traditionally relied on fossil fuel and energy-intensive reactions, has much to benefit from photochemical innovations. Through prioritizing research in fields such as catalysis and electrochemical systems in reaction engineering, industries can facilitate the transition towards decarbonization and reaction chemistry electrification. Specialists' recommendations are centered around taking advantage of photochemical processes to bridge the gap between batch and continuous processes, reducing wastage and enhancing overall sustainability. When industries adopt these new approaches, they will not only improve their environmental footprint but also strengthen economic sustainability, advancing world sustainable development goals through an improved and green chemical horizon.

10.6.3. Challenges and Opportunities Ahead

As photochemistry progresses, though, researchers are confronted by a variety of issues that can check technological development. One of the greatest challenges is the complexity entailed in developing highly efficient photocatalysts. While theoretical understanding of molecular organic photochemistry has reached maturity, its translation through experiential means is still beyond reach. What is crucial is the need for new materials possessing stability as well as catalytic activity in visible-light reactions. The construction of covalent organic frameworks (COFs) presents a promising future, but it is delayed by challenges in making them reproducible and scalable (Tang et al., 2024). While scientists can attempt to overcome these problems, the way to high-performance photochemical applications needs to consider not only the chemistry but also material engineering. Amidst these challenges are opportunities arising from the multidisciplinary nature of photochemistry research. The coupling of photochemistry principles with the advancement of material sciences could propel the next generation of photocatalysts towards revolutionizing energy production and chemical synthesis. For instance, the study of new redox processes in materials such as zinc-based batteries reveals the potential of energy storage and photochemical process coupling (Tang et al., 2024). As researchers seek the means of enhancing the electrochemical properties of these materials, they are also unlocking new possibilities for the applications of photochemistry in renewable energy systems. Thus, interdisciplinary collaboration does not only yield solutions to present issues but also creates possibilities for future solutions. In the future, photochemistry hinges on breaking the double dynamics of competition and cooperation in the research world. Although individual research groups may seek revolutionary discoveries, information sharing within collaborative networks can significantly help with overall development. COF design and synthesis as active visible-light photocatalysts are just one demonstration of how a collaborative effort is required to bridge material performance hurdles while extending their multidirectional uses (Alberto et al., 2022). Additionally, by combining knowledge in various research activities and open scientific discourse, the photochemistry community is likely to create a solid platform for innovation. Such collaborative efforts will not only optimize existing methodologies but also initiate innovative methods of photochemical applications, thereby making the field dynamic and responsive to upcoming challenges and opportunities.

SUMMARY

In concluding an inquiry into photochemistry, the prospect in the future for this science is huge and diverse, driven by advances that can address some of the world's most compelling issues. Developments in energy conversion and storage technologies, especially in the solar energy sector, have made photochemistry a core component of the transition towards sustainable conduct. As discussed in certain case studies, like the development in Zn-based batteries, new redox processes were invented that significantly enhance electrochemical performance. Not only are such developments making energy storage technologies more efficient but also basic studies that inspire follow-up innovation in renewable energy technologies and exemplify the key synergy between theoretical research and practical application in photochemistry (Tang et al., 2024). Moreover, the integration of nanomaterials into photochemical applications is a new wave in the field. The synthesis of novel nanomaterials has made it possible to develop complex biosensing platforms, as emphasized in the recent reviews on optical fiber biosensors. Nanomaterials leverage unique optical features that can increase sensitivity and detection limits, therefore enabling improved specificity and response time in biosensing processes. These innovations underscore the necessity for cross-disciplinary research where photochemistry and nanotechnology unite to create high-impact applications in healthcare, environmental monitoring, and beyond. By embracing these innovations, researchers and practitioners have unprecedented potential for the utilization of photochemistry in real-world applications (Li et al., 2022). In summary, the future of photochemistry lies at the nexus of innovation and application with a clear road map leading to an enhanced quality of life through an enhanced technology. The more the various industries realize the role played by photochemical processes in overcoming energy and environmental crises, the more the research and development in this area becomes a matter of utmost importance. From the momentum established by the recent breakthroughs in solar power, battery technology, and biosensing technology, it will be essential to maintain innovation in order to realize a sustainable future. The insights gained by case study and ongoing research form a solid basis for all forthcoming endeavors and guarantee that photochemistry remains a central area of exploration as it continues to meet challenges yet unknown. By pursuing these lines, the field will not just be progressing scientifically but will also be contributing to the progress of society.

The photochemistry science has revolutionary potential across many industries, particularly in energy storage devices. Given the increased demand for green power sources, breakthroughs in photochemical processes have emerged as competitive alternatives to traditional processes. Of specific interest in this regard are advancements in photoredox reactions, which utilize light to enable redox processes, allowing chemical conversions to become more efficient and selective. Recent research reveals the promise of the integration of such techniques into the design of cleaner energy systems, e.g., photoenergy conversion and novel battery technologies. In this context, systematical research into redox reactions, particularly in zinc-based batteries, is called for to ensure optimal electrochemical efficiency and advance the practical application of photochemistry to energy storage (Tang et al., 2024). The importance of ongoing research and each other's insights in the field cannot be overstated. Agreement from professionals in response engineering prescribes that a lot can be accomplished in photochemistry through concerted efforts in focused industries like pharmaceuticals and specialty chemicals. Overcoming barriers calls for an integrated understanding of catalytic processes along with electrochemical systems. With the application of cutting-edge reactor technology and enhanced manufacturing methods, researchers aim to optimize photochemical reactions to be industrially feasible. As witnessed in the agreement of industry and academic professionals, the overcoming of these technological hurdles will be central not just to improving existing processes but also to enabling global decarbonization and sustainability of chemical production (Bollini et al., 2023). Moreover, a study of photochemistry case studies demystifies the utilitarian applications and successes of new breakthroughs. For example, the application of photochemical methods in the design of more environmentally benign synthetic routes illustrates the double benefit of ecotoxicity and profitability. These case studies serve as concrete demonstrations of the translations of theoretical breakthroughs into applicatory utility, illustrating tangible outcomes that prove this field worthwhile in the future. As businesses increasingly adopt such new approaches, there is a strong necessity to assess their impacts on the overall production efficiency and sustainability programs. The ongoing deliberation at the scientific level regarding such advancements will not only determine modern practices but also determine the photochemical environment in the future, triggering even more innovations toward sustainable development (Tang et al., 2024).

REFERENCES

1. Abu Bakar, M. S., Asyraf, M. R. M., Syamsir, A., Najeeb, M. I., Alhayek, A., Al-Dala'ien, R. N., Manar, G., et al. (2024). Thermochromic polymer nanocomposites for the heat detection system: Recent progress on properties, applications, and challenges. *Polymers, 16*, 1545. https://doi.org/10.3390/polym16111545 (accessed on 28th September 2025).
2. Adamovich, I., Agarwal, S., Ahedo, E., Alves, L. L., Baalrud, S., Babaeva, N. Y., Bogaerts, A., et al. (2022). The 2022 plasma roadmap: Low temperature plasma science and technology. *Journal of Physics D: Applied Physics, 55*, 373001. https://doi.org/10.1088/1361-6463/ac5e1c (accessed on 28th September 2025).
3. Allman, R., Cowburn, P., & Mason, M. (2025). *Effect of Photodynamic Therapy in Combination with Ionizing Radiation on Human Squamous Cell Carcinoma Cell Lines of the Head and Neck*. https://core.ac.uk/download/pdf/8038363.pdf (accessed on 28th September 2025).
4. Bollini, P., Diwan, M., Gautam, P. K., Hartman, R. L., Hickman, D. A., Johnson, M. D., Kawase, M., et al. (2023). Vision 2050: Reaction engineering roadmap. *ACS Engineering Au, 3*, 364–390. https://doi.org/10.1021/acsengineeringau.3c00023 (accessed on 28th September 2025).
5. Bragina, O. A., Gekaluyk, A. S., Goltsov, A., et al. (2017). *Laser-Induced Generation of Singlet Oxygen and Its Role in the Cerebrovascular Physiology*. https://core.ac.uk/download/228178032.pdf (accessed on 28th September 2025).
6. Buragohain, D., Chaudhary, S., Punpeng, G., Sharma, A., Am-in, N., & Wuttisittikulkij, L. (2023). Analyzing the impact and prospects of metaverse in learning environments through systematic and case study research. *IEEE Access, 11*, 141261–141276. https://www.semanticscholar.org/paper/3aab687520ae071c9bbd84fae668351e726912e6 (accessed on 28th September 2025).
7. Corrigan, N., Yeow, J., Judzewitsch, P. R., Xu, J., & Boyer, C. (2018). Seeing the light: Advancing materials chemistry through photopolymerization. *Angewandte Chemie International Edition, 58*, 5170–5189. https://doi.org/10.1002/anie.201805473 (accessed on 28th September 2025).
8. Dutta, T., Chaturvedi, P., Llamas-Garro, I., Velázquez-González, J. S., Dubey, R., & Mishra, S. K. (2024). Smart materials for flexible electronics and devices: Hydrogel. *RSC Advances, 14*, 12984–13004. https://doi.org/10.1039/d4ra01168f (accessed on 28th September 2025).

9. Eleweuwa, H. O., Udele, N. O., Bakare-Abidola, T., Ezeifegbu, C. G., Okejevwa, G. D., & Olaoye, J. (2025). Analysis of energy efficiency in water treatment plants: Present conditions and future directions. *World Journal of Advanced Research and Reviews*. https://www.semanticscholar.org/paper/06d71b40933c859da9e3603e7b3f149cb4c2a806 (accessed on 28th September 2025).
10. Fan, Y., Xiao, H., Wang, Y., Wang, S., & Sun, H. (2024). Global research on nanomaterials for liver cancer from 2004 to 2023: A bibliometric and visual analysis. *Discover Oncology, 15*. https://doi.org/10.1007/s12672-024-01735-1 (accessed on 28th September 2025).
11. Fenstermacher, M., Abbate, J., Abe, S., Abrams, T., Adams, M., Adamson, B., Aiba, N., et al. (2021). DIII-D research advancing the physics basis for optimizing the tokamak approach to fusion energy. *Nuclear Fusion*. https://www.semanticscholar.org/paper/8425996e23b9cab8852e30dada2b0d1333acc95f (accessed on 28th September 2025).
12. Ferreira, M. F. S., Brambilla, G., Thévenaz, L., Feng, X., Zhang, L., Sumetsky, M., Jones, C., et al. (2023). Roadmap on optical sensors. *Journal of Optics, 26*, 013001. https://doi.org/10.1088/2040-8986/ad0e85 (accessed on 28th September 2025).
13. Grama, S. B., Liu, Z., & Li, J. (2022). Emerging trends in genetic engineering of microalgae for commercial applications. *Marine Drugs, 20*, 285. https://doi.org/10.3390/md20050285 (accessed on 28th September 2025).
14. Herrera Luna, J. C. (2022). *Production of New Carbon-Heteroatom Bonds Induced by Visible Light*. https://doi.org/10.4995/thesis/10251/191051 (accessed on 28th September 2025).
15. Hossain, M. A., Coe-Sessions, K., Ault, J. W., Gboyero, F. O., Wenzel, M. J., Dhokale, B., Davies, A. E., et al. (2024). Covalent integration of polymers and porous organic frameworks. *Frontiers in Chemistry, 12*. https://doi.org/10.3389/fchem.2024.1502401 (accessed on 28th September 2025).
16. Jia, Z., Gao, L., Qin, L., & Yin, J. (2023). Chemical recycling of PET to value-added products. *RSC Sustainability, 1*, 2135–2147. https://doi.org/10.1039/d3su00311f (accessed on 28th September 2025).
17. Kaufman, J. D., Elkind, M. S. V., Bhatnagar, A., Koehler, K., Balmes, J. R., Sidney, S., Burroughs Peña, M. S., et al. (2020). Guidance to reduce the cardiovascular burden of ambient air pollutants: A policy statement from the American Heart Association. *Circulation, 142*. https://doi.org/10.1161/cir.0000000000000930 (accessed on 28th September 2025).

18. Li, M., Singh, R., Wang, Y., Marques, C., Zhang, B., & Kumar, S. (2022). Advances in novel nanomaterial-based optical fiber biosensors—A review. *Biosensors, 12*, 843. https://doi.org/10.3390/bios12100843 (accessed on 28th September 2025).

19. López-Magano, A., Daliran, S., Oveisi, A. R., Mas-Ballesté, R., Dhakshinamoorthy, A., Alemán, J., García, H., et al. (2022). Recent advances in the use of covalent organic frameworks as heterogenous photocatalysts in organic synthesis. *Advanced Materials, 35*. https://doi.org/10.1002/adma.202209475 (accessed on 28th September 2025).

20. Manyoundou, J. R. (2025). Global assessment of energy consumption in conventional drinking water treatment plants. *European Journal of Theoretical and Applied Sciences*. https://www.semanticscholar.org/paper/4566ef5a9fd53d35b0216ecee86a87e2809e3b90 (accessed on 28th September 2025).

21. N/A. (2023). *Nano Plus Science and Technology of Nanomaterials, 7*. https://doi.org/10.48187/stnanomat.2023.7.001 (accessed on 28th September 2025).

22. Nandi, S., Adimule, V., Kerur, S. S., Dhanalakshmi, S., & Barmavatu, P. (2024). Recent advances in metal-organic framework derived nanocomposites in gas and chemical sensors: A review. *Sensing Technology, 2*. https://doi.org/10.1080/28361466.2024.2317385 (accessed on 28th September 2025).

23. Quijano Velasco, P., Hippalgaonkar, K., & Ramalingam, B. (2025). Emerging trends in the optimization of organic synthesis through high-throughput tools and machine learning. *Beilstein Journal of Organic Chemistry, 21*, 10–38. https://doi.org/10.3762/bjoc.21.3 (accessed on 28th September 2025).

24. Rao, A. V., & Rao, L. G. (2023). Dietary carotenoids–Sources, properties, and role in human health. *Physiology*. https://doi.org/10.5772/intechopen.111008 (accessed on 28th September 2025).

25. Shang, Y., Hasan, M. K., Ahammed, G. J., Li, M., Yin, H., & Zhou, J. (2019). Applications of nanotechnology in plant growth and crop protection: A review. *Molecules, 24*, 2558. https://doi.org/10.3390/molecules24142558 (accessed on 28th September 2025).

26. Sun, K., Wen, Z., & Wang, T. (2025). Integration of AI into pharmaceutical engineering education: Achievements, impact, and future prospects. *Advances in Education, Humanities and Social Science Research*. https://www.semanticscholar.org/paper/c1ffaa8e458a754704f22a66d47e19529f091c9d (accessed on 28th September 2025).

27. Tang, L., Peng, H., Kang, J., Chen, H., Zhang, M., Liu, Y., Kim, D. H., et al. (2024). Zn-based batteries for sustainable energy storage: Strategies and mechanisms. *Chemical Society Reviews, 53*, 4877–4925. https://doi.org/10.1039/d3cs00295k (accessed on 28th September 2025).

28. Uddin, S. M. I., Sumon, R. I., Mozumder, M. A. I., Chowdhury, M. K. H., Armand, T. P. T., & Kim, H. C. (2025). Innovations and challenges of AI in film: A methodological framework for future exploration. *ACM Transactions on Multimedia Computing, Communications and Applications*. https://www.semanticscholar.org/paper/4686a729c684adaceeb5ca9b8812674abba48fd7 (accessed on 28th September 2025).

29. Xin, Z., Zhang, M., Cui, H., Ding, X., Zhang, T., Wu, L., Cui, H., et al. (2023). Algae: A robust living material against cancer. *International Journal of Nanomedicine, 18*, 5243–5264. https://doi.org/10.2147/ijn.s423412 (accessed on 28th September 2025).

30. Yang, Z., Liu, Y., Cao, K., Zhang, X., Jiang, H., & Li, J. (2021). Synthetic reactions driven by electron-donor–acceptor (EDA) complexes. *Beilstein Journal of Organic Chemistry, 17*, 771–799. https://doi.org/10.3762/bjoc.17.67 (accessed on 28th September 2025).

31. Zhang, Y., Xue, L., Carter, W., Pei, C., Chen, T., Mu, J., Wang, Y., et al. (2021). Development of ozone reactivity scales for volatile organic compounds in a Chinese megacity. *Atmospheric Chemistry and Physics, 21*, 11053–11068. https://doi.org/10.5194/acp-21-11053-2021 (accessed on 28th September 2025).

INDEX

Symbols

3D printing technology 98

A

Advanced Oxidation Processes (AOPs) 133, 135, 136, 137, 138, 140, 143, 144, 147, 148, 150, 152, 153, 154, 156, 157, 159, 161, 162, 163, 166, 167, 178, 193, 195, 196
Aggregation-induced emission (AIE) 53, 65
Agricultural practices 275
AI-aided modeling 253
Applications of photochemical reactors 337
Artificial intelligence (AI) 232, 427

B

Besides formal education 382
Biochar-based systems 165
Biochemistry 5
bioelectrochemical systems (BES) 279
Biofilm recovery and ecosystem 36
Biogenic volatile organic compounds (BVOCs) 229, 232
Biological and chemical systems 52
Biomass burning (BB) 2
Broader framework 365
Build sophisticated models 222

C

Carbon-based compounds 269
Carbon-free fuels 431
Carbon nanocomposites (CCNs) 107
Catalytic ozonation process (COP) 271
Chemical oxygen demand (COD) 154, 167
Chemical vapor deposition (CVD) 29
clofibric acid (CA) 303
Closed Loop Advanced Sanitation System (CLASS) 153
COF-based photocatalysts 403
Co-metabolic activities 280
Compound similarity networks (CSN) 304
Comprehending photochemistry 1
Computational chemistry developments 48
Computational fluid dynamics (CFD) 335
Condensable particulate matter (CPM) 226
Contaminants of concern (CECs) 165
Continuous-flow photocatalysis 192
Controlled ecological life support system (CELSS) 319
Cost-benefit study 354
Cost-effectiveness analysis 244
Cost-effective sensing 399

Covalent organic frameworks (COFs) 47, 49, 62, 80, 105, 397, 403, 406, 423, 424, 432
Cross-fertilization between 270
Current environmental problems 18
Cutting-edge materials 56

D

Data-driven policies 224
Degradation-induced loss 117
Degradation of pollutants 298
Density functional theory (DFT) 11
Dependency-based cities 386
Diagnostic imaging technology 421
Dipole-dipole interaction mechanism 59
Dissolved organic carbon (DOC) 24
Dissolved organic matter (DOM) 24
Dodecylbenzene sulfonate (DBS) 134, 140, 146, 151
Double-barreled thrust 257

E

Ecological life support 319, 327, 350
Ecological support systems 207
Effluent Dominated Hydrosystems (EDHs) 339
Electron-hole pairs 88, 89, 91, 93, 94
Emissions trading scheme (ETS) 389
Encompass environmental 353
Energy-intensive reactions 431
Energy storage systems (ESS) 375
Energy-transfer reactions 64
Engineered nanomaterials (ENMs) 179, 181, 184, 185, 201
Environmental pollutant degradation 424
Ever-present need 186
External quantum efficiency (EQE) 60

F

First-order kinetics 268, 271, 289, 290, 291, 292, 298, 302
First-row transition 55, 58, 71, 86
Flat-plate collectors 315, 316, 317
Fluorescence and resonance energy transfer (FRET 66
Fluorescence spectroscopy 66, 68, 69
Foreign direct investment (FDI) 387
Förster resonance energy transfer (FRET), 406
Fossil fuel dependence 359
Fourier Transform Infrared Spectroscopy (FTIR) 296
Full-scale projects 195
Fundamental principles of photochemistry 6
Future technologies 404, 429

G

Global energy infrastructure 357
Goal-specific applications 287
Green infrastructure 232
Ground-level ozone formation 223

H

Hard-Soft Multivariate Curve Resolution-Alternating Least Squares (HS-MCR-ALS) 190
Heterogeneous wastewaters 153
High-efficiency production 317
High-energy radiation 329
High-level quantum chemical 268
High-performance liquid chromatography (HPLC) 293, 294
Hole-transporting layer (HTL) 79
Human health 267
Hybrid photocatalysts 101
Hybrid semiconductors 102
Hydrogen-bonded organic frameworks (HOFs) 406
Hydroxyapatite (HAp) 93

I

Including ecosystems and biodiversity 354
Intergovernmental Panel on Climate Change (IPCC) 357

K

Kinetic modeling 271, 288

L

Larger-scale environmental 96
Life Cycle Assessment (LCA) 24
Light-activated transitions 54
Light-controlled release 31
Light-emitting diodes (LEDs) 330
Light-induced cleavage 182
Light-induced photocatalytic 119
Light-matter interaction 318, 404
Light-sensitive biomaterials 73, 75
Light-sensitive systems 73
Liquid chromatography coupled with mass spectrometry (LC-MS) 294
Lithium-ion batteries 377
Local institutions typically 371
Localized surface plasmon resonance (LSPR) 97, 104
Long-lasting organic pollutants 177
Long-term activation 143
Long-term compliance 255
Long-term costs 205, 206
Long-term liabilities 163
Low-cost wireless sensor 245, 260

M

Machine learning 409, 410, 419, 423, 427, 428, 437
Magnetic resonance imaging (MRI) 421
Mass spectrometry (MS) 296, 297
Matter-light interactions 45
Metal-based photocatalysts 99, 100
Metal-doped metal 253
Metal halide perovskites (MHPs) 118
Metal-organic frameworks (MOFs) 105, 291
methyl hydroperoxide (MHP) 78
Methyl tert-butyl ether (MTBE). 137
Model Selection (MS) 183

N

Nanoscale material synthesis 121
National renewable energy policies 369
Natural organic matter (NOM) 25, 184, 187, 213
Non-carcinogenic effects 179
Nuclear magnetic resonance (NMR) 66, 277, 293

O

Operational parameters 332, 333
Organic light-emitting diodes (OLEDs) 53, 59

P

Paris Agreement emphasizes 369
Patient-focused therapies 419
Perovskite solar cells (PSCs) 79
Peroxyacetyl nitrate (PAN) 12
Pharmaceutical degradation in wastewater 302
Pharmaceutical synthesis 338
Photocatalysis 87, 88, 89, 90, 91, 95, 107, 108, 109, 111, 112, 113, 116, 117, 119, 187, 188, 189, 191, 192, 233
Photocatalytic materials 89, 90, 92, 93, 94, 95, 98, 106, 107, 108, 112, 116, 123, 124
Photochemical-based systems 421
Photochemical ozone formation potentials (POCPs), 223
Photochemical processes 48, 49, 62, 413, 414, 415
Photochemical reactions 61, 77, 313, 314, 315, 316, 317, 319, 320, 322, 323, 325, 327, 329, 331, 332, 334, 335, 336, 338, 340, 342, 345
Photochemical reactions harness 31
Photochemical reactor 315, 316, 317, 318, 319, 320, 326, 329, 330, 334, 338, 339, 342, 343, 344, 345, 349
Photochemical reactor design and opera-

tion 342
Photochemical research 251
Photochemical systems 10
Photochemical technologies 179, 183, 196, 205, 206, 208, 209, 210, 211, 214, 408, 413
Photochemical technology 323, 334, 346, 347
Photochemical water treatment 181
Photochemistry 228
Photochemistry development 417
Photochemistry-mediated photolysis 8
Photodynamic therapy (PDT) 418, 419
Photo-Fenton process 194, 197
Photophysical and photochemical 45, 81
Photophysics and photochemistry 49
Photopolymerization processes 16
Polyethylene terephthalate (PET), 426
Polyfluoroalkyl substances (PFASs) 138
Polymeric carbon nitride (PCN), 118
Principles of photochemistry 182, 183
Prioritize environmental integrity 356
Process-based solutions 247
Process-controlling systems 235
Public health 3, 20, 21, 22, 25, 50, 69
Public health and local ecosystems 241
Public perception 379, 380
Public-private partnerships 256, 373

Q

Quantum theory of light 54, 55

R

Reactive oxygen species (ROS) 37, 107, 143, 179, 185, 190, 197, 236
Real-time visualizations 58
Real-world applications 9, 103, 107, 117, 120, 124
Real-world potential 31
Real-world scenarios 159
Real-world technologies 73
Relating kinetic models 290
Renewable energy 361, 362, 363, 365, 367, 392, 395
Renewable energy technology 355, 367, 368, 378, 379, 383
Return on investment (ROI) 346
Rural locations 381

S

Secondary organic aerosol (SOA) 76
Selective non-catalytic reduction (SNCR) 239
Shared-Socioeconomic Pathway (SSP) 363
Short-term economic 250
Silicon-based organic 66
Single-atom catalysts (SACs) 114, 116
Small-scale reactions 322
Smart grid technology 375, 376
So-called renewable 358
Society life-cycle costing (S-LCC) 355
Socio-economic benefits 361, 386
Solar disinfection (SODIS) 198
Solar energy collection 116
Solar energy conversion 399, 411, 412
Solar-induced chlorophyll fluorescence (SIF) 56, 79
Solar-powered photochemical 255
Solar-powered photochemistry 200
Solar-powered reaction 202
solar water treatment 198, 201, 202
Solution technologies 345
Sonophotocatalysis processes 109
Spectroscopic methods 295, 296
Spectroscopy techniques 66, 67
Spin-orbit coupling 59
Stand-alone science 400
Sunlight-activated photocatalyst 189
Sunlight-powered reactions 18
Sun-powered photochemical processes 200
Super-redox processes 68
Sustainable Development Goals (SDGs) 367

T

Technological advancements 231

Temperature-degradation efficiency 284
Textile industry wastewater (TIWW) 167
textile wastewater industry (TIWW) 154
Theoretical calculations dictate 321
Theory-application gap 427
Theory-practice divides 100
Time-dependent behavior 271
Time-resolved spectroscopy 69, 70
Time-resolved techniques 71
Titanium dioxide (TiO2) 11, 15, 25, 26, 29, 88, 97, 108, 109, 119, 181, 187, 189, 191, 198, 212, 214
Top-line economic indicators 361
Total organic carbon (TOC) 141, 146, 150, 154, 159, 167
Trace organic contaminants (TOCs) 280
Transition metal dichalcogenides (TMDCs) 301
Triphenyl phosphate (TPHP) 282
Tris(2-chloroisopropyl) phosphate (TCPP) 278
Tropospheric Ozone Assessment Report (TOAR) 4, 20
Two-dimensional materials 98
Two-way approach 26
Typically titanium dioxide (TiO2) 147

U

Ultrahigh-performance liquid 278
Ultraperformance liquid chromatography–mass spectrometry (UPLC/MS) 95
United Nations Framework Convention on Climate Change (UNFCCC) 368
Utility-scale solar 365

V

Visible-light reactions 432
Volatile organic compounds (VOCs) 12, 16, 18, 19, 21, 23, 107, 111, 221, 224, 225, 228, 229, 230, 233, 236, 238, 240, 241, 249, 254, 416

W

Waste management programs 355
Water-based systems 58
Water treatment chemical management 211
Wavelength-sensitive materials 61
Well-delineated parameters 344
Wireless sensor networks (WSNs) 242

Z

Zero-order principle 289
Zero-valent iron (ZVI) 141, 145, 146, 150
Zn-based batteries 413, 423, 432, 438